Implementing New Technologies:
Choice, Decision and Change
in Manufacturing

Implementing New Technologies:
Choice, Decision and Change in Manufacturing

*Edited by Ed Rhodes
and David Wield
at the Open University*

Basil Blackwell
in association with
The Open University

First published 1985

Basil Blackwell Ltd
108 Cowley Road, Oxford OX4 1JF, UK

Basil Blackwell Inc.
432 Park Avenue South, Suite 1505,
New York, NY 10016, USA

British Library Cataloguing in Publication Data

Implementing new technologies: choice, decision
and change in manufacturing.
1. Technological innovations
I. Rhodes, Ed II. Wield, David III. Open
University
670.42'7   T173.8

ISBN 0-631-14379-3
ISBN 0-631-14381-5 Pbk

Library of Congress Cataloging in Publication Data

Implementing new technologies.

Includes index.
1. Production management—Addresses, essays, lectures.
2. Production engineering—Addresses, essays, lectures.
I. Rhodes, Ed.   II. Wield, David.
TS148.147 1985   658.5   85-11172
ISBN 0-631-14379-3
ISBN 0-631-14381-5 (pbk.)

Typeset by Photo-Graphics, Honiton, Devon
Printed in Great Britain by T.J. Press Ltd, Padstow, Cornwall

Contents

Part 4   Corporate Strategies for New Technology

Part 5   Financing Innovation

Part 6   New Technology and the Organization of Work

Preface

This reader has been produced in conjunction with an Open University course on the Implementation of New Technologies, part of a Master's degree in Manufacturing funded by the Science and Engineering Research Council (SERC). The course is primarily intended for people with a scientific, technical or other professional background in manufacturing. However, in common with this book, it will also be of interest to a wider group.

Whilst the book does include material on innovation in the sense of 'first commercial transaction involving new products or processes', it is more concerned with the consequences of technology diffusion and technology transfer, concentrating on the adoption within production units of technologies which are novel to the unit, even though not necessarily inherently 'new'.

The course and the collection of readings is directed towards a concept of 'implementing new technologies' that includes study of: decision forming and decision taking (including areas like feasibility studies and pilot projects); project planning; the 'conversion' or application of plans; and consolidating the change after insertion of new technologies. The readings are also concerned with the product market, financial and other environments within which implementation takes place and with the mediating influence of corporate strategy between these environments and individual projects.

Thus the course and this reader adopt a multidisciplinary approach derived from production engineering and management, technology policy, financial management, industrial relations and organizational behaviour.

It includes for example, the opinions and analysis of senior managers, engineering consultants and financial strategists, as well as academics from the fields of technology policy, industrial relations, management and production engineering.

In teaching the Open University course, this reader, consisting of edited extracts and commissioned articles, will be used in conjunction with four other teaching texts, video cassette film material, audio cassettes and other teaching aids. Further information on the course and its teaching materials may be obtained from the authors at the Faculty of Technology.

We would like to acknowledge the assistance of all those who helped with the book's preparation: John Bessant, Ernest Braun, Alan Chatterton, Peter Senker and Brian Small for suggesting and commenting on articles; other members of the course team for their comments, particularly Suzanne Brown, Margaret Bruce and Stephen Brown; the managers of the companies Babcock Power Ltd, Marks and Spencer, Courtaulds, Abbey Hosiery and Snackco, who kindly allowed us access, often at very short notice, both to their factories and to their time; John Taylor of the Open University Publishing Division; Tricia O'Doherty and Nessie Tait for typing the drafts of the introductions.

PART 1
Introduction

1.0

Arrangement of this Reader

This reader is divided into six parts. In part 1 we consider some of the relevant themes and issues. Part 2 raises questions about the nature and boundaries of technological change within the production unit. Part 3 takes the subject of technological change and innovation more firmly down to the level of manufacturing. Part 4 examines some of the technical and operational difficulties of implementing technological changes at production unit level, together with strategies for improving implementation. Part 5 deals with issues of resource allocation and particularly financial appraisal and control and their crucial relationship with decision making under conditions of extreme uncertainty. Finally, Part 6 concentrates on the work-force-related issues of implementation, using work force in the general sense of referring to all employees in a production unit. This general introduction deals very briefly with a number of issues, aiming to present a general appreciation of the context within which the reader has been prepared.

1.1

The Implementation of New Technologies: An Introduction

Ed Rhodes and David Wield

New Technologies

Discussion of the implementation of new technologies within the focus of much current debate is likely to be perceived in terms of the application of microelectronic technologies. However, at least within the context of manufacturing, such focus is misleading. Some forms of microelectronic technology are now scarcely 'new' by any definition nor are they the only 'new' technologies, particularly at the level of the individual firms. Studies of the diffusion of technological innovations (see Nabseth and Ray 1974, Ray 1983 for example) have emphasized that diffusion is frequently a prolonged process. This is not simply a matter of tardiness in adoption but a reflection of the very considerable obstacles that may have to be overcome to adapt them to the particular needs of the individual firm and the product market niche within which it competes. The process of diffusion is hardly ever simply one of acquiring 'off the shelf' packages of new technology, but involves progressive development and evaluation, involving both adaption within workplaces, and cooperative relationships with equipment suppliers. This process of adaptation emphasizes that established technologies can nonetheless be 'new' in terms of the issues and problems that are provided by the particular circumstances (such as technical, organizational and social) of the production unit. The term 'new' technology is thus not only used for the establishment of complete new production facilities. It also relates to large- and small-scale changes within established production systems which are a powerful though often neglected source of technological development.

However, our focus is narrowed in at least two respects. Firstly, we have concentrated on the application of new technologies to production processes rather than their incorporation within products although we emphasize the artificiality of this distinction. Secondly, our primary concern is with specific production operations in manufacturing. Since many of the issues and problems associated with the introduction of new technologies are similar across broad areas of economic activity, the separation of manufacturing processes from those of other industrial and service sectors is somewhat arbitrary. This is emphasized by the example of microelectronic technology with its broad impact and its potential to integrate many different kinds of production process. Even within manufacturing this can include, for example, integrating production operations with component handling and movement, and other manufacturing operations with aspects of purchasing, scheduling and design.

Implementing New Technologies

We use the word 'implementation' to include a wider range of activities than is generally meant by the term. In general, the term implementation has usually been taken to relate to the process of putting policy intentions – 'decisions' – into action. In the case of new technologies, the relevant activities may include: the acquisition of new equipment, consumables and so on; the undertaking of associated construction work – increasing floor loading, installing services etc.; equipment installation; consultation; training; cost control; commissioning; and handover. It is this level of activity that Braun in reading 3.2 describes as 'implementation'. As he points out, the activities involved can be highly complex, depending upon the nature of the project, and can involve high levels of uncertainty. This is, of course, likely to be the case where major re-equipment programmes are involved, as in the example of the food industry (Williams, reading 6.1) and that of power generation equipment (Wield, reading 2.1). However, even in comparatively small scale and limited examples of change such as the initial ventures into CAD described by Senker in reading 5.1, the complexities and difficulties can be considerable. This is particularly true in smaller organizations with a more limited range of inside expertise and support. Not surprisingly therefore, there is a considerable range of things that can go wrong with new technology projects – although the precise extent to which they occur is uncertain. Hussey (1984, p. 143) has pointed out that managers appear 'to have a propensity to take personal credit for things that go right but to avoid taking personal blame for things that go wrong'.

Among the manifestation of problems are delays in project initiation that give competitors a crucial time advantage. Unit costs of production may be inflated by overspecification of project requirements. Underspecification may constrain future development and flexibility. For a variety of reasons new technology may fail to provide all the predicted benefits or perform at specified levels and may have unforeseen adverse effects at other points in the production system. Completion may be achieved late and only after significant additional costs have been incurred. There is also the problem of non-initiation of new technology projects whether because new technology is not perceived as a means by which a firm's problems may be remedied or because proposals for change are rejected.

Problems like these raise a number of issues. There is the question of why it is that things go wrong in these ways even in seemingly competent, successful and well-managed organizations. In seeking to establish the reasons it is clear that one needs to look beyond the types of activities that we referred to above in terms of implementation just being the 'putting of policy decisions into action'. Thus, no amount of skills or experience in this 'putting into action phase' is likely to be able to successfully overcome the effects of policy decisions that are based upon erroneous assumptions or inadequate analysis. For example, Sciberras' study of the television industry (reading 4.4) indicates the dangers of basing decisions about introducing new technology on a misreading of product market trends. Similarly it is difficult to overcome errors in the complex processes of planning prior to the 'putting into action' of decisions. Additionally, it is important to look at events after the active stages of application. It is evident from a number of studies (see Gold 1980 and Teubal 1983) that quite apart from the benefits derived from growing familiarity with new technology (the learning curve effect) there are frequently subsequent benefits from further adaptations and developments within the production unit. In order to integrate these kinds of issues the concept of implementation needs to be broad enough to be able to assist in the analysis and understanding of these kinds of problems. Thus, in the course we emphasize the importance of viewing implementation as a broad change process. We divide the implementation process into four phases: an *initiating* phase which embraces the initial stimuli for technological change and the many elements of the decision-forming processes such as the development and assessment of policy alternatives; a *planning* phase; the *application* phase referred to above (which is similar to Braun's implementation phase in reading 3.2); and a *consolidation* phase which is concerned with full completion of the many elements of a project and with the stimulation and formalization of the post-application phase.

These activities of initiation and decision forming–planning–application– and later consolidation of change, are closely related. This is not solely in a linear sense, although there obviously is a progression of emphasis from one set of activities to another – but in a dynamic, interactive sense since, as case studies tend to show, new technology projects, particularly when complex and extending over a long time period, are subject to some reappraisal and adaptation as new possibilities or problems become apparent. The implementation process is located within an environment which has two main dimensions. The first of these is the *internal* environment which involves the particular circumstances of the individual firm such as the extent of organizational differentiation, the characteristics of production and associated technologies – the degree of integration for example; the nature of the financial appraisal and control systems; the character of industrial relations. The *external environment* includes the characteristics of the sector(s) of the product market aimed at by the production unit; the technological context – the rate of change and diffusion of knowledge, supplier characteristics etc.; the broad political context, including levels of state intervention and support relevant to technology. We emphasized above that the phases of the implementation process are interactive. Part of the interrelationship stems from the continuities provided by different groups of factors throughout the process of change. Firstly, we can trace the broad range of issues which embrace the work force at all levels, from management to the shop floor. Whether they are involved in the initiation of proposals to change or in responding to these, ultimately there has to be a dialogue, even if this is no more than a dialogue of the deaf and it embraces much wider issues than those of consultation or training we referred to earlier. Secondly, the technical and operational issues can be traced through, from the issues involved in initial selection of equipment to decisions on the methods of utilization or adaptation during say the commissioning or posthandover stages. Finally, there are the broad organizational and resource issues – particularly financial – which are likely to regulate the overall approach to change. These broad sets of factors provide an organizing framework for the latter three parts of this book (parts 4, 5 and 6).

However, the identification of a general framework for conceptualizing the implementation process raises the important and fundamental issue of whether it is ultimately possible to try and understand the issues and the nature of problems at such a level. It may be felt that the factors involved in specific areas of manufacturing and within specific workplaces are at a level of uniqueness that makes this impossible to achieve meaningfully. Certainly it is important to be aware of the potential significance of such factors. At a general level one can do no more than to point to their possible location and potential effects.

Thus in the study of Marks and Spencer (Braham, reading 4.1) there are unusual elements both in the technological orientation of the policies adopted by M & S as a customer and in the broader elements of the symbiotic relationship between M & S and its suppliers. But a number of broadly valid lessons and conclusions can be drawn from the example and these may be relevant to the specific circumstances of others. Yet pleas of 'uniqueness' should surely be resisted. In most cases genuine elements of uniqueness are likely to be significantly outweighed by the factors that are, more or less, shared with other firms. Additionally elements of supposed uniqueness often do not bear close scrutiny and may be found to have elements in common with other examples. Thus, the example of M & S has a considerable amount in common with the relationship between the National Coal Board and its suppliers as described by Townsend (1980), and with other close customer–supplier relationships. Reading 2.1 includes discussion of the important relationship of Babcock Power Ltd with the CEGB. Braun takes up the issue of the diversity of circumstances of change (reading 3.2) but regards it, as we do, as nonetheless part of a common phenomenon, which he terms manufacturing innovation. However, he emphasizes that it is difficult to understand the problems 'without some form of theoretical framework for analysis' (p.88). On the other hand it is still important to be aware of issues which arise rarely or uniquely since these sources of genuine uniqueness and abnormality can be highly instructive.

Differences in the precise location of concern and in the use of terminology between the approach of Braun and that which we outlined above should not obscure similarities of approach or purpose. Like him we are concerned to try and establish which factors and circumstances may lead both to decisions to adopt new technology and to successes and failures in doing so.

In this pursuit Braun, like Macdonald (reading 2.2) and others (e.g. Gold 1980, and Wilkinson 1983), is critical of much work on innovation. This often focuses on the events leading up to the first commercial application of an innovation and has tended to ignore subsequent innovativeness within manufacturing organizations whose effort at improvement and adaptation to their specific needs 'constitutes a system of continuously changing potentials and limitations' (Gold 1980). As we pointed out above, the nature of the environment of implementation is important and perhaps above all else its importance lies in fostering or limiting the extent of innovativeness within organizations. But what are the elements of this environment?

National Manufacturing Performance

In part they include factors we referred to above such as the general political environment and whether or not it supports innovativeness at firm level. In addition, factors within the financial system are likely to be of critical importance – in reading 5.4 for example Hayes and Garvin relate the preferred systems of financial appraisal in the USA to declining capital investment and R & D investment. The same analysis can be applied to the UK. However in considering the particular problems of the United Kingdom, it is arguable that the most significant element of the environment of technological change is located at a very general and fundamental level.

Briefly, it is most obviously reflected in comparative trade performance, a matter not only of price competition but product quality and other nonprice factors (Pavitt 1980). There has been a marked rise in the percentage of the domestic market accounted for by imports – now about one-third of sales compared with one-sixth in 1970. In the world market the British share has fallen from approximately 25 per cent in 1950 to about 9 per cent in the early 1980s, a change that is in some contrast to the stable or rising pattern of the other EEC industrial countries and Japan. Perhaps most disturbingly for the long term, those shares of the world market that have been retained have increasingly been of less technically sophisticated products: 'although British manufacturers specialized in broadly the right product lines, they received a low price per ton or per unit exported; technically inferior products have to be sold at the bottom end of the market' – which, it should be added, is also the slower growing end of the market (Williams, *et al.* 1983, p. 14). While there are some notable exceptions to this, as a tendency it has long been observable. Writing of the mid-nineteenth century, Hobsbawn refers to 'the traditional "underdeveloped" slant of the British economy … a steady flight from the modern, resistant and competitive market' (1968, pp. 145–6).

The possession of the Empire is undoubtedly a strong factor in explaining this 'underdeveloped' slant for not only did it provide protected markets but it contributed to the comparatively large level of GNP devoted to expenditure on armaments which also helps to explain the heavy bias in R & D expenditure. R & D expenditure in Britain was relatively high until fairly recently but in real terms is now diminishing. It has been and remains concentrated in sectors such as aircraft, aerospace research and defence-related electronics. Although academically prestigious, R & D expenditure has not often offered significant commercial returns. This contrasts with the USA which while having a similar R & D concentration, experiences a greater commercial yield because of larger market size and is possibly supported by a greater propensity to innovate. The other EEC industrial nations appear to have deployed R & D expenditure more effectively by greater concentration on areas such as chemicals, electrical machinery and instruments which offered better commercial returns (Channon 1973). But the Imperial legacy alone does not provide an adequate explanation of Britain's indifferent manufacturing performance.

Low Level of Innovation

The range of possible reasons is considerable and cannot be fully considered here but some stand out in relation to our particular concerns. The first of these, contrasting with levels of R & D expenditure, is the apparently low propensity to innovate which has been evident since at least the mid-nineteenth century. Concern at the implications of the slow rate of application of scientific knowledge to industry was expressed as early as 1835 by Richard Cobden (Mant 1979). Walker (1980) has shown how the UK increasingly lagged behind the Federal Republic of Germany and the USA in exploiting innovations even, in some cases, where these originated in Britain. 'The gloomy picture that emerges from the UK's economic performance after 1883 is by all standards rather familiar: a low ratio of domestic investment with, as a consequence, a declining rate of growth of productivity, slow growth of exports with rapid growth of imports of manufactures, and slow adoption of new technologies relative to competitor countries' (p. 170). This was particularly evident in the newer industries of the period such as chemicals or many branches of electrical engineering.

That investment in these industries was often both limited and late was not a consequence of lack of capital for investment although the owners of capital in the UK were evidently more ready than most to look for more profitable opportunities in the world outside. That many entrepreneurs did not meet the challenge of new industrial competitors and took refuge in the protected but less sophisticated markets of the Empire may be felt to be ultimately a reflection of characteristics within the work force as a whole.

Work Force

For many people, reference to the work force is likely to conjure up first an image of a work force that is manual, and then one of a group that is overpaid (as recent Government statements emphasize), strike prone and resistant to change. The 'work force' of course, embraces a much wider group of the population but the persistence of the caricature, despite all the prevailing evidence, probably points to one of the underlying problems – i.e. managerial beliefs about the nature of work force behaviour rather than of inherent characteristics of the work force. Apart from the highly publicized and often atypical examples that fuel the mythology, the general experience and research does point to a level of potential work force adaptability and flexibility both in the normal processes of work and where change is a prospect. Failure to realize this potential is more likely to be a matter of management approach to both the importance of consensus-based change and to the choices offered by new technologies. This is emphasized by case studies such as those of readings 2.1, 6.1, 6.2, 6.3 and 6.4. It is also borne out by what appears to be the rather different experience of Japanese and other multinational companies using the British work force. The point is well put by Mant (1979) when he suggests that there is either 'something quite distinctive about the foreign style and practices which liberates something special in British employees, and/or there is some special population of Britons drawn to the foreign based multinational in the first place' (p. 83). Thus, as Channon concludes in his study of enterprise structure and strategy, 'in large measure management got the industrial relations it deserved' (1973, p. 227).

The issue of strike proneness is more susceptible to statistical investigation and does not stand up to scrutiny outside a few highly concentrated areas, particularly when international comparison is made. Similarly, UK labour costs have been shown by regular surveys undertaken by the Swedish Employers Confederation, to be in the middle to lower end of a distribution that also includes Japan and the countries of North America and Western Europe. When social changes paid by the employer are included, providing figures of total hourly labour costs, Britain is consistently at or close to the bottom of the distribution. Thus by the 1970s Britain had become and remains 'a country of "cheap labour" within this group of advanced countries' (NIESR 1984). Yet cheapness of labour does not translate into low labour costs per unit of production, for levels of UK labour productivity are markedly lower than those of the USA for example (Smith *et al.* 1982).

There is a possible link between low costs of labour and low labour productivity because the latter reduces the returns on investment in labour-saving technology. Such evidence as there is (e.g. Caves *et al.* 1968; Blackaby 1979) indicates that investment per employee is lower in the UK than in most other industrialized countries (see reading 4.4 which considers the example of the television industry). This again is not a new phenomenon. Habakkuk (1962) has examined the much greater readiness of nineteenth-century American employers to invest in labour-saving technologies and relates this to factor costs. But Walker (1980) points out that there was a similar readiness to adopt labour-saving technologies in Continental Europe even though labour costs were generally lower.

Work Force Skills

In part, Walker relates this to the low level of skill in the UK work force and to management responsiveness to new opportunities. As a consequence of its pioneering role in industrialization, Britain in the early nineteenth century had a considerable reservoir of skill and experience at all levels in the work force. Much of this advantage was lost as other countries industrialized and Britain failed to follow their lead in the development of a formalized system of technical and other education. Not only was Britain late in following suit but 'a stigma was attached to scientific and technical education – or for that matter any formal education with an industrial leaning – in sharp contrast with Europe and North America where it was in high demand' (Walker 1980, p. 25).

Albu (1980) has documented this in the case of professional engineers. While the British universities belatedly gave greater attention to their training, engineering has tended to remain a poorly paid and low-status occupation in comparison with other professional and managerial groups and in comparison with Europe. Within manufacturing and particularly in the context of new investment this has probably contributed to the UK's poor innovative performance. Production engineers, for example, often do not carry the weight that their counterparts in the Federal Republic of Germany or France are able to carry. This is a problem increasingly shared by the USA which also has manufacturing problems. Increasingly 'the production men' have been ignored (see Hayes and Abernathy, reading 4.3). Like Mant (1979) and others in Britain they see power as being increasingly exercised by generalist or 'professional' management, the first doctrine of which is 'that neither industry experience nor hands-on technological expertise counts for very much. It encourages the faithful to make decisions about technological matters simply as if they were adjuncts to finance or marketing decisions'. Yet while marketing and financial considerations must be taken into account, technological issues cannot 'be resolved with the same methodologies applied to these other fields' (p. 170).

The issue of skill within the work force extends, however, well-beyond the level of training of engineers. Prais (1981) has identified a large proportion of the British labour force who have no vocational qualifications at all – about two-thirds of the work force compared with one-third in the Federal Republic of Germany. Later work (Prais and Wagner 1983) suggests that the German work force gaining vocational qualifications did so 'at standards which are generally as high as, and on the whole a little higher than, those attained by the smaller proportion in Britain' (p. 63). Commenting on this shortage of intermediate-level skills in the UK population, Katrak (1982) draws attention to the work of Keesing (1968) on the ratios of skilled to unskilled workers in a working population. Applied to Britain this would predict that 'the UK's exports to the advanced industrialized countries would be intensive in unskilled labour while its imports from those countries would be intensive in skilled labour' (Katrak p. 39). During the period 1968–78 he found clear evidence of an adverse shift in this direction. This tends to bear out what he refers to as a 'low technology syndrome' in Britain which 'has been a less successful innovator, in relation ᵗᴊ aι ieast some of its major foreign competitors' (Katrak, p. 39).

Taken together, factors such as those referred to above present a somewhat dismal picture and it is worth emphasizing that nonetheless, many firms do succeed despite them. But in general, the picture is still well summarized by Pavitt's analogy (1980) of industrial Britain perceiving itself to be a First Division team while in reality it has been relegated to the Second Division in which it 'will increasingly be challenged by the newly industrializing countries recently promoted from the Third Division' (p. 13). Factors such as those we have referred to, represent a considerable hurdle to be overcome in responding to this challenge, let alone seeking restoration to the First Division. Yet they prompt the further question of why it is that the UK in particular has experienced such problems. As we suggested above, the legacy of the Empire must provide part of any underlying explanation since it apparently touches so many issues, from the export of capital to the location of R & D and the historic pattern of trade.

Also at a fundamental level, and possibly related to the impact of the Empire, is the nature of British culture. The association of cultural factors with industrial performance is one that many find hard to accept yet it is an explanation that in another context, that of Japan, seems to be readily accepted. Various writers such as Hobsbawn (1968), Lewis (1978) and Wiener (1981) have in different ways pointed to what are in effect anti-industrial values within the prevailing pattern of beliefs and values. While lip service is often paid in political circles to the need for economic progress and technological change, writers like those referred to above have compellingly demonstrated that in most facets of society, educational – whether school or university – financial, administrative and managerial, the prevailing values do not generally favour activities associated with production. The outcome has been 'the spectacle (not necessarily all for the bad) of an industrial society led by men with "mind forg'd manacles" restraining their concepts and their actions' (Wiener, p. 10). A reflection of this is provided by Bessant and Grunt's (1985) comparison of manufacturing innovation in Britain and the Federal Republic of Germany which found that it was in cultural factors that the main differences between the two countries could be found. 'Whereas the German culture encourages many of the characteristics of technical progressiveness, the dominant UK culture tends to inhibit their development and to emphasize the stable/ short-term view.' Against this background it is perhaps not surprising that Dahrendorf was led to suggest that 'an effective economic strategy for Britain will probably have to begin in the cultural sphere' (quoted in Wiener p. 4).

We have considered these issues at some length in the introduction because they emphasize the underlying nature of some of the problems and the scale of the difficulties that have to be overcome if performance in implementation and manufacturing innovation is to be improved. Ultimately some of the issues are well outside the parameters of the individual production unit but they are, nonetheless, of great and direct relevance. What they imply is that 'for managers in the UK to be technically progressive, they must, to some extent swim against the cultural tide' (Bessant and Grunt 1985). This may require attention to some of the more technical issues of management and control of projects which, in part, are dealt with in our Open University course. But between these micro issues and the 'mega–macro' ones we have referred to above, there is a broad area of approach and conceptualization of the issues that needs to be considered. It is this area of identification of opportunities and choices, of the factors shaping both technological decision making and the subsequent attempts to realize and/or modify these decisions, that we concentrate on in this reader.

References

Albu, A., 1980 British attitudes to engineering education: a historical perspective, in Pavitt 1980

Bessant, J., Grunt, M., 1985 *Management and Manufacturing Innovation in the UK and West Germany*, Gower Press, Aldershot

Blackaby, F. (ed.), 1979 *De-industrialization*, Heinemann

Caves, R.E. and associates, 1968 *Britain's Economic Prospects*, Brookings, Washington DC

Channon, D.F., 1973 *The Strategy and Structure of British Enterprise*, Macmillan

Dahrendorf, R., 1976 Europe: Some are more equal, *Listener*, October 14

Gold, B., 1980 On the adoption of technological decision making in industry – superficial models and complex decision processes. *Omega*, Vol. 8, No. 5

Habakkuk, H.J., 1962 *American and British Technology in the 19th Century*, Cambridge UP

Hobsbawn, E., 1968 *Industry and Empire*, Pelican

Hussey, D.E., 1984 Strategic management: Lessons from success and failure, *Long Range Planning*, Vol. 17, No. 1

Katrak, H., 1982 Labour skills, R & D and capital requirements in the international trade and investment of the United Kingdom, 1968–78, *NIESR Economic Review*, No. 101

Keesing, D.B., 1968 Labor skills and the structure of trade in manufactures, in Kenan, P.B., Lawrence, R., (eds.), *The Open Economy*, Columbia UP

Lewis, W.A., 1978 *Growth and Fluctuations, 1870–1913*, Allen and Unwin

Mant, A., 1979 *The Rise and Fall of the British Manager*, Pan Books

Nasbeth, L., Ray, G.F., 1974 *The Diffusion of Process Innovations*, NIESR

NIESR, 1984 Industrial labour costs, 1971–83, *Economic Review*, No. 110

Pavitt, K., ed. 1980 *Technical Innovation and British Economic Performance*, Macmillan

Prais, S.J., 1981 Vocational qualifications of the labour force in Britain and Germany, *NIESR Economic Review*, No. 98

Prais, S.J., Wagner, K., 1983 Some practical aspects of human capital investment: training standards in five occupations in Britain and Germany. *NIESR Economic Review*, No. 105

Ray, G.F., 1983 The diffusion of mature technologies, *NIESR Economic Review*, No. 106

Smith, A.D., Hitchens, D., Davies, S.W., 1982 International industrial productivity: A comparison of Britain, America and Germany. NIESR *Economic Review*, No. 101.

Teubal, M., 1983 The accumulation of intangibles by high-technology firms, in Macdonald, S., McLamberton, D., Manderville, T., *The Trouble with Technology*, Francis Pinter, London

Townsend, J., 1980 Innovation in coal-mining machinery: the case of the Anderton Shearer Loader, in Pavitt 1980

Walker, W.B., 1980 Britain's industrial performance, 1850–1950: a failure to adjust, in Pavitt 1980

Wiener, M.J., 1981 *English Culture and the Decline of the Industrial Spirit, 1850–1980*, Cambridge UP

Wilkinson, B., 1983 *The Shopfloor Politics of New Technology*, Heinemann, London

Williams, K., Williams, J., Thomas, D., 1983 *Why are the British Bad at Manufacturing*, Routledge and Kegan Paul

PART 2
The Problematic Nature of Technological Change

2.0

Introduction

The purpose of the readings in Part 2 is principally to introduce important concepts of technological change and technological innovation. The case study Exercise '81 (reading 2.1) raises related themes Indeed, it also raises a broad range of other relevant issues, demonstrating the complex interrelationships between technical, industrial relations, organizational and financial factors that may be involved in implementing new technologies. It also indicates the very long time scales that may be involved in the achievement of major technological change. In the case of Babcock Power Ltd., the timescale is likely to extend over some 15 years from the initiation of the first stage – Exercise '81 in 1978. However, the breadth of change in this example raises the question of which elements of Exercise '81 are technological and which are simply matters of organizational development.

The answer is pointed to by Macdonald (in reading 2.2) who argues that we tend to perceive technology in a narrow way that is essentially rooted in machines, with a consequent lack of understanding of technological processes. He argues for a much broader view of technology which does not necessarily include machines but which considers the body of knowledge which is compulsory for 'getting things done'. This directs our attention towards some of the more nebulous aspects of technological change such as relationships between technical experts and commercial experts. Because of their 'difficult' nature, these aspects are often ignored despite their possible significance for success or failure. This, as Macdonald says, also indicates that the entire environment of technological change needs careful examination since many of the less tangible factors are ultimately located in the product market and elsewhere.

The reading also considers the nature of innovation as a process which is often treated as being linear, from invention to application by the firm, with descending levels of prestige and original contribution as one progresses through. Macdonald criticizes such treatment as being in sharp contrast to reality where much of the innovative effort is to be found at the level of the individual production unit. This issue is also taken up by Sahal in reading 2.3 although his concern is primarily with the general process of innovation from initial invention through to the first commercial application. With the development of scientific method and with the formalization of research and development work within companies and elsewhere, this is often assumed to now be 'a rational process subject to prediction, regulation, and control'. Drawing upon specific examples and studies of various invention–innovation processes he demonstrates that they remain highly dependent upon chance factors and that inventions may be separated from innovations by lengthy but varied periods of time. This underlines the importance of recognizing the multiple causes that may shape both research and development. Multi-causality and the element of chance suggest that R & D policy should be approached in terms of seeking selectivity and distrusting it. Sahal also suggests that there should be a more even balance between the element of research and that of development – a somewhat 'unholy' view that runs counter to established cultural and educational values.

Sahal also looks at the concept of 'long waves' or Kondratieff cycles and their relationship to what he terms generic innovations such as steam power or electrical power which 'not only lay the basis of new endeavours but also cut across several new and old industries'. These give rise to a subsequent clustering of interrelated innovations which, he says, are an important element in the cyclical nature of economic change. This issue is also considered by Godet in reading 2.4 although he approaches it from a very different direction. Godet, who is Scientific Advisor to the French Ministry of Industry and Research, criticizes the notion that new technologies by themselves will get us out of the economic depression and suggests that we should beware of information technology given the possibilities it holds for communication 'Concordes' in the future. He asserts that technical and economic change is taking place faster than social and organizational change so that there is a crisis of socioinstitutional regulation. Therefore it is the 'social' rules of the game that must be examined and adapted if change is to be accommodated without catastrophes. He does not mean by this that technical change should be slowed, but that it should be subject to increasing scrutiny, related to carefully considered needs and should be 'mastered, stimulated and oriented in a direction which will prove more appropriate to social aspirations and real economic constraints'. At the level of application to industrial processes, this means examination of the social conditions involved in the installation of new technology – an issue taken up in some of the readings in the final part.

2.1

Exercise '81: The Introduction of New Manufacturing Technologies into Babcock Power Ltd

David Wield

Introduction

This case study focuses on major changes in a large company whose main product is steam generating equipment for electricity production. The study begins by situating the company in the context of the British steam generating industry and its markets. It goes on to describe Exercise '81, a cost-reduction programme which followed a major £20 million investment in new equipment. Its principal corporate goal was, given the large decline in the home market, to cut costs by 25 per cent to make the company internationally competitive in an increasingly cut-throat market environment. This included a major reorganization of working practices in the company. Babcock management set out with a view that management and unions would need to identify a joint goal if changes radical enough to sustain Babcock's competitiveness were to be achieved. The case study describes senior management perceptions of, and actions to gain acceptance of, the changes in all sections of the work force situated in the highly industrialized and unionized Clydeside. Technical, operational and financial aspects of Exercise '81 will also be detailed and evaluated. Finally, events will be summarized since the implementation of this new technology.

Company History and Background

Babcock Power Ltd is a large subsidiary of the British-based manufacturing conglomerate, Babcock International plc. The company (prior to 1902 a part of Babcock and Wilcox of the USA) has a long history of boiler production. The first Babcock and Wilcox boiler was patented in 1867. In 1882 the first power station in the world to give a public supply of electricity was equipped with Babcock and Wilcox boilers at Holborn Viaduct.

Babcock Power now manufactures steam raising equipment fired both by fossil and nuclear fuels, pressure vessels for process plant, and undertakes fabrication and machining of components for the general engineering industry. It has continued to be profitable through the current recession, bringing in 38 per cent of Babcock International's total profits of £38 million in 1983.

Babcock and Wilcox began production at its Renfrew site in 1896. The site has been gradually extended since to occupy a 180 acre site, eight miles west of Glasgow. The Production Division is one of five operating divisions of Babcock Power, and Renfrew is its main manufacturing facility. Employment figures on site were 5,300 in mid-1976. At the time of Exercise '81 they were 3,500 and they had shrunk in early 1985 to 2,600. Output at Renfrew has been at about 20,000 tons per annum over this period. (Babcock finds output measurement by weight to be relatively accurate.) Annual turnover has been around £70 million.

The Technology of, and Market for, Generating Boilers

Babcock Power is now one of only two large manufacturers of steam raising equipment in Britain. The other is part of Northern Engineering Industries (NEI) at Gateshead (previously part of the Clarke–Chapman group). Babcock is the bigger of the two. Babcock's principal market until the late 1970s was the home electricity generating Boards' coal-fired power-station boilers and associated machinery such as coal-handling equipment and pipework, leading up to Drax 'B', the last 2640 MW coal-fired station to be ordered by the CEGB in 1978. Babcock also produced similar sized oil-fired boilers for Grain and Peterhead in 1973, and has made gas-cooled nuclear reactor boilers since the early 1950s. Babcock's research design and manufacturing development was thus strongly influenced by the CEGB, which generated seven-eighths of the UK electricity demand, and by the domestic market (Smith *et al*. 1984).

Boiler manufacture involves the cutting, bending and welding of metals to very high specifications (see figure 1). Since many of the products are large and heavy, space and material handling are important constraints on manufacture. Boiler size and steam pressure have significantly increased since the 1950s through a combination of increased thickness and strength of steels, better machining and welding technologies and improved nondestructive quality control systems like X-ray and ultrasonic testing systems. In the 1950s and 1960s the capacity of generating sets grew from 100 MW to 660 MW. The largest power stations have three or four generating sets, each with a boiler and turbine generator. Thus considerable advances in manufacturing and erection procedures were needed.

More recently there has been a concentration on reducing capital cost and total project time by shortening manufacturing time cycles and simplifying commissioning procedures. Materials' utilization has also been substantially reduced, lowering weight and cost. Exercise '81 was an attempt by Babock to cut costs by 25 per cent to allow it to compete more

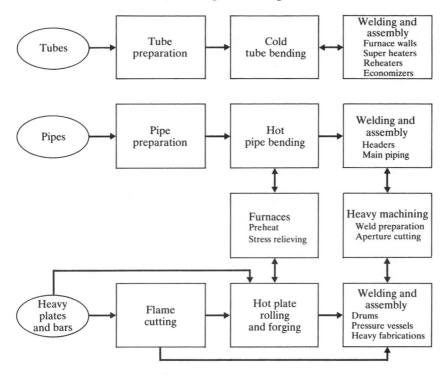

Figure 1   The boiler-making production process
Source: *CPRS 1977*

successfully internationally. Although not all of the advantages could be applied in a British power station, Hong Kong's Castle Peak 'B' 660 MW boiler, ordered in 1981, cost only about 70 per cent of the cost of Drax 'B', the last coal-fired boiler ordered by the CEGB in 1978.

Until the late 1970s Babcock was highly dependent on the home generating boards, particularly the CEGB, for its business. As various other companies supplying large utilities have found (British Airways, the National Coal Board and British Telecom), dependence on a single customer can be extremely dangerous. In the case of CEGB and the power industry this was exacerbated by a number of factors. Overestimation of the growth of electricity demand led to a heavy power station ordering programme in the early 1960s.

Orders came unevenly even in the good years. CEGB has had a strong influence on design, quality assurance and delivery monitoring. The Board is said to have greater influence than any other utility in Europe and North America on its suppliers, both before and after the development of a firm project specification. The Board sets up standards with very wide scope, covering management control systems as well as material and manufacturing quality control. On-site inspection involves 'almost constant presence on the suppliers premises .... In the design field it has been suggested that

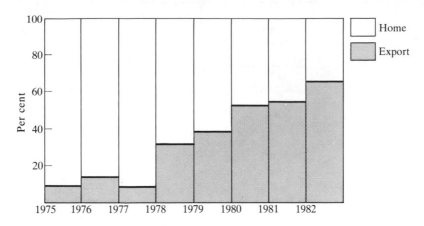

Figure 2   Distribution of work at Babcock Powers' Renfrew factory
Source: *McIntyre* et al. *1983*

the "special" relationship the Board enjoys with the industry has led in the past to a great deal of technical change after manufacturing has been initiated. The potentially disruptive impact of such change is considerable, and is not limited to CEGB work only, but can affect any work in progress at that time, including export orders' (CPRS, p. 30–1).

By the late 1960s the home generating Board's order rate had dropped from about 7,000 MW (7 GW) a year to 2,000 MW (2 GW) a year and to even less in the 1970s as overcapacity reached 35 per cent on top of the 20 per cent margin needed for repairs, maintenance and unexpected cold spells. The power industry began to tender more aggressively for export orders. About 20 per cent of the boiler makers orders were for export in the mid-1970s.

Since then, Babcock has successfully obtained large orders in such 'traditional' markets for British companies as Hong Kong, South Africa, Zimbabwe and India, pulling up Babcock's export work load from 10 per cent to 65 per cent between 1975 and 1982 (see figure 2). But competition is intensifying as domestic markets are shrinking in the countries where the world boiler manufacturers are located. In the early 1980s there were reckoned to be at least 12 major boiler makers worldwide based in Britain, the US, Japan, France and Germany, and another ten or so were able to obtain at least some export orders. At the same time there has been a trend for some countries to increase local manufacture while also increasing their capacity to make large equipment. As a result, the export market has also become increasingly competitive. By the late 1970s Babcock knew that to continue in business until more orders were forthcoming to rebuild British power stations, the company would need to substantially improve productivity and cut operating costs.

An Unsuccessful Merger, and an Investment Plan

It was the policy of industrial rationalization in the mid-1970s that produced the germ of the idea for modernizing the Renfrew plant.

In 1977 the Government's Central Policy Review Staff published an investigation of boiler makers and turbine makers for the electrical supply industry. It recommended merging the two boiler makers, Babcock and Clarke Chapman, and the two turbine makers, GEC and Parsons. Babcock and Chapman began plans to merge, getting as far as appointing a Chief Executive designate from the nuclear industry. It was in the context of the merger that he encouraged an upgrading and modernization of manufacturing on the Renfrew site. Babcock Power management drew up a three-phase plan to upgrade the site. The first and second phase were to upgrade without significantly increasing capacity. The third phase was to increase capacity.

The three-phase plan was to gradually replace all the old buildings by three-bay production units, each unit becoming a semi-autonomous factory having separate accountability for quality, delivery and costs. The first phase was to build one such block – the machine and assembly factory. The second phase (which is under discussion for completion in 1988 at the time of writing in early 1985) will increase the heavy fabrication unit. The final phase would substantially increase production capacity in the event of orders to rebuild British power stations.

The overall plan was to provide 'total' manufacturing flexibility meaning that any bay of any shop could be used for any purpose. Babcock had bays of all sorts of sizes, strengths and ages. In the recent past there had been many examples of constraints on manufacture – of doors not being big enough to get completed jobs out, and of only having one bay suitable for a particular job and finding it already being used for another purpose. In phase 1, for example, the structure of the new block allows the two biggest 100 ton cranes to be moved for use in any of the three bays, so that around 200 tons can be lifted at once.

The alternative to the plan was seen by management as accepting gradual closure of older parts of the Renfrew works, including the outdated machine shop. It was felt that this would further reduce the competitiveness of the rest of the works. It would thus be unlikely that a viable manufacturing unit would exist to meet future demand for home power stations.

In the event the proposed merger between Babcock and Clarke Chapman didn't go ahead. Serious problems arose between GEC and Parsons. GEC, much the larger of the two turbine makers, insisted on a 'slimming down' at Parsons which was unacceptable to Parsons' management and work force. In early 1978 all four companies were involved in complex merger negotiations at the same time as large public campaigns were underway to get the last big CEGB power station order, that for Drax 'B'. Clarke Chapman joined with Parsons to form Northern Engineering Industries and eventually pulled out of merger negotiations with Babcock. Successful in its joint management–union campaign to bring the early ordering of Drax 'B', Babcock went ahead with its modernization plans. Since then Babcock has worked for GEC in its largest turnkey orders. In the mid-1970s all the major turbine makers except the British had significant experience in carrying out turnkey orders. In general it is the turbine manufacturer that acts as main turnkey contractor. In a turnkey the customer contracts with a single supplier who is responsible for the provision and commissioning of the complete power station. About half of all power station export contracts in open markets have recently been negotiated on a turnkey basis. GEC now has considerable turnkey expertise, subcontracting boiler manufacture to Babcock at Castle Peak in Hong Kong, and Balco in India, but has also worked with other boiler makers.

The Plan for Phase One of the Modernization Programme

The decision to include, in the first phase, a new general machine shop resulted from the knowledge that this would have the greatest effect on

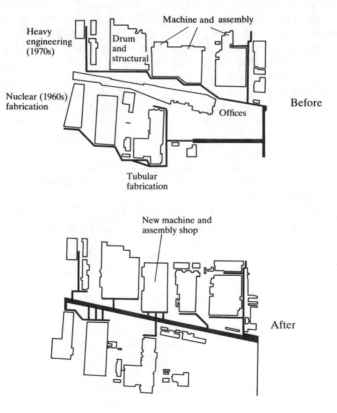

Figure 3   The Renfrew factory before and after Phase One
Source: *McIntyre* et al. *1983*

business as a whole (see figure 3). It was the oldest part of the factory, dating back to 1895, and each of the other factories on the site were to a degree dependent on its performance. Upgrading of the machine shop was on both counts a top priority to improve the sites' overall performance and thus generate funds for future phases of modernization.

The new building of 180,000 square feet would be divided into three bays. One bay would house the general machine shop and the other two would be used for fabrication and assembly of products with a high proportion of machined components.

The plan had three elements:

(1) updating the plant and working practices by capital investment;
(2) improving the effectiveness of management systems;
(3) increasing the effectiveness of people.

The plan was combined with a cost reduction programme, called Exercise '81, to reduce Production Division costs by 25 per cent below those of 1979. This, with similar exercises in the other two divisions of Babcock Power, it was hoped would be the key to improving the company's place in the highly competitive overseas market.

The Use of Consultants

Babcock decided early on that they should bring in production engineering consultants Ingersoll Engineers. Babcock did not have expertise in some areas of production engineering, particularly in the area of computer numerically controlled machine tools. Ingersoll had strong expertise in this area which supplemented their skills and backed-up Babcock's case for project approval to its Board of Directors.

But equally Babcock were concerned not to lose ownership of the project. Babcock wanted its own engineers to 'own' their ideas, and not be in the position to start blaming the consultants if anything went wrong. So they restricted Ingersoll to production engineering, and to the first capital investment element of the plan. But Babcock are clear that they could not have done without Ingersoll even though they did have differences. For example, Ingersoll thought that Babcock's ceiling for the machine shop had too much height. But Babcock felt that their shops had been too custom built in the past and needed to be more flexible for adaptation to equipment and tasks in decades to come, when no-one knew what would be produced in them. Babcock management point to a 100 ton crane in this shop that has already allowed them to do things not thought of at the design stage.

Team Composition

A special project team for the investment programme consisting entirely of in-house senior management was formed at the beginning. The team worked together so that it was in control and owned its own ideas. The project itself was proposed between May 1978 and October 1978. At the beginning in May only three or four people knew about it. The team was formed in Spring 1979 and remained together until the end of the project in late 1981. It was chaired by the Managing Director and consisted of seven other managers. The Ingersoll consultants working full time reported on a day to day basis to the production engineering manager who was the only manager working full time on this project. At higher levels they reported up to the Managing Director. The project team met weekly or fortnightly to keep the project under control.

The Content of the Investment Programme 1980–81

The total investment was £20 million over two years: £8 million on plant, £8 million on the new building. A further £4 million was spent on computing facilities, computer-aided drafting, a welding development facility and a new central road. The funds came from retained profits and a government grant. Investment was around 15 per cent of turnover per annum.

The relatively high cost of the building was because it might be in continuous use for 50 to 75 years and thus would be required to house several generations of plant and equipment. For example, to achieve maximum availability of floor space for production most of the fixed plant was located in the machine shop to provide flexibility in the other bays. Offices, stores and service functions were housed in annexes. To minimize energy costs, insulated cladding and semi-automatic heating and lighting systems were installed.

Three major factors were established when considering the machine tool requirements:

(1) Components for current and projected markets indicated a need for a wide range of dimensions and batch sizes.
(2) Components in the current and projected work mix could be grouped into families.
(3) The capacity of the new facility required the potential of doubling that of the existing machine stop.

Two separate investigations were carried out. One by a team of Babcock production engineers and the other by Ingersoll Engineers. Each team assessed the machine tools needed to service the range and volume of components and the comparison of the assessments indicated similar requirements. Differences were reviewed and consultation led to agreement on the size, type and numbers of machines. It was then agreed that there was a requirement for 64 machines, 16 of which would be computer numerically controlled (CNC), ranging from a £100,000 bar lathe to a £1.6 million machining centre capable of boring, milling and drilling products up to 7m diameter, 5m high, up to a maximum weight of 125 tons with eight axis CNC control. The total value of CNC machines was to be in excess of £5 million. One clear principle was to standarize machinery suppliers as far as possible. A policy was adopted to purchase a range of machines from one supplier to minimize the types of systems, to standardize programming and to facilitate maintenance and skilled staff movement from machine to machine. Sources for machine tools were investigated and prospective suppliers were required to identify reference installations, carry out machining tests and demonstrate training and after-sales service capabilities. Babcock also decided to go for well-tried machinery and not for development machines with the associated risk of commissioning delays.

A three-dimensional scale model was used in developing the shop layout. This took seven months. There were daily half-hour sessions with

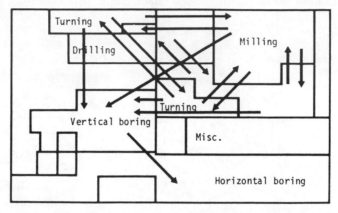

As it was: Babcock machine shop layout. No foreman had more than 30 per cent of responsibility for parts, control was difficult, lead times were 3 to 6 weeks in theory (far more in practice), and workload suffered peaks and troughs. Work movement pattern appears most awkward.

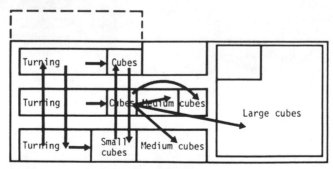

As it is: Renfrew's new machines rearranged in new factory permit a logical movement of components in production. Foremen now have responsibility for some 98 per cent of part-work content. Control is better but parts/documentation still moves to different supervisors/areas. Improvements on this layout mean that the 'production centre' idea will soon be working, also that parts movement is much less and that lead times can be reduced. Nonetheless the machine-shop still suffers from peaks and troughs – which Babcock's positive move into subcontracting should be able to even out.

Figure 4   Changes to machine shop layout
Source: *Holland 1982*

all the project team discussing possible changes in layout using a 'lego' type three-dimensional scale model. Discussions were held weekly with the shop management. Union representatives and operators were encouraged to view the model and comment on the proposals. The decision in favour of a production centre layout (a grouping of different types of machines) rather than a functional layout (based upon grouping of similar type machines), was the result of many weeks of analysis and debate. A key factor emerging from the analysis was that more than 90 per cent of each range of components could be controlled by a single supervisor (foreman) from the raw material to the finished state. Each supervisor would be responsible for a group of machines comprising lathes, grinding, drilling and milling machines. They would be grouped so that 90 per cent of components coming into each production centre could be finished there. Figure 4 shows details of the layout change. The products were grouped

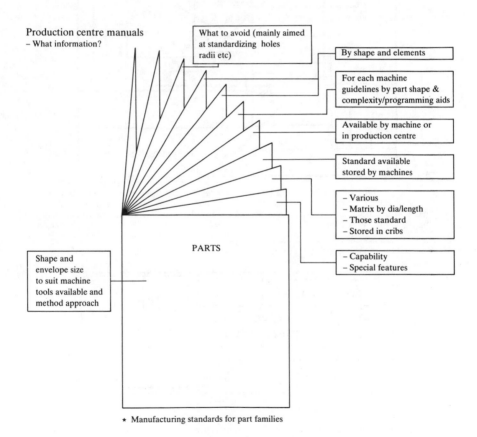

Figure 5   Example of new production organization and control
Source: *Holland 1982*

under six headings. First turned parts: small, medium and large; second, cube-shaped parts: small, medium and large. Babcock was producing over 600 varieties of components and these were rationalized into families for production in specified production centres. Manuals were produced showing machine capability, tooling available and production time indications for parts (see figure 5).

This approach helped to rationalize machine tool use. Seventy machines were installed in comparison with the old shops' 190. Thirty machines were new and 17 of those had CNC. The final decision on machinery was to buy the CNC lathes from Swedturn (a Swedish company), the milling machines from Fritz Werner (Federal Republic of Germany), boring machines from Giddings and Lewis Fraser (Britain) and a Scheiss-Froriep machining centre at an installed cost of £2 million (from the Federal Republic of Germany).

The introduction of high performance machine tools would reduce cycle times and therefore improved materials handling would be necessary. The requirements were analysed using typical parts on existing machines. Relating these to the new shop and applying the anticipated productivity improvements, component movement patterns were established which led to decisions on overhead cranes and forklift trucks. Mechanical lifting would be required for 40 per cent of the components and various types of jibs were selected after technical and user evaluation.

'Spindles cutting' – targets for production centres			
PC1	Turning machines	– If bar work	– 70-80%
		– If chucking work	– 50-70%
	Others	– Occasionally (end and centering) (mill and drill)	
PC2	Turning machines	– Chucking work	– 70%
	Others	– Occasionally (mill and drill)	
PC3	If slave tables being used		– 70%
	If work not suitable for slave tables		– 50%
PC4			– 80-90%
PC5	Borers and machining centres		– 80-90%
PC6			– 80-90%

Figure 6   'Spindle management' targets
Source: *Holland 1982*

Automatic swarf removal would help minimize operator involvement in nonproductive activities and the requirement for indirect operatives would be reduced by the installation of the jib cranes and overhead cranes with floor control. A kind of 'spindle management' was used to ensure that NC machines were cutting. Targets were set (see figure 6) for machine cutting time. The idea was that services and back-up (including raw materials, production control, labour rotation and back-up technical services) should be geared to keep spindles cutting. Audit sheets were prepared to monitor early difficulties (see figure 7). Four services which were essential to the effective operation of the new factory were housed within annexes of the building: quality control, a standards room, plant maintenance and a tool servicing and presetting area.

A policy of in-line inspection was adopted and inspection tables were located at each production centre. A standards room was provided with the necessary facilities for maintaining the accuracy of gauges and measuring equipment, and maintenance techniques such as condition monitoring and electronic diagnostics were introduced.

It was known that the number of CNC machines being introduced over a relatively short period of time would create problems especially since the company had no previous experience of CNC. To minimize the problems, CNC machines not requiring special foundations were ordered early and installed in the old shop six months before they were needed in the new shop. The supervisors and workers used the six months to learn how to use and maintain them, the object being to extend the learning period and build confidence in the new technology.

The selection and training of supervisors, operators, maintenance personnel and programmers were recognized as being crucial to the success of the project. Strathclyde University and Paisley College of Technology were commissioned to provide appreciation and selection courses on CNC. The machine tool suppliers undertook training at their works and on-the-job training after machine installation.

It was a requirement that output be maintained during the transfer of production from the old to the new building and this needed careful and detailed planning. It was essential given the dependence of the whole works on a continuous supply of machine components, that the transition from the old to the new buildings was completed in four weeks which included a holiday period. This task involved the transfer and installation of all but the largest machines. These were built in situ over a more extended period of several months. The coordination of civil engineering works, services installation and delivery and commissioning of machine tools and cranes was an extremely complex task for which the project team undertook a form of network analysis. For example, one crucial part of the schedule was the insertion of the large machining centre. The big overhead crane had to be installed first and to do this the roof had to be installed on one part of the building. The time period leading up to the operation of the machining centre was altogether about nine months.

Financial Appraisal

Simultaneously with the costing of the physical aspects of the investment, Babcock managers began to firm up the work force and other recurrent costs. These costs were built up on a standard hour basis, based on direct labour, which includes fixed costs. A series of costings were built up based on different values for major variables.

To give a general example of the variables which could be changed, one

Machine production audit.– Fritz Werner Milling
Distribution of available hours

Good prod			Delays													Monitor				
Week No	Set-up	Machine (total E.P.)	Good prod sub total	Initial tape prove-out	Await programme	Tooling	Training directs	Await maintenance	Additional set-up	Await supervision	Await inspection	Await work study	Await material	Await transport	Await crane	Union	Miscellaneous	Delay sub total	Total available hours	% Non productive	Standard hours	Operator performance	Effective performance

Figure 7 Example of audit sheet compiled by foremen and supervisors at Babcock identifying reasons for low productivity
Source: Holland 1982

could build in three assumptions of the grant level from the Department of Trade and Industry (DTI), say 10 per cent, 20 per cent or 30 per cent. One could also estimate the fixed and variable costs relating to three assumptions of output as a proportion of installed capacity, say 50 per cent, 60 per cent and 70 per cent. This produced a set of alternative calculations, profit, funding requirements and so on, with resulting estimates of return on investment. The Babcock management chose two of the most likely combinations of variables, and used a discounted cash flow (DCF) calculation as a basic indicator of the viability of the investment.

But this was linked with another important element. Careful costing of future expenditure was undertaken for various levels of output under cost headings like energy costs, rates, labour, training and supervision. These costs were used to build up initial budgets for managers. The final costings were used to monitor management performance in the first years of production. Therefore a firm budgetary control system was set up based on these agreed costings. The estimating process was in a sense a negotiating process from which commitments were gained from managers. 'If X made an estimate then one tried to pin them to that estimate as next year's target figure. They would back away if they weren't as confident as they tried to appear.' It was a kind of modelling. If the results weren't liked then the financial staff went back and looked at where the major problem areas appeared to lie and discussed them with the managers involved. For example, one could check how much electricity was actually used in a machine shop and from this check if estimates were out of line. So, for ongoing costs a 'what if' approach was used and a gradual refinement process which took six months. One looked at which parts of the equation gave the most dramatic effect. Further, on the benefits side it is again difficult to judge benefits on a narrow basis. One cannot judge benefits from the machine shop itself since the programme was a full site programme.

In terms of results on the basis of the whole site the figures accord more or less to the overall forecasts agreed but within a different scenario to that forecast. There has been a lower volume of production but at a lower cost than envisaged. Also, the investment was justified on the basis of a home generator market which has not materialized to the extent envisaged. But the new technology has allowed Babcock to do things that they did not know they would be able to do when the appraisals were made. These are discussed below.

Cash flow control was crucial. Money was saved by keeping careful control of communication from project team to finance division. The project team notified finance division well in advance when payments became due and when they could be delayed. Dates were juggled to find periods when suppliers would give bigger discounts. Some machinery was leased and some of the terms were particularly advantageous.

Management Systems

Management systems needed to be enhanced to deal with the new methods of work. The two major requirements were that any new system had to be developed and implemented before production was transferred to the new shop, and maximum use had to be made of existing systems. The systems developed were computer-aided process planning, on-line job recording and production control. All these were developed in-house.

Computer-aided process planning was an enhancement of an existing system. The new version produced a process plan covering all machining operations, and also a detail sheet giving speeds, feeds and tooling requirements for individual operations. Production control and on-line job recording were already being developed when the project was launched and could be applied to any manufacturing department in the works.

In addition a Kongsberg PC200 part programming centre with four interactive graphics screens was installed for programming the 16 CNC machines. A significant number of programmers were recruited from the shop floor. A computer-aided drafting system was also introduced which has reduced the lead time from receipt of an order to provision of manufacturing drawings.

Babcock management has found that implementing all these systems has significantly reduced work-in-progress, and has improved deliveries by reducing inter-operation times from one week to three days. There have also been reductions in the relative number of staff servicing the systems.

Work Force Issues

The Renfrew works is situated in the heart of Clydeside, which is a traditional stronghold of cohesive, strategic trade unionism and has often been the scene of intense industrial relations conflict. A piecework payment system was in use and much of the time of management and supervision was taken up by disputes on payment. In 1970 a progressive changeover to a measured day work system based on workstudy times was under way and by 1973 incentive payments had been completely phased out. This was contributed to by developing new relationships between management, unions and employees, based on more open styles of management involving strong consultation with the work force and with the long-established unions. In 1978 a plant-wide productivity bonus system was introduced which operated for six years.

In 1977, when a 30 per cent redundancy was inevitable given the slow down in the home generating Boards' ordering programme, the relationships between management, unions and employees had developed to a sufficient level that the redundancy was carried out without any union sanctions being applied. By that time, joint management and union projects were being introduced, the first being a campaign to bring about the early ordering of the Drax 'B' Power Station, necessary to avoid closure of the Renfrew works. The campaign was successful and demonstrated what could be achieved when management and unions have an objective they can share. This lesson has been learned with a series of joint objectives including Exercise '81, that we describe below.

When the planning of the new machine and assembly factory was started it was evident that this was a major project requiring the active support of all employees. The aim was for everyone on the site to have the same level of understanding as the most senior managers and union representatives. One medium of communication used – a radical step in 1978 – was in-house video. The first video programme set out management's plan and management's assessment of the key issues. It emphasized that there was no question of differential payments, a successful measured day work system having been negotiated with the unions. It stated the need for commitment and the need for change. The programme was shown to small groups over 2½ days in 13 viewing centres and 145 viewing sessions. The use of small groups made question and answer sessions possible after each showing. The sessions were crowded into a short period of time so that all employees would hear of the plans at about the same time. At the question and answer sessions over 1,000 reactions were noted down, gone through with the six union convenors, and summarized into nine significant issues. This was done in late January 1979 because the full commitment of the work force to change was needed before a final decision on investment.

Another video programme was then made on these nine issues, which included: equipment and services to be provided, buildings and layout, how natural wastage would be controlled, payment systems, job flexibility, markets and Japanese competition. The six convenors, with a senior manager as chairman of the group, now 'owned' these ideas. A news sheet was also produced to keep people informed, and a third video programme was produced in November 1979 to bring the work force up to date with the fact that a final decision to go ahead had been made involving commitments from management and unions.

The first two video programmes were very important in paving the way for the commitments which needed to be negotiated before modernization could take place. Before the final decision was made in November 1979, a written commitment from the unions was asked for and obtained to six issues of working practices (see figure 8).

The negotiations for these took about four months in mid-1979. One example of the new flexibility of work practices was that the jobs of separate crane operators with their assistants were eliminated. The operators on the machine shop floor could use cranes if they needed them.

Special consultative committees in addition to the previous consultative committees were also set up to implement the changes by representing all employees moving into the new machine shop. No senior shop stewards were involved in these committees because, by chance, there were no senior shop stewards in the machine and assembly sections. Examples of the types of issues they dealt with were: the working environment; what was needed by employees in terms of washrooms, change rooms and so on; and the acceptance of machinery. Employees were sent to look at the new machinery working in other factories.

Factory Rebuild—Phase 1

Memorandum of commitment

1 Unions are prepared to accept with Management the commitment to achieve a return on the investment during Phase 1, sufficient to finalise further investment at Renfrew site to enable the Company to remain competitive and therefore ensure future orders.

2 All parties accept the aim of maintaining a steady total level of employees on the site while recognising that the mix between staff and hourly paid employees, direct and support employees, will change in line with load and the need to increase output per man employed.

3 Unions will not press for unnecessary job retention and Management will carry out changes without redundancy, relying on non replacement of leavers and redeployment. Employees will accept redeployment with re-training where required.

4 Unions will actively support Management by maximizing every reasonable opportunity to reduce manufacturing costs, to improve deliveries and to meet quality requirements.

5 Unions are committed with Management to filling the New Factory to capacity. This will require transfer of work and or employees from other areas to the New Factory. Due attention will be paid to the work load in existing manufacturing areas during the transitional period.

6 Where new manning or job allocation issues arise, these will be resolved by combined Union/Management discussion as per domestic Agreements.
  To assist resolution, Management will ensure that early discussions take place before the event and this will involve the supply of as much information as is available including observations and visits where appropriate. Unions accept that it is essential to put new equipment and procedures into production as quickly as possible. Both parties accept that current procedures may on occasion be inadequate for maintaining efficient production when manning or job allocation issues arise. Management and Unions are committed to the development of a resolution procedure which will ensure that operations are effectively manned while the procedure is being followed.

Figure 8   The six commitments
Source: *Babcock Power Ltd*

Exercise '81

The unions were intimately involved in Exercise '81, of which the rebuilt machine shop was an integral part. Exercise '81 had as its goal cutting costs by 25 per cent. Senior management decided that this was necessary to remain internationally competitive. This was a clearly recognizable goal for all employees. Although unit costs of production would be reduced by 7 per cent as a result of the new investment it was clear that all functions within Babcock Power would have to make reductions in operating and contract costs. When the goal to reduce costs by 25 per cent was put to the

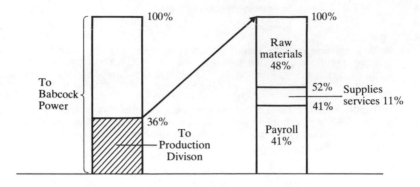

Figure 9   Cost breakdown of a typical contract
Source: *Babcock Power Ltd*

unions it was explained that cutting costs was not just cutting labour costs. The example used was that even if all employees in Babcock Production Division worked for nothing it would still not cut 25 per cent off costs.

The explanation of cost cutting was carried out using a typical boiler contract and its cost analysis (see figure 9).

The idea was to show that to cut 25 per cent was not an impossible task. A communication exercise was held. All functional groups within the plant were asked what was a reasonable cut that they thought was possible to achieve by considering each of the major cost headings (raw material, supplies and services and payroll) in turn. All functions were asked how they might reduce their costs. For example, the design function felt that they could save 10 per cent of raw materials by reducing quantity and specification of materials. Those involved in materials utilization suggested that 5 per cent could be saved and so on (see figure 10). The idea was to demonstrate that a series of relatively small savings could add up to 25 per cent.

	Design	Method	Materials utilization	Purchasing	Labour mix	Labour productivity
Raw materials	10		5	10		
Supplies (e.g. welding electrodes)	3	5	5	5	3	7
Payroll	5	5			5	10

Figure 10   Typical percentage cost cutting exercise
Source: *Babcock Power Ltd*

The results above were a typical example. When they were added together they showed a 23 per cent decrease in costs, very close to the 25 per cent target. So a case was made to all employees that if everyone became involved and the problem was approached methodically, such a target was attainable over a period of two years. A joint management and union steering committee, chaired by the Managing Director, was set up to determine strategy and monitor results and in each area of the organization, planning committees were set up to identify potential savings. Twenty-four such cost saving committees were set up. The savings were investigated and evaluated by employee action groups and this resulted in managers initiating revisions in their budgets to reflect the agreed level of saving.

The steering committee monitored management objectives and style and had the power to call in managers if they gave unsatisfactory answers to complaints from employee action committees. The 24 planning committees and employee action groups altogether involved more than 1,000 employees (see figure 11).

Figure 11   Flow chart for Exercise '81
Source: *McIntyre* et al. *1983*

These groups worked extremely well on production cost and overhead reduction. On design it was more difficult to monitor effectiveness. But at the end of the first year, 1981, the targets had been achieved and only 10 per cent remained to be achieved in 1982 to meet the overall target of a 25 per cent reduction in the 1979 costs (see figure 12). From the outset of Exercise '81, it was assumed in tendering for all overseas contracts that the targeted cost reductions would be achieved and, as a direct result of the lower bids which were thus possible, contracts worth some £256 million were secured in Hong Kong and Zimbabwe. This assumption has been justified by the savings achieved so far. However by 1983 it had become clear that a 25 per cent reduction in costs wasn't enough to remain competitive as international markets contracted. But that is another story.

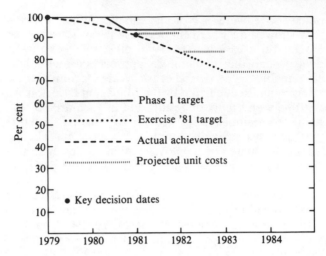

Figure 12   Target and achieved unit costs of production
Source: *McIntyre* et al. *1983*

Results

By modernizing the key facilities for engineering, development and production and systematically reducing costs it has been possible to extend capability in the markets in which Babcock has traditionally operated and to diversify into other markets. The new engineering facilities enable the company to react more quickly to market requirements and to reduce lead times.

The modernization of the machining, fabrication and assembly facilities has reduced manufacturing unit costs and reduced time cycles for machined components sometimes to a quarter of those that were previously possible. It is also now possible to undertake complex machining operations at the competitive prices that are required to provide access to new markets. Because Babcock selected general purpose machinery for traditional markets they have been able to move into new markets like aerospace and North Sea construction work. Boiler work in 1985 only accounts for 50 per cent of work carried out at Babcock whereas three years earlier it accounted for 70 per cent. Babcock has a much larger product range now and the management like to see it as a more diverse jobbing shop. It gets more contracts in the £1 million to £10 million range – jobs where very high quality machining is required, like producing accurate dies for satellite tracking equipment.

Throughput in the new machine shop has been increased by 30 per cent over the old shops with 45 per cent less floor space, 42 per cent fewer machine tools and 35 per cent fewer indirect shop floor personnel. There are also 9 per cent fewer support staff in relation to the number of direct operatives. One question arises. Why did Babcock increase capacity at a time when economic conditions were so adverse? Firstly, some of the manufacturing facilities were becoming very uncompetitive. New capacity was needed to remain competitive in a business they felt they knew well with an excellent chance of continuing success. Secondly, it was incorrectly anticipated by the CEGB, Babcock and others in 1978 that the home generating boards would be ordering substantial new generating capacity by 1985. Babcock management were also keen to avoid enforced redundancies at the time in order to show their commitment to keep a skilled work force together.

The programme involved introducing some 700 employees to advanced technology in a traditional, heavy engineering company and adopting new work practices. Babcock management believes that this degree of change has been achieved by defining objectives, by detailed planning and by building on management and union relationships which have been developed progressively since the mid-1970s. The managers involved consider that the change 'wasn't easy but it was remarkably smooth'. They believe that they had a massive advantage in having continuity in the management team and the union team since 1974. A relationship of trust has built up from this stability.

The Future

Even after all these changes it became clear in 1983 that a 25 per cent reduction in cost wasn't enough to keep things afloat as competition increased and markets decreased. In April 1983 some redundancies were declared to reduce costs by 33 per cent from 1979. That was called 'Operation Overhead', and has been followed with a programme of decentralization within the Renfrew works. In 1984 an £18 million proposal for the second phase was under discussion.

The proposal for a second phase of investment (1985 to 1988) would once again make investment the catalyst for changes. Babcock management are aware that things will have to be changed even more if Babcock is to continue to compete. There will need to be further changes in work practices so that Babcock is treated as if it is a greenfield site. A senior manager said in May 1984 'The only way we will really have a future is to start by putting down where ideally we want to be... . And this will include not only the union structure and retraining but the management structure too. No practice will be sacrosanct. We have 12 weeks to produce a model of how this site should be operated as if it were totally new – as if there were no management, social or union constraints' (Smith, 1984).

References

Central Policy Review Staff (CPRS) 1977 *The Future of the UK Power Plant Manufacturing Industry.* HMSO, London

Holland, T., 1982 How a little help can smooth the way for productivity. *Metalworking Production*, pp. 120–2

McIntyre, J., Greenwood, J.H., Harper, J.L., 1983 Increased manufacturing effectiveness in the production of steam generating equipment. *Proceedings of the Institution of Mechanical Engineers*, 197 (B): 117–22

Smith, A.K., Kellett, G.M., Harper, J.L., Upton, D.E., 1984 The development of design and technology of boilers. *Lloyds Register of Shipping*

Works Management 1984 May, p. 5

2.2

Technology Beyond Machines

Stuart Macdonald

Introduction

The sort of games psychiatrists play would reveal the close association of technology with machines which is widely perceived but which is not evident in official dictionary definitions of either word. The *Concise Oxford Dictionary* makes no mention of technology under machine, nor of machine in its rather cryptic definition of technology as the 'science of the industrial arts'. Yet it is unusual to imagine technology without machines – not necessarily huge iron monsters crushing objects in factories, but machines in the wider sense of the inanimate, shaped to perform useful tasks. Perhaps the main distinction between science and technology is that science is allowed to be abstract while technology is pre-eminently practical. Notions of technology divorced from physical means of application are difficult and uncomfortable. Similarly, machines – in the broad sense – are seen as technology, the means by which science is reduced to practice. The more modern and intricate the machine, the more it is technology and the less a fortuitous assembly of basic elements. Thus, while a capstan is definitely technology, a numerically controlled capstan is much more so. The height of high technology industry is measured chiefly by the sophistication of the machinery it produces; the proximity to science of product, rather than of process or application, is apparently critical. The discipline of economics has been outstanding in its confident confusion of technology with machines: technology has traditionally been subsumed within capital – within the 'buildings and machinery' category, in fact – and technological change measured by changes in capital input. Consequently, a new computer has been assumed to have the same effect on production as a new factory shed of similar cost. This reading is concerning with that part

Source: from S. Macdonald, D. McL. Lamberton and T. Manderville, *The Trouble with Technology*, Frances Pinter, London, 1983.

of technology beyond the machine, a part whose importance remains largely unappreciated.

Machines in Perspective

It is, of course, quite possible for technology to involve no machines at all. For example, a crop rotation is technology, and new crop rotations are decidedly technological change; indeed, that very change in technology was probably more significant than any other in agricultural improvement in Europe during the eighteenth and nineteenth centuries. Similarly, it is quite possible to view change in administrative or managerial practices as technological change once perceived associations with machinery are removed. Technology can be regarded simply as the way things are done, and technological change as the adoption of what are thought to be better ways of doing things – a definition not too dissimilar to that provided by Schumpeter long ago.[1] Obviously machines are likely to be involved in this process, but it is very difficult to envisage technology which is completely embodied in a machine. Even total automation requires technology beyond its hardware; the hardware is useless without software (and what is sometimes called 'peopleware') to direct and control its operation, and the whole lot is pointless without organization of raw materials entering the factory, and of finished products leaving the factory.

If, then, machines do not fully embrace technology, it is obviously necessary to establish an understanding of just what does. Technology is really the sum of knowledge – of received information – which allow things to be done, a role which frequently requires the use of machines, and the information they incorporate, but conceivably may not. Such a definition is not as catholic as it may at first appear; it is reserved for practical achievement and excludes aims, ideals and philosophies. However, the definition extends far beyond techniques, which are really just the tools of technology and, as Nelson has put it, are bereft of the 'logy' – the theory – part of technology.[2] The history of regional input–output analysis has provided a fine example of the distinction between technique and technology, and of the inadequacy of the former when applied without the corpus of knowledge encompassed by the latter.[3]

Technological change, then, may be defined as the addition of new knowledge to old knowledge, usually to allow things to be done in what are thought to be better ways, and sometimes to do new things altogether. Technology being inherently practical, experience is supposed to establish whether the new ways actually are superior to the old – or even to none at all. A major problem in such assessment is that it is much easier to gauge the performance of that part of the technology closely associated with the machine; the technology of the rest of the process is more nebulous and much harder to assess, which may result in the simplistic assumption that if the machine is technically efficient, the technology in which it operates is likely to be efficient too. That new machines are indicative of efficient technological systems and old ones of inefficient systems has been disproved in many developing countries.

The Linear Innovation Process

Clearly, if technological change requires new knowledge, the process by which information is created for technological change is worth consideration. The initial adoption of the new is innovation, and the traditional model of technological innovation is of a process inaugurated by research, and involving the trauma of development before the climax of innovation. Such a model assumes a convenient linearity which is almost certainly unjustified, but which is also assumed in much government science and technology policy, presumably for the same convenience. Consequently, the main element in the process is judged to be research, and policy to stimulate research is justified on the grounds that it leads to innovation. Revelations that development is much more expensive, time-consuming and uncertain than research have done little to qualify enthusiasm for what is imagined to be the seminal part of the innovation process. What appears after innovation is seen as another process altogether – that of diffusion. Geographers are especially fond of plotting rates of diffusion over space and time, and their approach typifies the assumption that innovation is the end of one process and diffusion the start of another, as if innovation marked the point at which technological crystals had formed and were ready for immutable distribution. Until fairly recently it was fashionable to measure technology lag in terms of the time taken for an innovation to spread from its origin,[4] implying that an innovation can be regarded simply as a technological module requiring only to be plugged in to other organizations, other economies and other cultures.

Yet technology is the totality of information which allows things to be done, and total information is unlikely to arrive in a crystallized package from the conventional research and development process. All that can reasonably be expected to emerge from that process is information which must be supplemented by other information before things will be done. Some of this other information will be embodied in the hardware of associated technologies if what is termed 'technological compatibility' is to be achieved. Babbage's computing engine failed because the technology of mechanics was less advanced than that of mathematics,[5] but current telecommunications networks provide examples aplenty of the problems of technological incompatibility.[6] Some of the additional information is analogous to the software necessary to render computer hardware operational, but much of it is information required to effect complementary organizational change. For example, a point-of-sale computer system adopted by a supermarket cannot work at all without software information to supplement the hardware information embodied in the machine, but its usefulness remains limited to fairly basic check-out functions unless there is organizational change to allow the supermarket to take advantage of, say, the new range of management information made available. Clearly this is part of a post-innovation process – though it may provide the basis for further innovation – rather than part of the traditional innovation process, but equally clearly it cannot be entirely separated from that traditional innovation process. Innovation is an integral part of what is in reality a much larger and more complicated innovative process.

The Innovative Process

A major advantage of the use of a model venturing well-beyond innovation is that it loosens the perceived connection between technology and machinery. The existing quasi-linear model, extending from pure basic research to innovation, becomes increasingly machine-centred as it progresses. But beyond innovation. machine orientation diminishes rapidly and technological change is revealed as a broad information process of which the encapsulation of information in machinery can be only a part. Currently, R & D statistics are obsessed with the creation of technological information in the secondary sector. Although R & D activity in the tertiary sector is acknowledged, it fits uneasily with the preconception that technology is machine-orientated. Thus, for example, the Australian government gives industrial R & D grants for work on computer hardware, but only exceptionally for work on computer software. Of the seven activities distinguished in the definitive Frascati Manual as part of the scientific and technological innovation process – R & D, new product marketing, patent work, final product or design engineering, tooling and industrial engineering, manufacturing start-up and financial and organizational changes – all, even the last, are seen as tributary to hardware.[7] There is little attempt to disaggregate R & D statistics in the tertiary sector; in Australia the business enterprise category with the second largest R & D expenditure is the 'other not elsewhere classified' category of the 'other industries' section, a virtually anonymous section responsible for about a third of Australian industrial R & D performance. Though the tertiary sector in developed economies is now responsible for most employment and wealth creation, and is the location of most information activities, it is still not acceptable to consider it a major source of technological information – presumably not because it does not create this information, but because it does so without also creating machines. Such a distinction is not constructive and is discouraged by the notion of an innovative, rather than an innovation, process.

Participants in a Process

This new concept might also affect the kudos afforded the participants in technological change. At present, high status is associated with basic scientific research, and status declines as R & D becomes more applied, involving engineers rather than scientists. After innovation, status plummets and that afforded the salesman is scarcely worth measuring. Yet the characteristically British tradition that a salesman need know nothing about the product, that if a man can sell one thing he can sell anything, contrasts with philosophy elsewhere which insists that a salesman must know the product well if he is to channel information both to and from the market. The Texas Instruments procedure of encouraging scientists to accompany major developments to the market would be anathema to those who espouse the British tradition, and yet it is merely practical acknowledgement of the existence of a single information process extending well beyond innovation.[8]

Inasmuch as technological change is viewed as a process of information flow as well as information creation, that flow is imagined to be a simple process involving pushing promising research results into industry for development. Even that, though, may not be easy when research has been conducted in organizations distant in every way from industry. Better formal channels, such as on-line computer networks, to facilitate this flow may prove counterproductive if they are constructed by demolishing the irregular and informal channels along which much technological information seems to flow. Informal, personal contact is likely to be discouraged when individuals are not only in different sorts of organizations, but also perceive themselves to be performing discrete functions – with appropriate status – in a process that ceases with innovation. A key element of success in new high-technology firms is the working relationship between the technical expert and the commercial expert; both play essential information roles in the innovative process, but the coordination of information embodied in individuals with very different characteristics is not easily achieved. Unless there is some success in coordination, scientists will continue to talk mainly with other scientists, engineers will talk mainly with other engineers and neither will see much profit in wasting time with salesmen.

Beyond Innovation

Although a staunch defence of the linear model of innovation – invention leading to innovation, basic research to applied research and then to development, science producing technology – would be rare, it is very convenient when dealing with such an uncertain process as technological change to assume that everything hangs from research. As Gannicott has pointed out, even when the importance of 'reverse linearity' would suggest that government industrial R & D incentives have actually been counterproductive, convenience is a more fundamental goal of policy than efficiency.[9] The linear model permits justification of policy incentives for research and permits use of measured resource input to research as an indicator of output. While there probably is some linearity in the innovation process, it is far from clear, even in that limited process, just where the impetus starts and which direction it takes. In the full innovative process, the origin and direction of information flow are even less certain. Invention, it seems, may spring from innovation, development from marketing information, applied research from production problems.[10] Thus, consideration of an innovative, rather than an innovation, process makes much less tenable the assumption, discernible in much literature on patents, that invention is the key element which automatically initiates a chain reaction. Much of the usefulness of new technology depends on the way it is used, on information derived from, and applied by, the users themselves. The information supplied by even nineteenth-century agricultural labourers seems to have been critical to the usefulness of the new machinery they operated. So crucial is this user information reckoned to be that a process of horizontal innovation has been postulated – by the doyen of traditional diffusion studies, as it happens – in which users themselves are largely responsible for the creation and dissemination of technological information.[11]

The extension of the process required for technological change beyond innovation means, of course, that much discussion of the effects of technological change acquires new parameters. Technological change ceases to have an impact on, say, employment or productivity because change in such areas is itself regarded as part of the innovative process. Thus organizational change affecting employment or productivity is part of greater organizational change needed to complement the product of innovation to create a total technological information package. Unemployment should not be seen as a potential and unfortunate consequence of technological change, but as a possible ingredient of the total package of change as fundamental as any change in machinery. The problem of technological determinism is really a problem caused by the assumption that new information associated with machinery is superior to other information required for technological change. Associated with that supposed superiority is the elevated status of those responsible for creating information to make machinery rather than to make machinery useful. Because the technocrat does not recognize the importance, and perhaps even the existence, of other information than that which he or she possesses, he or she naturally assumes that only technical solutions are relevant to technological problems. General reluctance to challenge this assumption has suffered those most expert in the technical information of technology to expound on other parts of the innovative process with which they have little familiarity, and in which they are often ignorant of their lack of information. For example, in 1975 the Chairman of the Australian Atomic Energy Commission declared:

> Our technological civilisation produces a continuing stream of problems of a most complex technical character. In many cases judgement must be made on issues which cannot be proved absolutely one way or the other. Only a small proportion of the population is capable of understanding issues of this sort, even if they were to make the effort. Many elected representatives, though not all, are in the same situation. The experts must in the end be trusted.[12]

The scientist's contribution to science policy often provides evidence of this situation: in 1977, one of Australia's leading research chemists, and then Chairman of the country's major science policy advisory body, blithely ignored the endeavours of generations of historians:

> [The] social effects [of the stirrup] over the years were enormous. It led to the development of a specialised corps of mounted soldiers who

needed considerable support not only from foot-soldiers, but also from other men to feed and care for the horses. The mounted soldiers took to wearing armour and special methods had to be used to get them into the saddle. It was these mounted soldiers who became the squires and aristocrats and the medieval society was born—through the stirrup.[13]

Further, exponential growth in the quantity of information involved in the innovative process has increased and intensified specialization – most evident in the exclusive ritual language of many experts – and has helped to isolate each speciality from the rest of the process. Peer group assessment within existing disciplines struggling with new masses of information has discouraged serious interdisciplinary work and the formation of new disciplines. Thus, the study of technological change, for example, is fragmented, with little concern for the total process.[14] Vocational education has also discouraged information flow among even existing disciplines, and instead of alleviating specific manpower problems inherent in the process of technological change, has probably exacerbated them by sacrificing the flexibility that is a necessary accompaniment to all change.

Information and the Innovative Process

Perhaps the greatest penalty for associating technology with machines is exacted through the use of information technology itself. The peculiar characteristics of information as a good – for example, that information remains with the seller even after the buyer has taken possession, and that the buyer cannot be allowed to know what the information is before purchase – make information difficult to classify alongside other economic goods, and may be partly responsible for its neglect by economists. Yet the importance of information in a developed economy is manifest, and much new technology, especially that associated with microelectronics, computing and telecommunications, is designed to cater for information. New machinery offers extraordinary improvements in ability to assemble, store and process information. As might have been predicted, though, plummeting hardware costs have been accompanied by soaring software costs, reflecting the growing problem of providing machinery with information about what it should do with information. There is a much more serious problem. Information is still a most unfamiliar good, and it is understandable that machines which are able to produce cheaply more information should be welcomed as warmly as machines to produce cheaply more of any other good. Oversupply, apparently, is not possible – the more information the better:

> Computer-based management information systems have facilitated more effective management techniques by improving the extent and availability of information on which decisions are based. It is axiomatic that the wider the range of relevant information available to management, and the more accurate and timely that information, the better is

management able to monitor and evaluate organisational progress and development.[15]

What has been ignored in the euphoria surrounding the declining production cost of information is the consumption cost of the information produced. Resources are required to use information, and the more information produced, the more resources are required. An obvious example would be the time taken to read, digest and decide what to do with the information on a computer print-out. More subtle would be the cost of making a better decision when half a dozen alternatives are increased to several thousand, or when sheer quantity of information exceeds even short-term retention capacity and blurs the distinction between relevant and irrelevant. New machines for the production of information have made more information available for use, but have not necessarily made more information more useful. Thus they satisfy the definition of technological change only in that what has been adopted is thought to be a better way of doing things. It can be thought to be so because the costs of change – which are largely information costs – are often not assessed (particularly if they are internal to an organization), and are anyway not considered to be part of technological change. Again, it is the perception of technology as machine-orientated that is mainly responsible for this situation. If technology and technological change are ever to be better understood, a new and grander perception altogether is required, a perception of an innovative process which extends beyond the limits of innovation, far from machines, to encompass the entire environment of technological change.

References

1. Schumpeter, J., 1939 *Business Cycles*, McGraw-Hill, New York, Vol. 1, p. 84
2. Nelson, R., 1982 The role of knowledge in R & D efficiency. *Quarterly Journal of Economics* 97 (3): 453–70
3. Jensen, R., Macdonald, S., 1982 Technique and technology in regional input–output. *Annals of Regional Science* 16 (2): 27–45
4. Tilton, J., 1971 *International Diffusion of Technology. The Case of Semiconductors*. Brookings Institution, Washington DC.
5. Hollingdale, S., Toothill, G., 1975 *Electronic Computers*. Penguin, London, pp. 15–62
6. Macdonald, S., Mandeville, T., Lamberton, D., 1981 Telecommunications in the Pacific Region—impact of a new regime. *Telecommunications Policy* 5 (4): 243–50. Braun, E., Macdonald, S., 1982 *Revolution in Miniature. The History and Impact of Semiconductor Electronics*. Cambridge University Press, Cambridge, pp. 200–2
7. OECD, 1981 *The Measurement of Scientific and Technical Activities (Frascati Manual)*. Paris, pp. 15–16
8. Braun, E., Macdonald, S., 1982 *Revolution in Miniature. The History and Impact of Semiconductor Electronics*. Cambridge University Press, Cambridge, p. 141

9. Gannicott, K., 1980 Simple economics and difficult policies: the case of public money for research and development. Paper delivered to 50th Australian and New Zealand Association for the Advancement of Science Congress, Adelaide
10. Gibbons, M., Johnston, R., 1974 The roles of science in technological innovation. *Research Policy*, 3: 220–42
11. Leonard-Barton, D., Rogers, E., 1981 *Horizontal Diffusion of Innovations: An Alternative Paradigm to the Classical Diffusion Model,* Working Paper No. 1214, Sloan School of Management, M.I.T.; Rogers, E., 1967 *Bibliography on the Diffusion of Innovations*. Diffusion of Innovations Research Report No. 6, Department of Communication, Michigan State University
12. Baxter, P., 1975 Some comments on Ann Mozley Moyal's 'The Australian Atomic Energy Commission: a case study in Australian science and government'. *Search* 6: 458
13. Badger, G., 1977 ASTEC: Planning for science and technology in Australia. Public lecture, Griffith University, Brisbane, 10 September
14. Nelson, R., Winter, S., 1977 In search of a useful theory of innovation. *Research Policy* 6: 36–76
15. Elliott, R., 1981 Technological advances and the Australian banking system. In Goldsworthy, A., (ed.) *Technological Change—Impact of Information Technology*. Canberra, AGPS, p. 175

2.3

Invention, Innovation and Economic Evolution

Devendra Sahal

Rational View of Discovery

In *Science and the Modern World*, Alfred Whitehead hails the emergence of systematic research and development activity.

> The greatest invention of the nineteenth century was the invention of the method of invention. A new method entered into life. In order to understand our epoch, we can neglect all the details of change, such as railways, telegraphs, radios, spinning machines, synthetic dyes. We must concentrate on the method in itself; that is the real novelty, which has broken up the foundations of the old civilization. (p. 141)[13]

His observation was to surface time and again, in one form or the other. As Joseph Schumpeter, one of the most influential economists of the twentieth century, put it years later:

> It is much easier now than it has been in the past to do things that lie outside familiar routine – innovation itself is being reduced to routine. Technological progress is increasingly becoming the business of teams of trained specialists who turn out what is required and make it work in predictable ways. The romance of earlier commercial adventure is rapidly wearing away, because so many more things can be strictly calculated that had of old to be visualized in a flash of genius. (p. 132)[12]

In a similar vein, Professor John Kenneth Galbraith observes:

> It is a commonplace of modern technology that there is a high measure of certainty that problems have solutions before there is knowledge of how they are to be solved. (p. 17)[3]

Source: from *Technological Forecasting and Social Change* 23 (3): 213–35 (1983). ©Elsevier Science Publishing Co., Inc.

To cap it all, Dalton states:

> Technological innovation is no longer the haphazard result of occasional discovery. It has become institutionalized through corporate, university and governmental research. (p. 145)[2]

We are thus led to believe that technical progress is a rational process subject to prediction, regulation and control.

In reality, however, one finds that this is very rarely the case. A close examination of the available evidence reveals that innovation is seldom a predictable and almost always an erratic process despite all attempts to manage its course through systematic research and development activity. The emergence of fiber metallurgy is a case in point. This important new technology was created accidentally when an employee of a plastic company happened to drop a bunch of steel wool in a hot plastic mix during the course of his frustrating efforts to find new epoxy resin plastic for automotive applications. He immediately noticed that the steel wool made it possible for the plastic to harden in a uniform manner by absorbing and conducting the heat away from the mix while also adding extra strength to the material. In this way an entirely new avenue of technical progress was discovered. Metal fibers were soon combined with other substances in an attempt to develop new lightweight materials of extra strength and toughness. However, the initial applications of these materials, particularly in the aircraft industry, were not wholly successful and, for a while, the future of the new technology seemed indeed bleak. Fortunately, the innovation found a totally unexpected new use: in the construction of skis, golf clubs and other sports items. This provided a fresh spurt to its development. The reason was simple enough: sporting goods constructed from fiber-reinforced material could make something of a professional out of an amateur. Thus the prospects of fiber metallurgy were once again revived. Today it is widely regarded as a most successful technical breakthrough. The example is illustrative of the very general proposition that innovation is neither merely a mechanical nor solely a goal-driven process.[7-10]

The traditional, rational view of innovation does simplify matters considerably. It is nevertheless an article of faith reminiscent of Alexander Pope's attempt to assume away any trace of uncertainty in the universe:

> All nature is but art unknown to thee;
> All chance, direction which thou
>   canst not see;
> All discord, harmony not understood;
> All partial evil, universal good.

In the following an alternative theoretical view of the innovation process is advanced. The essence of the proposed viewpoint is that innovative systems are *inherently untidy* systems. The most important clue to any possible uniformity in the behaviour of such systems is therefore to be found in their very multiformity. Our search for invariance must begin with the consideration of the variance itself.

Emergence of Order through Chaos in Innovation Processes

One of the most striking features of technical progress is that there is virtually no innovation of any significance that has been made wholly by design. Paradoxical as it may sound, the occurrence of innovations by chance is particularly apparent in the conduct of basic research. For example, it was a sheer accident that Galvani, a physiologist, discovered electricity. He had dissected a frog and happened to leave its legs hanging on a set of different metals. Later he noticed that the frog's legs contracted due to generation of an electric spark. His observation paved the way for the development of the electric battery. Several decades later, Oersted unexpectedly found the relationship between electricity and magnetism as he happened to hold an electric wire parallel to a needle during the course of a lecture. Much to his surprise, he noticed that the needle changed its position. His observation laid the basis for the development of the dynamo. Similarly, Roentgen discovered X-rays while experimenting with electrical discharges in high vacuums. By chance he noticed that a barium-platino cyanide plate that was left in an adjacent darkroom became fluorescent. He rightly suspected that this was caused by some invisible radiation that passed through the wall. Subsequent investigations confirmed his discovery of X-rays.

Chance permeates much industrial research as well. For example, the development of antiknocking gasoline had been a baffling problem to manufacturers of engines of automobiles and aircraft for several years. At one point the problem seemed to defy any solution. It was rightly suspected that the problem lay in the fuel but wrongly believed that the knocking of the engine somehow depended on the colour of gasoline. One of the employees of General Motors then decided to try out various chemicals and, quite by accident, mixed iodine into gasoline. Much to his surprise, the fluke had worked. The existing theory of knocking proved to be wrong. The success of the iodine in preventing the engine from knocking paved the discovery of antidetonants.

Frequently, important innovations have occurred as 'accidental by-products' of some main line of research. For example, a great many earlier attempts to develop inexpensive antifreeze for automobiles seemed to make little headway. However, during the course of these attempts, a physician succeeded in making one of the first practical forms of synthetic rubber quite by chance. Likewise, Teflon was accidentally developed by a chemist in an attempt to make an improved refrigerant for Du Pont Co. [...]

These examples are illustrative of the fact that chance plays a central role in discovery. The term 'discovery' is used in a neutral sense to include both invention (creation of a new device) and innovation (its commercial introduction). Equally important, the act of creation does not depend upon *pure* chance alone; invariably, it is also preceded and followed by systematic effort. Thus the origin of new techniques lies in a phenomenon well described by the saying that chance favours the prepared mind. We find therefore that technical progress is a cumulative process of learning to learn and unlearn in a *probabilistic* manner. Furthermore, it can be shown that any phenomenon that depends upon a cumulative mechanism involving chance is governed by the negative binomial law. [...] A detailed specification of the stochastic scheme underlying the negative binomial distribution of innovations can be found in the earlier works of the author.[7-9] There are two key features of the proposed scheme. First, origin of new techniques lies in a wide *variety* of technical possibilities. Second, the emergence of new techniques moreover depends upon selection at the level of homogenous fields of R & D activity (e.g. the digital computer and scientific instrument fields). In essence, innovations arise out of selectivity in diversity. Selectivity has its basis in the accumulation of past experience, diversity in the acquisition of novel experience. Thus innovation is a process of learning to learn and unlearn that is governed by the negative binomial distribution. [The argument is further developed in the article from which this reading is taken.]

One main implication of the binomial law is that technical discoveries do not occur at a constant rate; rather, they tend to cluster. This implication of the theory is further examined here by computing nine-year moving totals of inventions as well as innovations over the entire period of 121 years. These are plotted on a vertical axis of a graph, with the horizontal axis being the centre of the nine-year intervals (figure 1). It can be seen that neither major invention nor major innovation has occurred at a steady pace; rather, both the creation of a new device and its commercial application have occurred in a wavelike fashion. Specifically, several prominent peaks are evident in each case: e.g. around years 1902 and 1924 in case of major inventions and around years 1908–13 and 1934–39 in case of major innovations. A further discussion of the phenomenon of clustering in fundamental technical breakthroughs is deferred until a later section in this reading. Suffice it here to note that the observed distribution of fundamental advances in technology is in keeping with the proposed theoretical scheme.

In sum, a certain order is evident in the process of discovery. Significantly, the origin of this order lies in the very chaos in research and development activity. There is unity in the deep structure underlying the apparent multiplicity of factors surrounding the process of technical breakthroughs.

The Phenomenon of Equivocation in the Transfer and Utilization of Technical Knowledge

One of the most important forms of technology transfer is the incorporation of new engineering knowledge into commercial products or the transformation of inventions into innovations. Hitherto, it was assumed

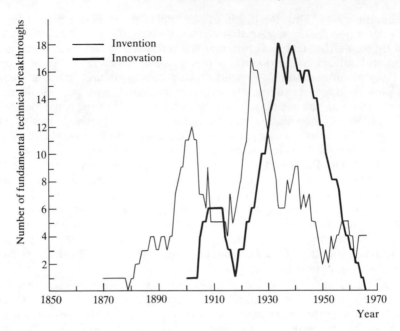

Figure 1   The advent of major technical inventions and innovations, 1850–1970

that inventions and innovations occur independently of one another. The flow of technical knowledge is never a wholly directed process.

The proposed viewpoint is justified by the fact that very often inventions and innovations are outcomes of markedly dissimilar activities undertaken by different individuals as well as different institutions. As Joseph Schumpeter[12] put it,

> Although most innovations can be traced to some conquest in the realm of either theoretical or practical knowledge that has occurred in the immediate or the remote past, there are many which cannot. Innovation is possible without anything we should identify as invention, and invention does not necessarily induce innovation, but produces of itself … no economically relevant effect at all. (p. 84)

Thus, a priori, it is plausible that the process of transformation of invention into innovation cannot be unequivocally specified in advance.

Disparity Principle of Innovation

It is a commonplace of observations that the time interval between invention and innovation varies considerably from one case to another. The gestation period – the time taken from the first patent to commercial application – was some 55 years for the float glass but a mere four years in the case of the transistor. Penicillin required nearly 15 years of effort prior to its commercial introduction in 1943 but streptomycin took only one year before it was successfully marketed in 1944. Is there any system to the observed time lags between invention and innovation?

There are two noteworthy aspects of the observed gestation periods in the advent of fundamental technical breakthroughs. First, the amount of time required in bringing an invention to the market place depends upon a host of factors ranging from debugging of the product to obtaining the necessary resource commitment for the venture. Thus the time lag between invention and innovation is likely to vary randomly from one case to another. Second, a close examination of the evidence reveals that a small number of technical breakthroughs demand extremely large gestation periods though many breakthroughs require gestation periods of only moderate length. There is a fundamental disparity in the way advances in technology come to pass.

Our investigation reveals the existence of a systematic pattern in the gestation periods of fundamental technical breakthroughs *because* (and not in spite) of the great variety in the circumstances surrounding the trans-formation of invention into innovation. The pattern can be very simply expressed as follows. Within any given industry, if we rank all fundamental technical breakthroughs according to the gestation time required in each case, then the product of that rank and the length of gestation period is approximately constant. (See table 1 for an illustrative example.) Thus the second longest gestation period will be nearly one-half of the longest, whereas the third longest gestation period will be nearly one-third the longest. The observed pattern has the important strategic implication that it is essential to guard against excessive investment in the commercial introduction of a *few* technical breakthroughs that require extremely large gestation periods. Rather, it is imperative to focus on *many* technical

Table 1   Illustrative application of the rank-time rule in the case of telecommunications industry

Innovation	Gestation (years)	Rank	Product of gestation time and rank
1. Magnetic tape recorder	39	1	39
2. A.M. radio	18	2	36
3. Television	13	3	39
4. Radar	9	4	36
5. Wireless telegraph	8	5.5	44
6. Wireless telephone	8	5.5	44
7. Triode vacuum tube	7	7	49
8. Magnetic recording	5	8	40
9. Transistor	4	9	36
10. Long-playing record	3	10	30

breakthroughs that demand gestation periods of only moderate length. In essence, one must not put all one's eggs in one basket.

Clustering of Fundamental Technical Breakthroughs and Long Waves of Economic Development

As noted earlier in this paper, fundamental technical breakthroughs do not occur singly; instead, they tend to cluster. In this respect, technical progress is well-described by the saying that it never rains but it pours. The nature of clustering in the occurrence of technical breakthroughs is further investigated here by means of spectral analysis of data. The results of our analysis reveal the existence of two cycles of 48 and 24 years of duration in the present time series on occurrence of major innovations. It is interesting to note that our analysis of fundamental technical breakthroughs derived from very different sources of data also discloses two innovation cycles of exactly the same duration.[9] The two series cover the same time period except that the earlier series is broader in its focus of technical advances than the present series. Together, they confirm the existence of innovation peaks in 1888 and 1936 associated with the 48-year cycle, and in 1863, 1882, 1909, 1936 and 1957 associated with the 24-year cycle. As pointed out elsewhere by the author, the peaks of these two major cycles *coincided* in 1936 thereby giving rise to an unusually large cluster of innovations at the time.[9] What, if any, significance is there to the observed clusters of fundamental innovations during certain periods of time?

In 1913, J. Van Gelderen advanced the thesis that economic development occurred in the form of long waves of about half a century consisting of 'spring-tides' of prosperity and 'ebb-tides' of relative scarcity. About a decade later, N.D. Kondratieff reached an essentially similar conclusion.[4] A controversy ensued that has lasted to this day: Are long cycles facts or artifacts of our measurement? Given their existence, what accounts for their origin? In his monumental work, Joseph Schumpeter contended that the origin of such long waves (Kondratieff cycles) lay in the clustering of fundamental innovations during certain periods of time.[11] Specifically, the first Kondratieff cycle (1787–1842) was driven mainly by innovations resulting from the development of steam power, the second (1843–97) by innovations centred around railroads and the third (1898–1953) by innovations made possible through the development of electric power and the internal combustion engine. The importance of Schumpeter's pioneering work cannot be overemphasized; however, it requires further verification.

It seems appropriate here to make a distinction between two categories of fundamental innovations: (1) *characteristic innovations*, which give rise to new socioeconomic activities or industries, and (2) *generic innovations*, which not only lay the basis of new socioeconomic endeavours but also cut across several new and old industries. For example, both the catalytic cracking process and the internal combustion engine were fundamental technical breakthroughs, but the former was a characteristic innovation, whereas the latter was generic. Although these two categories of innovation clearly overlap, they are nevertheless conceptually distinct. What distinguishes the two is the magnitude of the ripple effect: generic innovations tend to have more of a global effect and characteristic innovations more of a local effect.

It goes without saying that the development and adoption of a *generic* innovation across broad areas of industrial applications give rise to a cluster of interrelated innovations. We find therefore that the advent of steam power was accompanied by a host of innovations made in quick succession to one another. Starting from the turn of the nineteenth century, the cluster of innovations included: Murdock's long slide valve; Trevithick's high-pressure self-moving engine, return-flue boiler and fusible plug; the Longdon carriage and portable engine; Evan's grasshopper beam engine; Woolf's introduction of compounding; Col. Stevens' twin screw vessel; and so on. Contrast this with another cluster of *characteristic* innovations made during the mid-1930s: the fluorescent lamp, automatic drive in vehicles, the helicopter, antiknocking gasoline, Plexiglass, catalytic cracking of petroleum, Kodachrome, television, cyclotron, polyethylene and so on. Clearly, there are marked differences in the nature and significance of these two clusters of innovations even through they are not wholly independent of one another.

More generally, it is apparent that technical progress exhibits two distinct but related types of clustering phenomena in the occurrence of fundamental technical breakthroughs. First, we have a cluster of *interrelated* innovations arising from the advent of a handful of generic breakthroughs. This points to what may be called *technology cycles*. Second, we have a cluster of innovations consisting of several largely *unrelated* characteristic breakthroughs, pointing to what may be called *innovation cycles*. Evidently, the origin of long waves of economic development lies in technology cycles, not in innovation cycles. As discussed elsewhere by the author, the existence of technology cycles of differing durations accounts for the fact that economic development entails not only Kondratieff cycles of 50 years but also Kuznets cycles of approximately 30 years in length and Juglar cycles of 9–10 years.[7] According to the proposed theory, it is to be expected that *technological development* of a *generic innovation*, such as steam power, engenders long waves of economic development, whereas the *technological development* of a *characteristic innovation* within any given industry – e.g. the farm tractor in agriculture and locomotive in the railroad industry – induces medium and short waves of economic development. In sum, different innovations give rise to cycles of different duration. To reiterate, in contrast with the clustering phenomenon in characteristic innovations, the clustering phenomenon in generic innovations feeds upon itself in a synergistic manner, thereby setting in motion a new wave of economic development.

These findings have important implications for the prospects of surmounting the current worldwide economic stagnation. Thus it is important to distinguish between empty slogans and deliberate policies that both call for innovation as a way out of current worldwide economic malaise. Superficially, the two seem alike. In reality, they point to very different prescriptions. In particular, it is evident that the prospects of world economic recovery do not depend on an arithmomorphic search for some undefined cluster of (characteristic) technical breakthroughs. Rather, as discussed below, economic rejuvenation requires a concrete programme of action involving development, diffusion and transfer of a few key (generic) technical breakthroughs. Specifically, it is suggested that a strategy of economic renovation ought to be based upon gradual, step-by-step *development* of a few, already available fundamental innovations, such as those in the microelectronics, bioengineering and solar energy fields; the *diffusion and transfer* of these techniques across broad areas of industrial application; and concomitant organizational and social innovations.

Further, a distinction should be drawn between the proximate and the ultimate causes of long waves of economic development. According to the theory advanced here, it is *developmental* and *incremental* innovations that are the proximate causes of long waves although ultimately their origins can be traced to certain *fundamental* (generic) innovations. This viewpoint is also supported by the results of a wide variety of studies indicating that the cumulative effect of many seemingly minor changes in technology on productivity growth often tends to be quite substantial.[7] Thus, contrary to popular opinion, what we need in the future is not unprecedented expenditure on basic research at the expense of technological development effort. Rather, it is of paramount importance to strike a proper correlation between various components of R & D activity.

This brings us to another point: that R & D activity has a dual character incorporating elements of both a problem-solving and a problem-posing nature. Relatedly, fundamental innovations may arise as outcomes of not only efforts to overcome various bottlenecks but also as consequences of attempts to exploit new opportunities. The implication, of course, is that there is no one single set of economic circumstances that is most auspicious to occurrence of technical breakthroughs. Rather, a distinction should be drawn between two broad categories of fundamental innovation depending upon the associated sets of circumstances. First, we have what may be called *sunshine innovations*, made possible by excess resources under conditions of economic well-being. Second, we have what may be called *sunset innovations*, made possible by problemistic search under conditions of comparative scarcity.

Both common sense and theory suggest that it is periods of economic recovery and recession rather than economic prosperity and depression that are most conducive to occurrence of technical breakthroughs. Fundamental innovations are unlikely to peak at the *height* of economic prosperity, inasmuch as 'all work and no play makes Jack a dull boy'. On the other hand, fundamental innovations are also unlikely to peak in the *midst* of economic depression inasmuch as 'all play and no work makes Jack a mere toy'. Rather, clustering of fundamental innovations requires an optimal utilization of production capacity high enough to ensure adequate rewards and low enough to allow the necessary experimentation.

A priori, therefore, fundamental innovations are expected to peak during two time periods: (1) toward the *end* of depression and anywhere

Table 2   Long waves of economic development

Prosperity	Recession	Depression	Revival
1787–1800	1801–1813	1814–1827	1828–1842
1843–1857	1858–1869	1870–1885	1886–1897
1898–1911	1912–1925	1926–1939	1940–1952
1953–1965			

Sources: The dates of the long waves during 1787–1939 have been obtained from Kuznets.[5] Thereafter the dates are based on world trade statistics

Table 3   Clustering of fundamental (characteristic) innovations in various phases of long waves of economic development

Innovation cycle	Innovation peak	Phase of Kondratieff cycle	Type of breakthroughs
48–year	1888	Toward the beginning of recovery	Sunset
	1936	Toward the end of depression	Sunset
24–year	1863	In the midst of recession	Sunshine
	1882	Toward the end of depression	Sunset
	1909	Toward the end of prosperity	Sunshine
	1936	Toward the end of depression	Sunset
	1962	Toward the end of prosperity	Sunshine

during recovery, and (2) toward the *end* of prosperity and anywhere *during* recession. As shown in tables 2 and 3, the agreement between theory and the data is excellent. Moreover, it is interesting to note that, over the past 121 years, the clusters of (characteristic) innovations include both sunshine and sunset innovations rather than one at the exclusion of the other. Thus, we have the important result that although the *occurrence* of fundamental (characteristic) innovations may well be a matter of chance, the *rate* of innovation is a function of economic forces.

In summary, the relationship between long waves of economic development and fundamental innovations is of a reciprocal nature. Long waves are a cause of *characteristic* innovations as well as a consequence of *generic* innovations. Thus, the process of economic evolution is a determinant of the process of technological evolution as much as the process of technological evolution is a determinant of the process of economic evolution. The economic evolution reflects, as much as molds, the technological evolution.

Conclusions and Policy Implications

This reading has presented an attempt to probe into the processes of invention, innovation and economic evolution. The results of our investigation dispel the popular myth of a heroic father figure as a major force behind technical progress. Moreover, they shed some new light on a phenomenon that has long baffled social scientists, namely that virtually every major discovery in modern times has been made independently by two or more workers in the field at about the same time. Thus Robert Merton, an eminent sociologist, reaches the apparently paradoxical conclusion that

> The pattern of independent multiple discoveries in a science is the dominant pattern, rather than a subsidiary one. It is the singletons–discoveries made only once in the history of science–that are the residual cases, requiring special explanation. (p. 306)[6]

According to the theory advanced here it is to be expected that a discovery is characterized by several independent origins. The reason is simple: technical breakthroughs depend not only upon a certain amount of preaspiration but also upon chancy inspiration.

The theory has several important implications for R & D management during the 1980s and beyond. To begin with, it is apparent that innovations depend upon a multitude of causes rather than on any one single factor at the exclusion of all others. Thus success in innovative activity critically depends upon pursuing several small-scale scientific and technical experiments at the same time. In essence, variety in innovative activity is best stabilized through variety in R & D activity.

It is therefore futile to promote orderliness in the conduct of R & D activity; rather, our attempt should be to capitalize on the innate uncertainty of innovation. Indeed, we should seek to turn chance into fortune through pursuing *multiple approaches* to any given problem in R & D activity. Thus, it should be pointed out that popular attempts to avoid duplications in R & D constitute what is really a penny-wise but pound-foolish policy. Far from being wasteful, duplication in R & D is both an essential condition and a *sine qua non* of making progress.

Second, there is little to be gained and much to be lost from rationalizing the process of discovery. Considerations of efficiency may well require a monolithic hierarchy. Innovativeness, however, calls for creative diversity. The technological stagnation of the 1970s – as evidenced by the decline in the number of patents awarded in virtually all industrialized countries – is a case in point. Viewed from the proposed standpoint, the observed slowdown in the pace of technical progress has been, in no small measure, caused by the increasing centralization of the R & D activity over the years. If the considerations advanced here are any guide, nothing is more important in the future than to promote greater autonomy in the conduct of R & D enterprise.

Third, any attempt to routinize the course of research and development activity is doomed to failure. We find it reassuring that the importance of this point is not lost to the truly successful managers in the industry. 'The way to run this place', says Arno Penzias, the research vice-president of Bell Laboratories, 'is to hire smart people, point them in the right direction and get out of the way'.[1] In our view, planning should seek to promote not some fixed set of objectives but the capability to cope with and seize upon accidental and unexpected events. The only essential condition of a dynamic organization is the property of self-organization.

Fourth, and finally, one way out of current worldwide economic stagnation is to be found in the development of a few already available fundamental innovations such as those in the microelectronics, solar energy and biotechnology fields rather than in an endless search for more and more technical breakthroughs. Thus, contrary to certain popular prescriptions, what we need in the future is not massive increase in the expenditure on *basic research* at the expense of *technological development* efforts. Rather, it is imperative to strive for a mutuality between various components of R & D activity.

The results of our investigation further reveal the existence of a number of significant regularities in the origin of new techniques, transfer of technical knowledge and productivity growth. The central implication of these regularities for policy making is that not all roads lead to Rome – at least not in the same amount of time. Some selectivity is essential. It should be pointed out, however, that these regularities are themselves steady states of certain stochastic processes at work. Thus the only true stability in nature is indeed one of constant change. Needless to say, variety is the spice of life. The sum total of these considerations is that the policy makers must necessarily tread on a few more or less selected routes. In so doing, however, they should avoid taking the proverbial long lanes that have no turning.

Thus, following Whitehead's terminology, our advice to policy planners is very simple: seek selectivity and distrust it.

References

1. Begley, S., 1982 Ma Bell's Dream Factory, *Newsweek* Jan 25: 68–69
2. Dalton, G., 1974 *Economic Systems and Society: Capitalism, Communism and the Third World*. Penguin Books, London
3. Galbraith, J.K., 1967 *The New Industrial State*, 3rd edn. Houghton Miffin, Boston, 1967
4. Kondratieff, N.D., 1935 The long waves in economic life. *Rev. Econ. Stat.* 105–115
5. Kuznets, S. 1940 Schumpeter's business cycles. *Am. Econ. Rev.* 30: 250–371 Reprinted in Kuznets, S., *Economic Change*, Heinemann, London, pp. 105–24
6. Merton, R.K., 1961 The role of genius in scientific advance. *New Sci.* 259: 306–8
7. Sahal, D., *Patterns of Technological Innovation*. Addison-Wesley, Reading
8. Sahal, D., 1982 The form of technology governs the scope of its transfer. In *The Transfer and Utilization of Technical Knowledge*. Sahal, D., (ed.) 1982 D.C. Heath, Lexington
9. Sahal, D., 1982 Chance and Opportunity in Technological Innovation, submitted for publication 1982

10. Sahal, D., 1982 Structure and self-organization. *Behav. Sci.* 27: 249–58
11. Schumpeter, J., 1939 *Business Cycles*. McGraw Hill, New York, Vols. 1 and 2
12. Schumpeter, J., 1942 *Capitalism, Socialism, and Democracy*. Harper, New York
13. Whitehead, A.N., 1925 *Science and the Modern World*. Macmillan, New York

2.4

The Technological Mirage

Michel Godet

Current infatuation with new technology is too fashionable not to be suspicious. We have learned from the history of forecasting errors that dominant and fashionable points of view on a given issue generally prove to be inaccurate. A correct forecast, when it exists at all, is in most cases an exception and for this reason is hardly ever accepted. The dominant analysis of the present crisis and its future prospects does not escape this constraint of global conformity. The stakes are high, since if one's diagnosis is erroneous, there is little hope of prescribing appropriate remedies. (There are no good answers to a bad question.)

The Myth of the Technological Godmother

The technological myth, viewed as a tool to recover from crises, will soon become an article of blind faith in our minds. Technological forecasting is fast becoming the modern substitute for divine prediction. In the new creeds of the technologICal (and other 'ic') revolutions, nothing is *a priori* impossible and every problem has its solution, even those stemming from technology itself.

Given this turn of events, we raise a fundamental question: is technology not merely a mirage, a collective illusion? It is more than likely that the myth of salvation through a technological godmother will be long lived, because it is consciously maintained by an entire intelligentsia. Powerful lobbying by technocratic and industrial pressure groups which have risked their reputations can only change direction if/when the mirage fades in the face of evidence.

Yet it is patently clear that the technological myth is no better founded than the straightforward energy interpretation of the 1973–74 oil crisis. It is

Source: from *Futures* 16 (2), 1984.

now recognized that the slowdown of growth and the increase in inflationary pressures began at the end of the 1960s, i.e. before the first oil crisis. It is striking that practically nobody, after 1974, talked about any technologies other than those pertaining to energy problems. You can always find a tree behind which to obscure the forest you don't want to see.

Energy source independence was not the remedy for all ills. Furthermore, with hindsight it is striking that those countries which have been economically the most dynamic in the North and in the South are also those highly dependent on external sources for their energy requirements (Japan, the Federal Republic of Germany, South Korea), while those countries which have had oil revenues at their disposal (OPEC members and the UK) have stagnating industries.

The crisis is structural and not conjunctural in origin. A monetary explanation (in terms of inflation, the role of the dollar and of exchange rates) is also not well-founded. One must not confuse the symptom with the cause – a crisis of 'leadership'. It is likely that the current US economic recovery will prove to be no more than a flash in the pan, which will undoubtedly lead to enormous fiscal and trade deficits, which have previously served to ignite and kindle the process.

Crises are Opportunities

There remains the thesis that we are experiencing a crisis of transition, between two technological waves. As the technologies and production processes which led to growth in the 1950s and 1960s peter out, we must await new technologies and ensuing processes in order to recover from crisis ... or .... In the meantime, it would appear that technological voluntarism is incapable of speeding up this transitional phase. In fact, it is paradoxical that the drop in productive investment and in productivity comes at the very moment when the microprocessor revolution is initiating a new era of rising yields in industry and services. Perhaps one should see in this paradox a proof of increasing obstacles to yields of technology – obstacles which are not only socioeconomic but also organizational (e.g. excessive centralization, the complexity of large-scale systems).

In conclusion, the crisis is not an energy-related nor monetary, nor technological crisis, but rather a crisis of socio-institutional *régulation*. In effect, since technico-economic change is taking place faster than social change, social structures and modes of behaviour are proving increasingly inadequate in the face of the world today. According to this analysis, *the crisis is a reflection of the nonadaptation of contemporary structures, and continuing technological change only tends to exacerbate the imbalance.* In a changing world, the rules of the game and modes of behaviour must also change. There can be no creation without destruction. It is necessary, certainly, to 'make the rich pay', but provided we admit that we are *all* privileged to some extent, we too will have to give up some of our acquired privileges to facilitate a smooth (that is, avoiding catastrophes) changeover to a world immensely different from that which we have known up to now.

It is necessary to unravel social structures, as columnist Pierre Drouin has reminded us.[1]

Clearly, a responsible government cannot encourage the search for flexibility at the expense of the least well-off (in terms of wealth or security). Where there is no possibility of a new social deal through legislation, crises and unemployment are, alas, necessary to overcome the rigidities accumulated in past years, if only to help us push through the necessary measures. This viewpoint is somewhat difficult to accept, and certainly does call for political austerity, but it also carries the embryonic hope that the social change of tomorrow will be born from the crises of today. This view also leads to the conclusion that the crisis is only just starting, contrary to the claim that '*L'après crise est commencé*'.

The Need for Social Monitoring

It seems that social change, essential for recovery from the crisis, is only marginally stimulated by technical change. As René Eksl and Gerard Metayer have correctly emphasized, in regard to information technology, 'the balance sheet of experiments in France is disquieting: nowhere have these applications brought about the slightest social or cultural dynamic ... these techniques have been grafted on to existing structures and have systematically eluded questions of the redistribution of power and of the creation of a new balance between economic and social groups'.[2]

Technological innovation resolves nothing directly; and if sometimes it has the merit of throwing light on certain contradictions in our social structures, it also raises new problems of modes of living, work organization and social relations. Thus the pervasive use of computers in teaching may generate new forms of social exclusion, where some become 'computer-illiterate' in a society where those who have been computerized become 'illiterate' in every sense of the term.

Technological decisions should be subjected to increased social scrutiny, and to an open debate in which the experts consulted are really independent of technocratic and industrial pressure groups. In France, Parliament should set up an assessment office to evaluate technical options. It is high time for such a project to see the light of day; otherwise, decisions in telecommunications technology will be no more democratic than those already taken in nuclear power, given that the social stakes are far higher.

Product and Process Innovation

It must be made clear that our proposal here is not that a brake should be put on technical change. On the contrary, technical change should be mastered, stimulated and oriented in a direction which will prove more appropriate to social aspirations and real economic constraints.

Process innovation (automation, the massive introduction of industrial robots) is expected to develop rapidly because of the constraints of international competition. This trend to innovation in industrial processes seems inevitable, and this makes it all the more relevant to ask questions about the social conditions of its installation in business enterprises.

On the other hand, most innovations in products and in service industries do not get beyond the experimental stage (i.e. they are not seen as profitable). Some of those responsible are tempted to say 'let's create the need, and provide the product and "out there" they are sure to come round to using and buying it'.

A 'Concorde' in Communications and Office Automation Technology

The supply-side policy can only last for a short while. Sooner or later comes the sanction of the marketplace, and sometimes nothing is left of the innovation but the financial overspending. The realms of technological possibility are not necessarily socially desirable or economically viable. We must be careful, especially in lean times, not to implement (with cable TV, videotex or other networks) new 'Concordes' in communications and office automation technology.

What is the point, for example, of setting up technical support systems like videotex if the contents are lacking or useless? For videotex, in particular, there is a race on between Europe and North America. On both sides of the Atlantic, the main argument (I have recently been able to confirm this) used to justify this race is that 'we must do it because they are doing it'. Farce will soon displace illusion.

Other Needs to be Satisfied

There are perhaps 'needs' which we don't really need (the electronic telephone directory?), and there are certainly important needs which are not yet satisfied, especially in the area of social communication:

(1) The need for space and calm: what is the point of cluttering up meagre and noisy accommodation with new electronic gadgets?
(2) The need for green spaces and play areas for children in cities: electronic games cannot replace trees or swings.
(3) The need for communication and direct human contact, not only through the new media: otherwise 'everyone will be closer but no-one will have neighbours', etc ...

Economic growth is not an unalterable stock, but an annually renewable phenomenon. In this light, the citizen–consumer sees the rate of growth of the annual budgetary deficit as being less important than its redistribution or its qualitative contents (standards of living and social relations). It is about time that we realized that our people do not necessarily want to consume more but would prefer, by far, to live better.

References

1. Drouin, P., 1983 La glu. *Le Monde*, 6 October
2. Eksl, R., Metayer, G., 1983 Modernisme technique et conservatisme social. *Futuribles 65*: April

PART 3
Manufacturing Innovations as Processes

3.0

Introduction

Part 3 takes our subject more firmly down to the level of manufacturing. It is 'preluded' by reading 3.1, an extract from James Bright's now classic work, 'Automation and Management'. The reading deals with developments in the production of electric light bulbs. It is shown how the production of this seemingly simple product raised complex manufacturing problems. Solutions were found in a fashion probably typical of most technological change – evolutionary development that depends upon managerial adaptations of the production process, rather than upon radically new technology. The case study also demonstrates the interrelationship of change in products and processes – at times change in light-bulb design was required to accommodate change in the production process. His book was produced at a time when a widespread and rapid shift to highly automated factories was often predicted in the American media and elsewhere. Bright's concluding caution about the nature and pace of change remains highly relevant and reflects some of the issues raised by Godet.

Braun's article (reading 3.2) takes up the concept of manufacturing innovation. The article reflects a number of the points made by Sahal and by Macdonald in its emphasis on multicausality, on the significance of factors within the environment and on the importance of the innovative process within the production unit. The main concern of the paper is to develop a systematic conceptual framework which will facilitate understanding and analysis of the complex phenomena that were involved in manufacturing innovation and which will help identify common problems and issues likely to be experienced in projects of very different scales and types. Braun suggests this can be accomplished through use of a three-phase 'constellation theory'. Like Macdonald, Braun argues that the problem with much work on innovation is that it gives little or no attention to events beyond initial commercial application, regarding these as part of a diffusion process. This overemphasizes the role of formal R & D activities and overlooks the essential originality of much of the task of implementation within the workplace.

The role of technology users is an issue that is also taken up by Abernathy and Townsend – reading 3.3 – in which they explain their influential model of technological change and reiterate the importance of looking at change in terms of multicausality and of responding accordingly. They explain technological change as relating to a model of progressive change in the production process. They see the production process progressing from one where an initial stage in the production unit is essentially uncoordinated to one with more coordinated but segmented production systems and then to those which are highly integrated and thus systemic in character. They refer to Bright's study of electric lamps as illustrating both this progression and a further point they make about product and process development. They see this innovation process as essentially evolutionary and heavily dependent upon the technology user rather than the technology manufacturer. However, the transition between stages is seen as neither inevitable nor irreversible. Essentially it is a matter of conscious choice according to changing technology, product market, work force and opportunities.

The relationship between factors like these, decision-making processes and technological change is the concern of Gold, Peirce and Rosegger in reading 3.4. They point out that decisions about technological innovation are not unique but fit within the general decision-making processes of an organization. In addition they remind us that decision making is generally not a 'climactic, once-and-for-all choice' but, like technological change, is an evolutionary process, shaped by favourable or unfavourable experience. This has a number of implications including that of endeavouring to establish whether technological change has met the expectations set by original decisions. They indicate further parallels between decision making and technological change in that the former, while often perceived as a logical/rational process, is rarely thus. Decisions reflect the variations that are to be found in organizational objectives and in managerial values. Consequently, when appraising specific technologies, firms often come to very different conclusions. Having identified the scale of the difficulties the authors proceed to develop a model which contributes to better understanding of technological decision making.

3.1

Evolution of Automation: Electric Lamps

James R. Bright

Electric Lamp Manufacturing

The lamp industry is an especially interesting example of mechanization development because the electric lamp presents a complex manufacturing problem. Widely different kinds of materials – metals, glass, gas – are involved; a number of the components are extremely fragile; several of them must be made to a high degree of precision; the product requires delicate assembly work; the testing of every unit for functional performance is essential; and the fragility of the product requires packaging at the moment the testing is completed. Thus, lamp making employs many manufacturing arts and, as a problem in automation, it is unusually involved.

To encourage demand for electric current the mass production of lamps at low cost has been an objective of the lamp industry for many years. Today, several thousand types of lamps are manufactured on a commercial basis. About 80 per cent of this demand is concentrated in fewer than 100 types of lamps. (This statistic is exclusive of the vacuum tube industry which creates a product looking somewhat like the lamp, but far different.) To illustrate the progress of automation in the lamp industry, the 60-watt incandescent lamp was chosen. It is a typical size and can be isolated as a product of many years' design and manufacturing study.

Phase 1, Prior to 1920

A typical layout of a Westinghouse Electric Corporation manufacturing unit of this period is shown in figure 1.[1] It is a functional arrangement (departmental setup) with a stem department, an inserting department, an

Source: from James R. Bright, *Automation and Management*, Boston, MA: Division of Research, Harvard Business School, 1958, pp. 20–30, 38. Reprinted by permission.

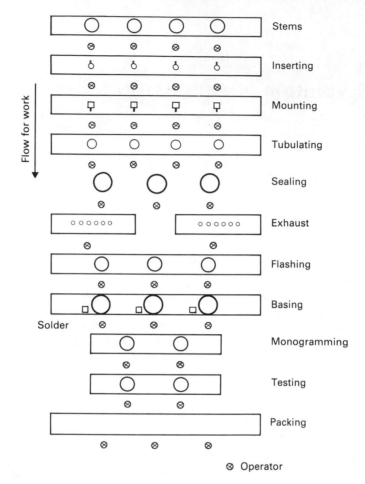

Figure 1   Electric lamp manufacturing by functional layout, Westinghouse Electric Corporation, 1908–20

exhaust department and a testing department, as shown. Some machinery was used, and in most cases an individual operator manned a single machine. Exhausting, however, was mechanized to the point where one operator manned a series of exhausting stations. There was a great deal of hand work in the assembly of the lamps.

The manufacturing areas in the earliest days were not laid out in sequence as shown in this illustration. Far more often each department was located in any convenient space without particular regard to being adjacent to the next operation in the production sequence. As each operator finished a batch of parts, he put them into a container which was hand carried or trucked to the next department. In general, this departmental layout was standard production technique prior to 1920. In some cases, it was carried on long after that, particularly for other than the common sizes of lamps. As of the period covered by figure 1 the standard operations (excluding the preparation of the parts) were:

(1) Stem making – combining of lead wires with glass, fibre and arbor.
(2) Inserting of stems – inserting wire with hook on one end in button or arbor.
(3) Mounting – mounting filament on supports and connecting with leads.
(4) Tubulating – combining exhaust tube to top of bulb.
(5) Sealing – combining tubulated bulb with mounted stem.
(6) Exhaust – exhausting air from bulb through exhaust tube and scaling tube.
(7) Flashing – initial lighting of lamp on predetermined schedule.
(8) Basing and soldering – affixing base to end of lamp and soldering connections.
(9) Monogramming – labelling of lamp.
(10) Test – final inspection.
(11) Packing – wrapping individual lamp, placing in outer container.

Approximately ten operators produced about 200 lamps per hour. There were two drawbacks to this production system: (a) large stocks of materials always were in process; (b) material-in-process was exposed to the atmosphere for a considerable time. Exposure had a deteriorating effect upon some of the components and thus affected quality.

Phase 2, 1920–25

About 1920–25 the manufacturing art advanced through the system illustrated in figure 2. This layout reflects the first successful grouping of equipment. In effect, it was a production line – a collection of machinery so arranged spatially and balanced in capacities that the product moved from one operation to the next with little waiting time, and the production rate of each machine grouping was roughly equivalent to that of its neighbours. This layout reduced the amount of material handling. It reflected a significant advance in the level of mechanization of some individual operations. In addition, more of the production operations were mechanized.

Possibly the important innovation from the point of view of manufacturing evolution was the pacing of operations by indexing machines.

'Indexing' means that materials are fed into the machine which then moves them from station to station (indexes), pausing momentarily so that successive production steps can be completed. A uniformity of flow – automatic, nonvarying timing – results. Indexing machines have the effect of contributing or forcing uniformity of pace to production operations on each side of themselves, in order to receive or provide parts at the indexing machines' rates.

Another advantage of indexing machinery is reduction of work-in-

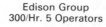

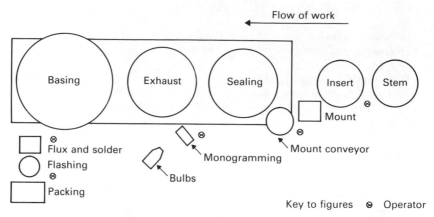

Figure 2   Evolution of lamp production line, Westinghouse Electric Corporation, 1920–25

process. The work does not sit on the workers' bench after he has finished an operation and until he completes the batch. In effect, there is only one piece in process at each work station (and sometimes a few more pieces in the handling system between stations). This saving in work-in-process further contributes to a gain in quality in lamp manufacturing, since the time in process is much shorter and less deterioration of parts is encountered.

A major innovation in this manufacturing system of 1920–25 was the rotating exhausting machine. The machine was connected to the exhausting system through a rotating valve, thus enabling continuous movement through the exhausting process. No pauses were required to engage and disengage the exhaust line to each bulb, which was, in effect, done automatically. Nor was there any waiting until the operator got around to disconnecting an exhausted bulb. The rotating table carried the lamps through the process and to the common unloading point at an unvarying pace.

The number of manufacturing operations was reduced from eleven to eight. The compounding of operations – that is, the performance of two or more operations at a single machine station – began to appear and is evident in the following list of operations required:

(1) Stem making, incorporating exhaust tube.
(2) Machine inserting and hook forming (one support at a time).
(3) Hand mounting of filament.
(4) Sealing and monogramming.
(5) Exhaust.

(6) Basing and soldering.
(7) Flashing.
(8) Packing.

These refinements were not due to improvement of machinery or mechanization of former hand jobs alone. A significant product design change was evident in this stage. The exhaust tube was incorporated in the stem of the lamp bulb, rather than in its apex. This had great advantages in simplifying the original sequence of production operations since (a) it eliminated one assembly job – No. 4 tubulating – and (b) it simplified No. 6, exhaust.

The indexing rate controlled the speed of production. It was such that the line produced approximately 300 lamps per hour using a crew of five operators.

Phase 3, 1925–36

During the period 1925–36 the lamp production line took the form schematically suggested in figure 3. Spreading of mechanization along the production sequence is evident. A considerable amount of labour was eliminated by mechanizing material movement between machines. Automatic transfer arms, automatic loading for the exhaust machine and similar work-feeding devices were part of the system.

Several other basic forms of mechanical evolution marked this major step: most important was the combining of sealing and exhausting into one machine. Compounding of several machines on one machine base produced important effects. Though the 'Sealex' machine was larger than either of the machines that it replaced, it required less total space. There was also a simplification of power supply, a reduction in space for servicing, space for operators and of course, the elimination of manual

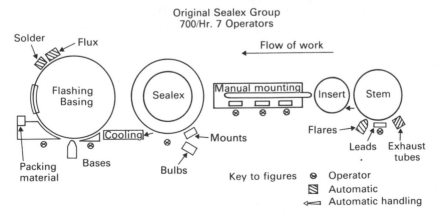

*Figure 3   Beginnings of integration and continuous flow in lamp production
Westinghouse Electric Corporation, 1925–36*

handling between the machines. These factors further reduced the work-in-process.

A technical process change at this stage forced the addition of another 'production process' – cooling. Because heat was applied to the bulb to facilitate exhausting, it became necessary to add a cooling conveyor. Thus, functional operation and movement were combined as the lamps cooled on their way to the basing machine.

Other significant steps were: the flashing test was integrated into the basing machine; fluxing and soldering were each made automatic. They were placed adjacent to the basing machine and were so closely integrated with it that they could almost be considered as part of it. The same was true of the exhaust tube operation and the flaring – both were made automatic and both were intimately linked with the stem-making machine.

This system produced approximately 700 lamps per hour. Seven persons were required to operate it, the two additional persons being needed for mounting.

Phase 4, 1936–55

Lamp manufacturing from 1936 through 1955 is approximately described by the layout of figure 4. Glass flares, exhaust tubing, lead wires and coils were automatically formed and delivered to the stem-making machine. The stem machine automatically assembled them and delivered the stems

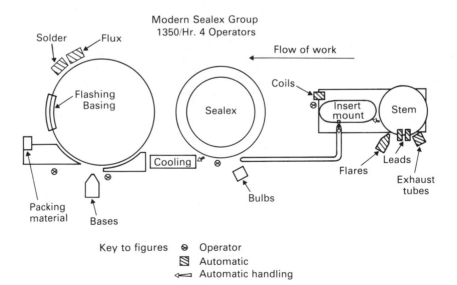

Figure 4   Spread of automaticity in lamp manufacturing, Westinghouse Electric Corporation, 1936–55

by conveyor to the next operation. The operator for this machine was a combined inspector, material supplier and, partially, a relief operator for the other positions. The mounting operation was made automatic. The mechanization of this assembly job eliminated three operators.

The compounding of functions in a single machine unit continued. The preparation of stem components, the assembling of these components and their mounting were combined in one machine.

This machine system produced 1,350 lamps per hour. A crew of four operators was required.

Westinghouse engineers have advised that it is entirely possible to perform, nonmanually, all the remaining operations of loading the sealing-exhaust machine, basing, testing and packing. However, they say:

> ... present design of machine does not lend itself readily to much higher speeds, which would have to be attained in order to justify the cost of conversion. With the progress already made there is a good possibility for more complete automation. Radical changes and improvements in equipment are envisioned by equipment designers for the not-too-distant future.

Parallel with 50 years of refinement of the production machinery and the mechanization of all the production operations, there were equally important, although perhaps less spectacular, design changes and developments that were essential to facilitate automatic production. Some of these are not pictured or charted since they include such things as changes in the gas content, in the hardness of metals, in the crystalline structure of metals or in electrical and other characteristics of the materials. For example, the concentration of wire into coils that could more easily be mounted or assembled mechanically was not possible without the development of wire that would resist sagging under the high temperatures that occurred when the lamp was lighted. Thus, the development of nonsag crystalline structure of tungsten was fully as important a step toward automation as was the machine for doing the job. This same improvement greatly increased the light output of the lamps.

Some of these changes in component parts, which contributed to progress in mechanization, were as follows.

The early machine-blown bulbs are illustrated in figure 5A. The light bulb neck and collet were not well-adapted for high-speed sealing. Furthermore the light bulb had to be sealed both at the base and at the exhaust tube. The key design change (figure 5B) was to machine-blow the bulb from a ribbon of glass, and to use a heavy neck and collet that were easily adapted to high-speed sealing. The exhaust tube was incorporated in the stem rather than in the bulb. Therefore, only one sealing job needed to be done after the bulb had been exhausted.

Refinements in the design of the stem also are shown in figure 5. The early conventional stem had hooks over which the straight wire was draped by hand, figure 5C. Anyone who has closely examined one of these light bulbs will appreciate the delicacy required to perform this difficult job manually.

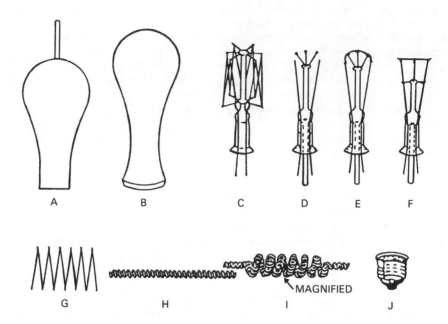

Figure 5   Component design changes that facilitated automatic forming and assembly in lamp manufacturing, Westinghouse Electric Corporation, 1908–55

The second step was to develop a conventional stem to mount a coiled filament, figure 5D. Clearly, fewer manipulations were required to mount the filament, and it would appear that the structure was stronger as well as simpler. Hooked leads with pigtail supports were provided for manual mounting. Notice the exhaust tube in place in the stem.

The automatically mounted stem for the coiled filament is shown in figure 5E. Here the leads were straight rather than hooked. To make the stem assembly, the filament, the glass flare, the leads, the exhaust tube and the support wires were automatically fed into the stem machine and automatically assembled. The stem itself was automatically mounted.

The final development is shown in figure 5F. This stem, too, was automatically assembled and mounted, but it was designed for the coiled coil filament. It can be seen in the figure that support requirements were reduced, which simplified the design of the stem.

Meanwhile, the coil itself underwent redesign. This light-producing filament is a very delicate piece of tungsten wire. In figure 5G we see the preformed wire that had to be draped over the supports of figure 5C. To drape such fine wire on flimsy supports by mechanical means was extremely difficult and was a manual operation. This filament, incidentally, had to operate in a vacuum, so creation of a suitable vacuum was essential.

To shorten the long filament and therefore simplify the draping operation, the filament was coiled, figure 5H. Since it was now of shorter length and was fairly rigid, it was adapted to automatic mounting. This made possible the stem construction in figure 5E. Further evolution of mechanization occurred even within this single part. Originally, the supports for this coil were secured around the filament by hand manipulation. Eventually this job was mechanized so that fixing the coil on the stem could be made completely automatic. This type of coil was used in gas-filled lamps.

To shorten the coil further and concentrate more light-producing element within the light bulb, the recoiled filament was created. Its appearance is shown in figure 5L. Because of its comparative rigidity it was well-adapted to automatic mounting. It is the standard in the 1955 electric lamps.

The type of base used in 1955 for automatic operations is shown in figure 5J. It, too, is adapted to prefilling, conveying through slide feeds, threading over lead wires and soldering operations. All these operations have been mechanized except for the threading of the lead wire. This job is well on the road toward mechanization, and then the entire basing operation will be automatic.

Thus, the entire mount and its component parts have evolved into a form of manufacturing and assembly that is highly automatic, as a result of progress in a number of areas.

Correlation of these changes in manufacturing technique that contributed to automation, is shown in Table 1. This table includes some detail not shown on the layouts. It further confirms that automatic manufacture of a complex, multipart product is not a matter of machinery alone. It requires parallel progress in materials and in product design.

In studying this and other automation systems it is evident that certain basic trends are inextricably woven into progress toward automaticity:

(1) mechanization of manually performed operations;
(2) arrangement of machinery into a production operation sequence in which all operations are done at approximately the same rate so that continuous flow can be achieved;
(3) combining of several functions into a single machine base (i.e. a compound machine);
(4) integration of all machines with automatic work feeding, work removal and material handling devices (between machines) so as to create a work movement system that is nonmanual;
(5) changes in the product design to permit mechanical manipulation, assembly, and other forms of nonmanual working in production operations;
(6) changes in material to permit either the use of a production technique that is more easily made automatic or a design form that is more easily mechanized.

A review of this accomplishment throws facts into contrast against general impressions and common statements about automation.

Table 1   Evolution and productivity of lamp manufacturing technique, Westinghouse Electric Corporation, 1908–55

Item	Phase 1 (figure 1) 1908–20	Phase 2 (figure 2) 1920–25	Phase 3 (figure 3) 1925–36	Phase 4 (figure 4) 1936–55
Type of layout and manufacturing system	Departmental	Semiproduction line sequence	Production line	Semicontinuous production line
Significant mechanization of operations	Very little, largely hand tools	Rotary indexing machines producing steady flow of parts on major operations	(1) Automatic manufacture of flares and exhaust tubes (2) Automatic soldering and fluxing (3) Automatic material handling: stem to insert; sealex to cooling; cooling to basing (4) Combination of sealing and exhaust; flashing and basing	(1) Automatic production of leads; coils (2) Automatic work feeding of exhaust tubes, leads, flares, coils (3) Automatic mount assembly (4) Automatic hand-ling; mount to sealex, sealex to cooling
Number of operators	10	5	7	4
Production/ hr/group	200	300	700	1,350
Lamps/operator/day	160	480	800	2,700

Lamp manufacturing has been undergoing constant mechanization for almost half a century. Although it is one of the most automatic multipart product manufacturing technologies in existence, it still is not 'fully automatic'. Automaticity varies in degree and character through the line.

Automaticity has not been a matter of machinery alone. It often has been delayed until material improvements and design changes have been created to make a new step in mechanization possible.

The operators are *not* required to be superskilled. On the contrary, their duties are lighter and are essentially those of patrolling, inspecting and workfeeding.

Net employment has not been reduced in the Westinghouse Lamp Division in spite of automation. The number of salaried and hourly paid employees in this division has grown from 2,762 in 1939 to 7,759 in March 1955. This is an increase of over 280 per cent in spite of a major manufacturing phase improvement during these years. (Earlier employment figures were not available, but are known to be lower than the 1939 figure. It should also be appreciated that hundreds of new types of lamps have been introduced, and many of these are produced in a far less mechanized manner than the 60-watt lamp which has been discussed.)

This accomplishment in highly automatic production is not the result of automatic control. Feedback control is not necessarily essential to automatic manufacturing.

The ultimate controlling factor in pacing the growth of automaticity has been machinery cost (or possibly policy) but definitely is not one of technical feasibility. Technically, the operation now could be far more automatic, and automaticity could have been achieved earlier.

What benefits have resulted from this effort?

The accomplishment of this lamp manufacturing programme is impressive. Results are charted in figure 6. The product is about 170 per cent more efficient in light-producing ability with the same amount of power consumption. The 1955 60-watt lamp sells for roughly one-tenth of the cost of the World War I 60-watt lamp. This price does not take into account the depreciation of the dollar over these years, which makes this cost reduction achievement even more striking. The individual worker of 1955 turns out about 17 times as many lamps daily as he did at the time of the First World War. Employment in the Westinghouse Lamp Division has climbed in spite of this steady progress toward automaticity and increased productivity.

[Bright also examined the shoe industry which then, as now, had experienced only a limited degree of mechanization let alone automation.] ...

Industry Requirements for Automation

In a comparison of electric lamp and shoe manufacturing, one can identify serious obstacles to automation that exist in many industries.

Like the manufacturer of shoes, manufacturers in many other industries do not have a free choice of materials. They do not have freedom to modify

84

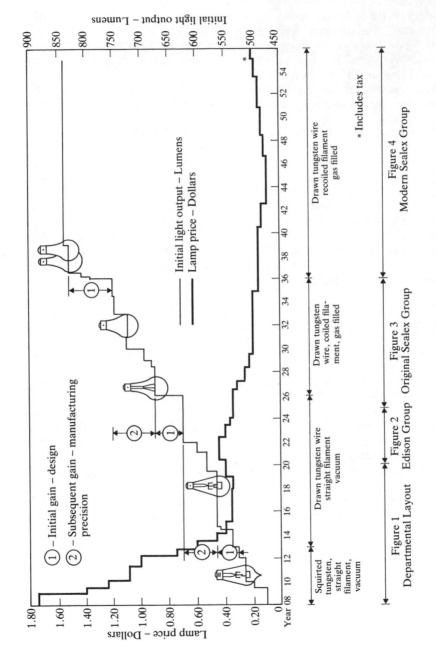

Figure 6 *Light output and price of 60-watt electric lamp related to design, material and manufacturing changes, Westinghouse Electric Corporation, 1908–55*

the shape of the product to simplify automatic manufacturing. They do not have the freedom to create or force uniformity. They do not have a sales environment that is predictable or that encourages manufacturing in large lots. They do not have personnel within their organizations who are conditioned to or capable of making an aggressive attack on mechanization.

Lack of uniformity throws difficulties in the way of automaticity. This uniformity may be missing in the design of the product, in the sales pattern, or even in the material itself. Any one of these obstacles can bar automation. The shoe industry faces all of them.

Similar examination of other industries will demonstrate that automatic production cannot be achieved overnight. We have seen that machinery, alone, is not the answer – it takes parallel progress in other production areas as discussed earlier. The shoe industry presents a new requirement: the need for a marketing environment that encourages uniformity and is compatible with inherent limitations in the machinery and process.

This comparative analysis seems to confirm that (1) automation is an evolutionary trend and not an absolute quality; (2) some industries are exceedingly difficult to automate or impractical to automate; (3) widespread 'fully automatic manufacturing' is in the far future. There is little doubt, however, that we shall continue to see gains in productivity as the production line is further refined and integrated into a harmonious machine-like whole, even though it is not completely automatic.

For these reasons, the managerial task for achieving automation is not simply that of keeping in touch with equipment developments. It is to pursue productivity improvements on a company-wide front, including design, materials, processes and marketing practices that will facilitate automaticity rather than in the machinery area alone.

The major problem for management is to perceive the direction and possibilities for constructing an environment that will support automaticity and to press forward at a rate and in a manner that are economically and technically desirable in their own particular firm, with due regard to easing the impact on the work force.

Note

The abrupt rises in light output were due initially to the change from a straight filament in a vacuum to a single filament coil in a gas filled bulb and later to the introduction of the coiled coil filament. The increasing automaticity resulted in more uniform lamps and less scrap. Better equipment provided higher precision processing which further increased quality and quantity of output.

Reference

1. Gero, W.B., 1955 *A Brief Summary of Automation in the Lamp Division*. Westinghouse Electric Corporation

3.2

Constellations for Manufacturing Innovation

Ernest Braun

Introduction

It has become generally recognized in recent years that the efficiency of manufacture of goods is a main determinant of economic success or failure in the internationally intensely competitive trading system of technologically advanced countries. It has also become recognized that technological innovation is a prime driving force in economic growth.[7] Because of the great importance of technological innovation, a great many attempts at a theoretical understanding of the process have been made. Full success eludes these attempts because of the great variety of innovations carried out in an even greater variety of settings. The present reading attempts therefore not so much to suggest a new theory of innovation, as a new heuristic method for fitting the complex phenomena into a systematic framework. In this way it is hoped to aid analysis of the multifaceted activity called technological innovation, to show that success or failure depend upon constellations of circumstances which can at least be understood if not foreseen, and to further our understanding by showing that certain sets of recognizable constellations exist.

Technological innovation is a term of great generality and in this reading we shall concentrate on that aspect of technological innovation which has the most direct bearing on manufacturing efficiency – manufacturing innovation. We define manufacturing innovation as a new method of producing an essentially established product by an essentially established process. Manufacturing innovation usually involves the installation of novel machinery and/or novel methods of controlling the manufacturing process. Very often the efficient use of the innovation requires a variety of organizational changes, but changes in organization alone do not constitute

Source: from *Omega* 9 (3), 1981.

manufacturing innovation. Some examples of manufacturing innovations might illustrate the meaning of the term. If a robot is introduced to spot-weld or to spray paint, this does not necessarily involve any change in the product or the process, yet it constitutes a major technological innovation. Similarly, a numerically controlled machine tool, a computer controlled automated warehouse, an automated feed to a machine; all these are examples of manufacturing innovation.

The most important feature of this type of innovation is that its sole objective is to improve the efficiency of manufacture. The improvement can be in labour productivity, capital productivity, production yield, savings in materials or energy, improvement in the quality of the product or in other parameters. Thus manufacturing innovation can be seen as one of the major weapons in the battle for world markets of established products. Having said that however, the influence of manufacturing innovation on product design and even product innovation cannot be denied. Just as a new product requires new production facilities, so the production facility influences the design of the product. The point is particularly clear in the case of microelectronics, where manufacturing procedures dominated the quality and characteristics of the product.[3] The case is equally valid in many other products. Manufacturing innovation is certainly a more important determinant of product innovation than it is often given credit for.

Existing innovation theory does not fully satisfy our requirements for understanding innovation in general and manufacturing innovation in particular on several grounds. Much of the best work, the work with the strongest empirical foundation, is based on analysis of factors important in successful as compared to unsuccessful innovations.[6] A great deal of useful insight was obtained in this way on the importance of factors such as innovation champions, market coupling and many more. By the very nature of this kind of investigation, however, the emphasis is on factors and not relationships between them or logical sequences of events. Some guidance to policy is obtained, but the statistical nature of the evidence and concentration on factors rather than on relationships and sequences, makes the guidance somewhat inadequate. Unfortunately, the repeatability of the statistical results is also far from satisfactory.[5]

From the point of view of manufacturing innovation, existing theory is deficient in two more respects. First it treats manufacturing innovation simply as a diffusion process. Although it is true that the introduction of, say, numerically controlled machine tools may be regarded as the diffusion of somebody else's product innovation, in doing so one overlooks much pioneering effort and truly innovative activity associated with the introduction of new manufacturing procedures into an organization. In many cases, and especially when electronic controls are added to existing machinery, diffusion is not the essential feature of the activity. Secondly, in concentrating on product and process innovation and disregarding manufacturing innovation, much existing work overemphasizes the role of deliberate R & D activities. Without wishing to underestimate the role of R & D or the importance of studying the coupling between R & D and innovative activity, this emphasis is unsuitable for the study of manufacturing innovation. For in this case formal R & D is often, though not always, transferred with the product from the machine tool or electronics manufacturer to the manufacturing innovator.[8]

In view of the importance of manufacturing innovation in the economic life of a technologically advanced country and in view of the enormous impetus given to this kind of innovation by the advent of microelectronics,[1] we have undertaken a series of case studies to further our understanding of the process of manufacturing innovation at the microeconomic level. The kind of questions we sought answers to were: the reasons for introducing the innovation; strategies of introduction; sources of supply; difficulties encountered; comparison between expectations and results; impacts on economic performance; balanced skill requirements.

Even in our preliminary results it became apparent that the diversity of firms, diversity of technologies and diversity of circumstances made it impossible to achieve meaningful generalizations without some form of theoretical framework for analysis.[4] The purpose of the present reading is to discuss such a theoretical framework – a constellation theory of manufacturing innovation.

Constellation Theory

It may be a truism to say that all complex phenomena – and manufacturing innovation is nothing if not a complex phenomenon – require a constellation of various circumstances to be propitious for the event to occur. The truism becomes useful when it helps us to turn our backs on the attractions of simple cause and effect relationships which are utterly inadequate to cope with the complexity of the real world, yet provide such comforting illusions of understanding. Constellation theory can never provide the simple comforts of delusion, but it can provide a framework for a deeper understanding of reality in a way which can prove useful in terms of policy actions.

For a manufacturing innovation to proceed, a constellation of circumstances must occur. Although the number of specific constellations can vary greatly from case to case, it is easy to identify certain phases in the constellations. These phases represent logical sequences, but do not necessarily correspond to time sequences.

In the best tradition of thermodynamics, we start off with the zeroth phase, for before we can begin to describe an innovation process in a given firm, we must describe the firm itself and its immediate surroundings. Thus the zeroth phase of a constellation for innovation is the cluster of circumstances in which and under which the firm operates. The parameters that matter here are the size of firm, industrial sector, state of the market, economic and political environment, technical, managerial and financial resources within the firm. Having described the firm and its environment, we proceed to the next phase, the first phase of the innovation process itself.

The first phase in any manufacturing innovation consists of the identification of a *weak link* in the existing manufacturing system. To identify such a link requires an actor (or group of actors) whose task it is to keep the production machinery as efficient as possible. The weak link may consist of a variety of individual circumstances, such as shortage of skills, inefficient flow of production, waste of energy, waste of materials, inadequate quality of product, unreliability of machinery, high maintenance costs, inadequate output, low productivity (capital or labour), low yield in production, contravention of safety regulations, unpleasant working conditions. The list is not exhaustive, but indicates the many flaws which may occur in a production system and which we have encountered in our case studies. It must be emphasized that weaknesses in a system are only relative to the best available practices and technologies. Weak links develop, they are not static. Indeed a weakness may only come into existence because of a new technology which had become available.

The first phase in a constellation leading to a manufacturing innovation, the identification of a weak link (or links) by an actor, is closely related to the second phase, which consists of the identification or development of a *possible solution* to the problem discovered in phase one. In the simplest case, the second phase may consist of the identification of commercially available equipment which will solve the problem. It is possible that the problem only exists in conjunction with an identifiable solution. The availability of new equipment may have made the old machinery obsolete and therefore uncompetitive.

The second phase may be very much more complex and more protracted and may contain constellations within the constellation. It may be necessary to develop modifications of commercially available equipment or it may even be necessary to carry out a complete research and development programme to resolve a production problem. Here again it needs to be emphasized that a production problem may only come into being as a result of new possibilities, perhaps discovered in a research programme. The analogy with 'market pull' and 'technology push', commonly identified with product innovation, exists in manufacturing innovation. There is, of course, a possibility of feedback between solution and problem; in other words, the problem may be modified in the course of its solution.

Examples in our case studies range from numerically controlled machine tools to solve bottlenecks in availability of skills or in accuracy and speed of machinery, to the addition of electronic monitors to machinery which was otherwise inadequately utilized by the work force.

As the first phase, so the second phase also requires actors. In the simplest case, the person who pinpointed the production problem may have done so in view of a solution that had been brought to his/her attention. In the most complex case, a whole range of actors may be required, including many of the figures familiar in theories of innovation: gate-keepers, research managers and researchers, development engineers, innovation champions.

The third phase is the *implementation of the innovation*. This can be a simple matter of installing and commissioning some commercially available equipment, with few repercussions on the manufacturing organization. It may, however, be a complex matter full of controversy, difficulty and far-reaching change. Among the many difficulties that may be encountered in the implementation stage of an innovation are: shortage of relevant skills, resistance to change by management and/or workers, shortage of capital, lack of space, need for extensive reorganization, need for change in working practices, renegotiation of manning and/or pay agreements, technical malfunctions in the manufacturing system caused by the introduction of the innovation, changes in power structures and/or hierarchies.

Clearly, problems of phase three feed back into phases one or two, i.e. both the initial problem and its initial solution may be modified during implementation. Sometimes the 'modification' may go as far as abandonment of the project.

The more complex the implementation stage proves to be, the more actors and actions will be required. Trade unions, production engineers, personnel managers, line management, management services – all kinds of people may become involved. The more people are involved, the more important it becomes that there should be a person or a small group in overall charge of the project. Again, the innovation champion becomes important.

The implementation stage has often been discussed in the literature on innovation and various strategies have been described, e.g. 'green field' approach, 'participation approach' and 'persuasion' approach. What needs to be emphasized here is the close relationship between all the stages of an innovation, implementation cannot be separated from the other phases.

The essential feature of this constellation theory of innovation is the assertion that an innovation can only proceed if a number of factors are in a suitable relationship with each other, i.e. the constellation of circumstances is right for the process to occur. The division of the innovation process into phases or stages serves the purpose of classification of what otherwise may appear a very diverse and unstructured maze of different innovations in different industries. Although the circumstances may differ greatly and factor analysis may give unstable results, the logical sequence described here will remain valid.

For the innovation to proceed and to yield results, the clusters of factors in the three phases must come into such relative 'positions' as to allow the 'flow' of change to occur – in analogy with the flow of current through a logic circuit with many switching elements. The analogy with electrical switches in a logic circuit leads to the concept of 'switching actions' to be undertaken by the various 'actors'. Only when the right actions are undertaken will the innovation take place. The actors 'operate switches' or 'set points' and the innovation flows if, and only if, they are correctly set.

Any policy aimed at smoothing the path of a particular innovation or of innovation in general, must aim to identify the 'switches' and help the 'actors' set them correctly.

Any analysis of a manufacturing innovation should aim to identify the clusters of factors in the various stages of innovation, their mutual relationships, the actions required to allow the 'flow' of the innovation, and the actors involved. The analysis must, of course, also include a comparison of the results of the innovation with the initial aims and thus, in a sense, close the loop of the complete process (see figure 1).

Applications of Constellation Theory

In principle, constellation theory can be applied to the analysis of a whole range of complex phenomena. As far as technological innovation is concerned, most certainly every kind of innovation requires favourable constellations. We shall, however, discuss only applications within the goals set in this reading and confine discussion to manufacturing innovation.

There are, in essence, three things that constellation theory can provide:

(1) It provides a framework for asking the right questions about an innovation process and a scheme for sorting, classifying and understanding the complex factual information that might be available from widely differing case studies. It is thus a heuristic device, a device aimed at classifying and understanding what otherwise might be a jungle of more or less well established facts with wide divergence for different cases.

The most important aspect of the heuristic role of constellation theory is that it enables comparisons to be made between widely separate sequences of events. At first sight it is hard to see that the installation of a multimillion pound tailor-made computer controlled manufacturing facility in a gigantic plant has any similarity with the purchase of an automatic feed for some machine in a tiny manufacturing firm. Yet the two events are aspects of the same phenomenon – manufacturing innovation – and have many common problems which can be understood with the aid of constellation theory.

(2) The second function of constellation theory is that of type-casting. We can recognize typical constellations such as:

(a) Very small firm living in symbiotic relationship as supplier of specialized items to large firm. Short of capital and management/technical sophistication. Innovates only under extreme pressure. Only path open to it is off-the-shelf purchase of modern equipment, with possibly slight modifications suggested by consultant.

(b) Small firm built on technical expertise in sophisticated market. Sophistication of product and competitive pressure force constant stream of manufacturing innovation, although mostly limited to purchase from outside and adaptation to needs.

(c) Large firm in competitive mass market. Manufacturing innovation vital aspect of survival. Devotes much management effort to this task and is able to mount own research and development as well as buy in the best machinery and best advice.

Admittedly, this kind of type-casting can never be complete and must always allow atypical exceptions. Nevertheless, it helps not only in the understanding of the innovative behaviour of the firm, but also sets limits to its possibilities. Simple rules can be derived, such as 'if you belong to

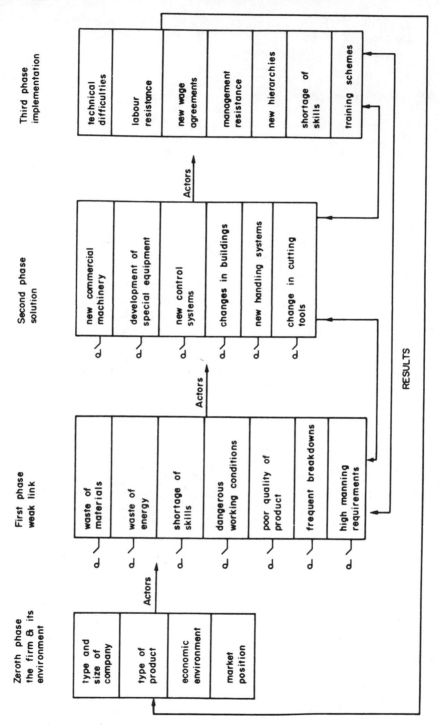

Figure 1   Analysis of a manufacturing innovation

type (a) of firms, do not attempt to get ahead of the competition by having sophisticated equipment specially developed for you'. Advice that smacks perhaps of the home-spun 'people in glasshouses should not throw stones'; but most observers are surprised at the failures that occur because of the absence of very simple rules of behaviour. We have seen cases of failed attempts to innovate into complex technology well-beyond the reach of the firm.

(3) Finally, constellation theory can help to guide and inform policy, both at management and government level. The analysis schematically shown in figure 1, pinpoints actions that need to be taken for the innovation to flow. Put in a different way, for a constellation to lead to successful innovation certain circumstances have to be manipulated. The point of constellation theory is to show where manipulations are possible and necessary.

For management, the implications are fairly clear. Management must decide on what sort of company it is managing and what sort of world it lives in and then decide on broad strategies for innovation. Thus the zeroth phase of constellation theory becomes a kind of 'know thyself' exercise. It may also be useful for management to recognize which, if any, typical innovative constellation the firm belongs to; know thy type as well as thyself.

Next management must see to it that all the key functions in innovation activity are allocated, looked after, and properly coordinated. It is necessary, for example, to have a person responsible for seeking out weaknesses in the production process and relating these weaknesses to solutions which might be available on the market. It is also necessary to allocate responsibility for implementation of innovations and, last but not least, for assessing the results of attempted innovations.

On the macroeconomic level there are many lessons to be learned. For example, it is possible that certain production problems may be common to a large number of small firms, none of whom are in a position to work out a solution to them. In a case like this, research associations or government schemes may provide an answer.

Obviously, these kinds of problem can be, and have been, identified and solved without the benefit of constellation theory. The theory does, however, help to look out for problems and solutions because it highlights points of action. Wherever there is an action point in a constellation diagram, there is a point to which micro- or macro-policy needs to address itself.

Thoughts on constellations have come through practical experience of empirical studies of manufacturing innovation. Some action points have been identified in this way, e.g. by Bessant,[2] who identified the need for a 'public actor' to help firms in phase two of the innovation constellation; the identification of a solution. In a forthcoming paper we shall apply this theory to analyse a number of further case studies.[4] Whether or not we succeed in achieving greater clarity will determine the success or failure of this proposed new way of looking at manufacturing innovation.

References

1. Bessant, J., *et al.* 1981 *The Impact of Microelectronics—Report and Bibliography*. Frances Pinter, London
2. Bessant, J., 1980 Factors influencing introduction of new manufacturing technology. Mimeographed TPU Report, to be published
3. Braun, E., MacDonald, S., 1978 *Revolution in Miniature*, Cambridge University Press, London
4. Braun, E., Moseley, R., Wilkinson, B., 1981 *Manufacturing Innovation in the West Midlands Materials Forming Industry*. To be published
5. Downs, G.W., Mohr, L.B., 1976 Conceptual issues in the study of innovation. *Admin. Sci. Q.* **21**: 700–14
6. Freeman, C., 1974 *The Economics of Industrial Innovation*. Penguin, Harmondsworth
7. Mensch, G., 1975 *Das Technologische Patt*. Umschau-Verlag, Breidenstein KG, Frankfurt
8. Nelson, R.R., Winter, S.G., 1977 In search of useful theory of innovation. *Res. Pol.* **6**: 36–76, p. 65

3.3

Technology, Productivity and Process Change

William J. Abernathy and Phillip L. Townsend

This reading develops a model which seeks to clarify the factors that facilitate or inhibit the successful application of new technological knowledge to improve productivity, where productivity is defined to include the effectiveness of the good or service that is produced as well as efficiency in the factors of production. The approach differs from much prior work in two important respects: the characteristics of technology users are seen as more important to the success of technological innovation than the characteristics of technology creators (i.e. the problem is more on the side of utilization and diffusion than in managing science and engineering), and that the factors which critically enable innovation are best described as patterns of conditions rather than in terms of single important variables.

The unit of analysis and the definition of the 'technology user' in this reading is the overall production process which is employed to create a product, whether the product is goods or a service. More specifically, the unit of analysis is defined to include the physical process (capital equipment, tasks, labour skills and process flow configuration), the product, the characteristics of input materials and the characteristics of product demand that are incident on the process. The term *productive segment* is used to describe these included elements which can otherwise be conceived as the vertical span of a production process that would typically be managed by the senior operating executive in an organization. The essential idea underlying the model is that a process, or productive segment, tends to evolve and change over time in a consistent and identifiable manner. That is, as a given productive segment develops over time, it follows a predictable profile that will be common among different industries.

The hypothesis is that the stage of development which a productive segment has reached along this profile will determine its propensity to host

Source: from *Technological Forecasting and Social Change*, 7, 1975.

particular types of innovation. The most likely form of successful innovation, the typical barriers and enabling conditions, and appropriate management skills all tend to depend upon the stage of development which the recipient process has reached. A model which explains these variations promises to reduce the number of variables and the complexity which the decision maker must face in managing technological innovation. In addition, it will provide a framework for transferring management expertise from one economic sector to another.

Understanding Process Change as a Pacing Element

To better understand the sources that pace innovation and diffusion, it is instructive to consider how the application of new technology has been related to process development in specific examples. Evidence concerning the course of evolutionary development in a number of diverse industries suggests that management's effort in improving the production process is itself a major pacing element that explains success over the long term. Technology cannot be applied until the recipient production process has reached an essential minimum level of definition and consistency. This is true whether the technological application is conceived as a process innovation from within the organization or as a new product innovation by some external organization.

One particularly interesting example is the development of the electric light bulb manufacturing process described in J. R. Bright's book [reading 3.1].[1] During a period from 1908 to 1955, the manufacturing process of electric light bulbs evolved from rather chaotic conditions, manual operations and rudimentary process equipment, almost to complete automation involving extensive process technology. Output per direct man-hour increased eighteenfold in this period. Although the application of much new potential technology was important to the increased productivity, Bright found that these applications were enabled by managerial changes in the process itself.

Bright describes a series of incremental changes which enabled the conversion of potential technology to operational practice. First, during the early phases a series of improvements were introduced that led the process from batch to continuous operation. These were achieved by adopting systematic layout, standardized operations and production sequence changes. Second, the product underwent redesign a number of times to facilitate these process changes and permit mechanical manipulation. Third, individual manual operations were replaced with direct mechanical analogies. Finally, automatic work feeding, removal and material handling devices were used to integrate all machines to form a highly automated process.

The important implication of this study is that management has a critical role in causing the process to evolve and in readying it for technological innovation. Initially, the process had to be *rational* before even simple technology could be effectively applied. Then individual manual operations could be first mechanically assisted and then automated. Later in the period (1936 forward in Bright's study) as process uncertainty or ambiguity was reduced, the process equipment began to be designed as an integrated whole – a systems development effort as it were.

Studies of other processes show that they have evolved with a strikingly similar pattern. Early gains in automobile manufacturing resulted from a similar evolution in the manufacturing process at Ford. The basic concepts here were not new. The use of interchangeable parts and the assembly line were accepted practices by the time Ford applied them to automobiles. (For example, line operations were employed by both the fifteenth century arsenal at Venice and early meatpacking operations in Chicago. Eli Whitney used interchangeable parts in manufacturing guns for John Adams. Ford made some contributions, apparently by combining and refining the two concepts.) They represented concepts of process organization rather than scientific advances. They became the basis for numerous changes in management practices and a stream of changes in many process operations, product design, material inputs etc., that made the automation of engine manufacture and automobile assembly possible.

This same pattern of process development is evident in nonindustrial sectors as diverse as medicine and health care delivery. As the physician's technique and knowledge in new fields like open-heart surgery are improved and the uncertainties and ambiguities in the process begin to be resolved, then the prospects for further significant improvement through innovation in capital-embodied technology grow. Important technological innovations here have followed practice, however, and attempts to promote progress in reverse order have not been notable for their success. In several instances highly automated, comprehensive systems, like automated multitest facilities and complete computer-based hospital information systems have been developed to promote substantial improvements in the current practice of health care delivery. The impact and acceptance of these technological innovations have been minimal, however, particularly when compared with development cost. These patterns, like those in light bulb manufacture, demonstrate an important role for new technology in supporting the development of important new medical fields, but a limited and risky role in pacing improvement.

The coordinated pattern of product and process change, developed by example above, is also evident in at least two independent fields of systematic research: the work done on 'operations technology' reported by Harvey,[4] Hickson *et al.*[5] and Woodward,[13] and studies of 'the product life cycle' of international trade by Vernon[11,13], Stobaugh[8] and others. The research on 'operations technology' based on studies of several hundred firms in the US and Great Britain shows that as a firm's production process becomes more highly integrated, and of a line flow nature, there are corresponding broad changes in product and the firm's organizational structure. For example, Harvey's[4] results show that with such changes the product becomes more standardized and the product line less diverse. Hickson's work suggests that the use of scientists and engineers (R & D input) also increases with increased integration of the production process and scale increase.[5]

Life cycle studies in international trade demonstrate a related pattern of change over time. As a product becomes more standardized, then more efficient, less flexible production technologies are chosen, and the basis of competition shifts increasingly from one that emphasizes innovation in product performance features to one of minimizing cost. These changes have important implications for the source and nature of technological innovation.

Observations that may be drawn from the preceding examples and literature lead to several hypotheses about the systematic properties of development in a productive segment. Five basic concepts comprise the underpinnings for a descriptive model. First, as costs are reduced and productivity is increased *production processes evolve toward greater predictability and more ordered or systematic relationships*. (While such progression is evident in many examples, it is systematically documented in Bright's study.[1]) A series of almost imperceptible changes are involved even if some major process alterations are readily apparent. Some productive segments remain essentially unchanged for long periods of time, but rapid productivity improvement will be associated with a movement toward the more systematic state.

Second, *development requires consistent progress in four* different factors: *process continuity and predictability, product improvement and standardization, process scale and improved material inputs*. Progress in each of these factors is apparent in light bulb manufacturing, automobiles, the results from studies of operations technology, technological change in international life cycle studies and less obviously in many service sector examples. All contribute to advancing productivity. This not only implies an interdependency between factors, but also as a third concept, that *a minimum degree of evenness in advance among productivity factors is needed*. The product standardization (any colour Ford as long as it is black), the use of the assembly line to rationalize production, division of labour and increase in scale cannot be separated out as sufficient conditions for Ford's results. Change in all of these aspects went hand in hand.

Fourth, *the innovation or utilization of product as well as process technology will be heavily influenced by the level of development* that has been attained within the productive segment. This will also influence the *origin and type of applicable innovation*.

In early stages of a productive segment's development, most technological applications are to new products, if new technology is a significant factor at all. The innovation in light bulbs was the light bulb itself. The important changes in the process at this stage are to make it predictable and rational and technology is of little importance. As the productive segment develops, technological inputs to process change are increasingly enabled but the product design necessarily stabilizes. This in turn reduces the input of new technology to the product. With further progression, process innovations and applications become more technologically complex, costly and ultimately become the vehicle for linking the process into an integrated system. In the extreme the disrupture cost of process change becomes the source of inhibition for product innovation. Many have argued that this is the state which the automotive industry has now achieved.

Fifth and finally, *the stream of changes within a productive segment builds cumulatively over time to significantly alter its aggregate characteristics*. Three definite stages can be identified that are distinctly different from one another but common to productive segments in the same phase of development, even though the segments are in different industries. These stages are described here as *unconnected, segmental* and *systemic*. These are named for the major structural properties of each stage, as summarized in figure 1. The rows in this figure represent stages, and entries under each column identify characteristics of the four productivity factors. The differences among stages are pervasive and beyond a purely technical structure have implications for the organization and the innovative climate of the organization that manages the productive segment. The three stages are discussed next and then the typical modes of change between stages are identified.

Uncoordinated

Early, in the life of process and product, technical advances and market expansion and redefinition result in frequent competitive improvements. The rates of product and process changes are high and there is great product diversity among competitors. The process itself typically is composed largely of unstandardized and manual operations or operations that rely upon general purpose equipment. During this state, the process is fluid, with loose and unsettled relationships between process elements. Such a system is 'organic' and responds easily to environmental change, but necessarily has 'slack' and is 'inefficient'.

Segmental

As an industry and its market matures, price competition becomes more intense. Production systems, designed increasingly for efficiency, become mechanistic and rigid. Tasks become more specialized and are subjected to more formal operating controls. In terms of process, the production system tends to become elaborated and tightly integrated through automation and process control. Some subprocesses may be highly automated with process-specific technology while others may still be essentially manual or rely upon general purpose equipment. As a result, production processes in this state will have a segmented quality. Such expensive development cannot occur until a productive segment is mature enough to have sufficient product volume and at least a few stable product designs.

Systemic

As a process becomes elaborated, increasingly integrated and investment in it becomes large, selective improvement of process elements become increasingly difficult. The process becomes so well integrated that changes become very costly since they must consider the whole processing system. Process redesign comes more slowly but may be spurred either by the development of a new technology or by a sudden or cumulative shift in the requirements of the market environment. If changes are resisted as market and technical forces continue to evolve, then the stage is set for revolutionary as opposed to evolutionary change within a productive segment.

Modes of Process Change

The right-hand side of figure 1 describes the type of process change activities that are associated with movement from one stage to another: process rationalization, systemic development and process product realignment. Change in the early stages is the easiest to achieve.

Process rationalization

The transition from the uncoordinated state is achieved after standard methods of production are adopted, process flows are routinized and (in general) part of the production process is rendered stable enough to justify greater expenditure for more automated and special equipment. A predictable and rationalized process will require that at least one product be produced in high volume with standardized specifications, which is normally associated with a middle stage of the product life cycle. The process will come to have islands of automation interspersed with manual operations. Innovations will increasingly result from the development of new technology. Suppliers develop a more continuous focus on the productive segment and on alternative technologies which might improve the process. Suppliers increasingly develop internal capability to innovate technology and increasingly initiate innovations in the processes of customers.

Systemic technological development

A combination of process rationalization, stabilization in product design and application of increasingly advanced technology must render the entire process much more stable before systemic development may occur. There is high emphasis on process engineering, including issues of inventory control, process balance and optimal equipment selection. Eventually, technological advances need to be treated as systems development problems. A specialized equipment supplier industry may form and be sustained in larger industries to conduct the development of process-specific equipment, since development costs are then amortized over more installations. This contrasts with the earlier stages where the process equipment suppliers will provide or adapt general-purpose (non-process-specific) equipment. Relationships with suppliers may also change. Unmanageable or difficult-to-automate process tasks will tend to be separated out of the main process and often turned over to feeder contractors. Considerable control is also gained over material suppliers so that inputs can be optimized to the needs of the process.

Process/product realignment

As a process evolves through its stages, productivity gains result from changes in the four factors – material inputs improvements, changed process technology and labour skills, larger process scale and a tailoring of product characteristics. The development will continue as long as the productive segment faces a similar environment. But what if a productive segment becomes stalled in any one stage? If no movement occurs because progression in one of the four productivity factors is barred or because management is ineffective, then productivity gains can be expected to diminish or even become negative.

A dilemma is apparent. Continued evolution toward a systemic state offers the benefits of high productivity but only at the cost of decreased flexibility and innovative capability. A productive segment places itself in a very vulnerable position when it achieves a systemic state and may then face limits to further development through blockage of one or more of the productivity factors. It must face competition from: innovative products that are produced by other flexible segments that are more capable of substituting products; foreign imports; competing products from other industries with high cross-elasticity of demand; or process changes by customers to eliminate the product directly. Once such a state is reached and further development becomes blocked then stagnation or death of the productive segment may be threatened. This is essentially the problem that faced Western Union, around the turn of the century, when as the giant of the communications industry it decided to shun the innovations in voice communications and specialize itself in efficient telegraphic operations. It is also reflected in Ford's position following the demise of the Model T in 1926, and in that of the manual typewriter producers in the 1940s and 1950s.

There are choices to be made other than for the management of a segment to accept economic demise. First, firms may elect to achieve lower costs by moving production to a foreign country. In fact, cases of foreign direct investment which are motivated by efficiency (most are not) may be viewed largely in just this way – as competitive reaction to approaching stagnation of a productive segment. For example, establishment of foreign shoe production operations by US companies at least partially reflects the inability of the shoe industry to move toward more productive stages of process development.

Alternatively, a decision may be made to backtrack along the traditional course of evolutionary process development to a more flexible state with greater capability for innovation but less process integration, efficiency and

Stage in this productive segment	Material and parts – inputs	Process characteristics		Scale	Product	Modes of process change
		Technology	Labour			
I Uncoordinated	– Raw material and parts used as available from supplier – Types and quality vary widely – Limited influence over supplier	– General purpose equipment and tools used as available from industry – Special adaptations to general purpose machines are made by user (jigs, fixtures etc.) Flow through process needs careful management control	Most workers have a broad range of performance skills Considerable flexibility exists in type of tasks each worker can and must perform Labour organization (if any) is along craft or skill (trade unionism)	– Capacity ill defined – Greater volume achieved by paralleling existing processes – Short-run economies of scale achieved through learning curve improvement of manual operations – Few scale barriers to entry into industry segment	– Great variety of products with different features and quality – Frequent design change – Market relatively insensitive to price and quality (imperfect market that is price inelastic)	(in transition from one stage to the next) Process rationalization: – Standardize tasks – Develop even flow through all process steps – Automate easy tasks – Introduce systematic or mechanized materials handling – Redesign product and process to automate difficult tasks →
II Segmental	Suppliers are strongly dependent Tailored material specifications imposed on supplier	– Process automation is evident for some process tasks and systematized work flow – Level of automation varies widely within process. Islands of highly automated equipment are linked by manual operations	– Manual tasks are highly structured and standardized – Labour is specialized with technical skills becoming more important – Overhead labour functions such as maintenance scheduling and control are a significant cost	– Capacity increased by equipment addition and advances to de-bottleneck particular operations – Minimum size process necessary to compete in industry segment	– Some segments of market sensitive to price and quality (encouraging standard products and scale economies) – Significant volume achieved in some few product lines	Systemic development: – Separate difficult-to-automate tasks from

III **Systemic**	– Input's characteristics are optimized to process needs – Supplier process integrated into overall process design – Tasks that cannot be automated are segregated from process and are often subcontracted or performed by suppliers	– Unique process equipment is designed for some tasks (often by outside firms) – Single units of equipment perform multiple process tasks and are integrated into automatic material handling equipment – Formal systems engineering is required for process change – Process equipment is designed as an integrated system, often by separate engineering groups or engineering companies – Licensed technology may dominate, depending on the industry	– Direct labour does monitoring and maintenance tasks – Most important skills concern technical process equipment operation – Labour classifications are rigid and are of primary concern to labour organization	– Complete new facilities designed to achieve economies through spread costs – Market growth and technological evolution pace scale increase – Antitrust laws, logistics or external factors eventually limit scale growth	– Product variability is low and volume is high – Standard products if price competition is prevalent or standard groups of products if product differentiation is prevalent – Co- and by-products play greater role

process or eliminate them
– Design products to have maximum common process elements
– Arrange administrative organization for congruence of control over process flow

Product and process realignment: to meet changing markets and technological advances. (May reset to earlier stage: 1 or 2) or stagnate during maturation

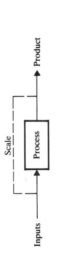

Inputs ⟶ Process ⟶ Product

Scale

Figure 1   Productivity factors in three process stages

capital intensity. This essentially is the course of action that major productive segments in the automotive industry followed after the end of the Ford Model T era in 1926. The assembly, engine and body plants at Ford reached their highest level of integration during the Model T era when the basic vehicle was standardized and the price was brought down to $890 (in 1958 dollars) and through-put time from iron ore receipt to vehicle delivery was reported to be as short as four days. Thereafter vehicle performance was increased together with price and the major process segments such as engine and assembly operations were more decoupled one from the other.

This choice is a major strategic long-range decision that management must make either implicitly or hopefully in an explicit manner that recognizes advantages and consequences. In effect, it is a trade-off decision between the capability for innovation and productivity improvement. Whether the firm chooses its response or is forced to react, environmental change will eventually cause most product segments to choose among: reverting to an earlier stage of the process life cycle, or 'escaping' by international migration or its own economic death.

Implications for Management and Research

The proposed model of process evolution should be viewed as a new theory about process change and its implication for innovation. The five concepts that were introduced in the section above represent the essential hypotheses of this model.

The basis for extending the essentially descriptive model of process evolution (as in figure 1) to one that is measurable and replicable, is demonstrated by the explicit methodological approach developed and applied by Harvey and Woodward.[4, 13] Such a model promises a method that can be used to assess the characteristics of a given productive segment in a way that can usefully clarify its capability for innovative activity and productivity improvement. Research is currently underway to develop the process model in this way and to test its implications for innovation and productivity improvement.

The implications for managers and policy makers fall into three major categories: managing technological innovation, improving the productivity within a productive segment itself and the use of the model as a framework for formulating policy in respect to both technology and productivity.

Several important issues in managing technological innovation are addressed by the model: the natural locus of innovation (or the most potentially fruitful source); the most appropriate type of innovation; and the array of barriers to innovation.

(1) The locus of innovation shifts with the stage of development. (A number of studies that have examined changes in the pattern of innovation over time indicate a dramatic shift in locus. Included among these are Knight's study of computers and Clarke's study of rocket engine

innovations.[6,2]) During the unconnected stage in the development of a process, innovative insight comes from those individuals or organizations that are intimately familiar with the recipient process, rather than those intimately familiar with new technologies. The critical input is not state-of-the-art technology but new insights about the need. Later, in the systemic stage, needs are well-defined, 'system like' and easily articulated. These needs lend themselves to complex technological solutions, and the innovator will frequently be one that brings new technological insights to the problem. (This point is the essential conclusion that is implied by numerous studies which show that 'need' is the principal stimulant of innovation.[7, 12]) This may be a formal engineering or R & D group, an equipment company or some other external source. In undertaking managerial action to stimulate innovation, it is important to appreciate these distinctions so that the most likely sources of innovation can be identified, nurtured and supported.

(2) The type of innovation that is likely to succeed, whether systems are technologically complex or simple, and whether applied to product or process, also depends upon the stage of development. During the unconnected state most technological applications are to the products that the productive segment will produce. Few are to process improvement, and those that do occur tend to be simple in application and to address single needs. Complex technological systems of process equipment do not 'take' well when the recipient process is ill defined, uncertain and unstructured. Systems technology has not been very successfully applied to solve ill defined process needs. The converse is true in the systemic stage. In a symmetric fashion, isolated radical innovations, of even major significance, seldom gain ready acceptance when the recipient productive segment is in the systemic stage. Like many of the theoretical concepts which seem to have application to pollution problems in automobile engines, most have effects that are too disruptive even though they promise high merit as a product improvement. The seemingly isolated innovation must in reality be incorporated as change throughout the systemic productive segment. A realistic assessment of the type of innovation that will be successful, and how it should be introduced, depends upon an understanding of the productive process that will receive it.

(3) The total array of barriers to an innovation, like the appropriate type of innovation, changes composition with the stage of development. In the unconnected stage, resistance centres around perceptions of irrelevance. Will the innovation work and meet a need? In the systemic state resistance stems from the disruptive nature of innovation. Will it displace vested interest and disrupt current practice? In both forms, resistance radiates into many different dimensions, labour practices and union positions, legislation, product acceptance etc., but the model helps to clarify the changing nature of these barriers.

The implications of the model for improving productivity follow rather directly from the five concepts that were defined earlier. Two are noted below to suggest its application.

First, the major factors in process change (product and product characteristics shown as columns in figure 1) need to advance with a minimum degree of evenness. To the extent that one or more factors are lagging, the risk increases and the return from improvement in other factors decreases. The model directs attention to a balanced consideration of all the productivity factors when change in any one is considered, thereby helping to uncover inconsistencies. The major study which provides the most direct evidence of this pacing is Project Hindsight.[7] Most breakthroughs resulted from the decision to build the system to which they contributed.

Second, a factor has an *order of progression*. A change which requires a reversal in the natural order of progression in any particular factor will have implications for the segment. For example, changes in the structure of manual tasks in a process may be sought to enrich job content and job satisfaction to workers. If such changes lead to less specialization, more

Table 1   Industry Productivity Patterns

High productivity industries	Some comparative low productivity industries
Unconnected stage	
Livestock and livestock products[a]	Forestry and fishery products[a]
Real estate and rental[a]	Medical and educational services[a]
Mining (ferrous ore)[b]	Mining (copper)[b]
Segmental stage	
Air transport[b]	New construction[a]
Radio and television receiver eq.[b]	Aircraft and parts[a]
Malt liquours[b]	Iron foundries[b]
Agricultural products[a]	
Hosiery[b]	Footwear[b]
Coal mining	
Systemic stage	
Electronic components[a]	
Office and computing machinery[a]	
Petroleum refining[b]	Glass and glass products[b]
Electric and gas utilities[b]	Steel[b]
Communications[a]	
Aluminium rolling and drawing[a]	Primary copper[b]
Product/process realignment	
Railroad transportation[b]	
Miscellaneous textiles and floor coverings[a]	

[a] Industry definitons correspond to the input–output model of the economy. Productivity improvement rates are projections for the 1965–80 period as reported by the US Dept. of Labor, *Patterns of Economic Growth*.[9] Industries with high rates of productivity improvement (left side) are those with annual improvements above 3.6 per cent. The low category (right side) corresponds to rates below 2.5 per cent.

[b] Industry definitions correspond to SIC standards. Productivity improvement is the annual average during the 1960–71 period. The high category represents a rate of improvement of 5½ per cent or better (left side). The low category is below 2.5 per cent.[10]

discretion and a general loss in the rationalization of major processes, then these changes may represent a general reversal in the direction of process evolution. As such they may thwart long-range productivity improvement. The model provides a basis for planning enrichment programmes so that they need not conflict with plans to improve productivity in the long run.

Finally, as a framework for policy evaluation, the model provides a new basis for strategy choice. It points out the course of action that can be pursued over both the short and long term as a strategy for continuing productivity improvement. At the same time it describes the consequences of following such a strategy to its ultimate conclusion. These consequences include the loss of adaptability to external changes in needs and a reduction in innovativeness. If these latter capabilities are important, then the best choice may be an intentional decision to *slow or reverse* evolutionary progress or to remain in that particular stage which offers the best trade-off between conflicting objectives. There is reason to believe that in practice progress may frequently be halted for long periods and then continued.

Some support for the hypothesized relationships between productivity improvement and production process structure is available from industry statistics. Although the elements included within a productive segment do not correspond well with the broad industry categories that are used in national economic statistics, there is some similarity among many segments in one industry. Table 1 lists several industries that are distinguished by their rates of productivity growth as defined by two different sources of industry statistics. The list includes some with very high and some with very low improvement rates. Those with high rates of productivity improvement are listed on the left-hand side of table 1. An effort has been made in this table to roughly and somewhat arbitrarily sort these into the three stages on the basis of the definition that would best describe the major productive segments included within the industry. A similar procedure was followed on the right-hand side with industries that have low rates of productivity growth. Although industry data are slippery, the resulting pattern suggests that industries with high rates of productivity improvement are those whose major included productive segments are progressing most rapidly from stage to stage. The industries with low rates of productivity improvement, on the other hand, seem to include segments that are best characterized as stalled in one stage of development, often for many years. The low productivity industries do not necessarily have the lowest rates of innovation. They do, however, apparently host a low rate of process innovation.

An interpretation of the data in table 1 is speculative, but it does suggest some important policy considerations. High rates of productivity are achieved in periods of rapid process change in an industry. The decision to pursue a strategy of productivity improvement need not be followed to its ultimate extreme, the systemic state. Each movement that is undertaken should be made in full recognition of the implicit trade-off that is involved – flexibility and innovative capability versus productivity improvement.

References

1. Bright, J.R., 1958 *Automation and Management*, Harvard University, Graduate School of Business Administration (Reading 3.1)
2. Clarke, R., 1968 Innovation in liquid propelled rocket engines, Ph.D. dissertation, Stanford University, Graduate School of Business
3. Gruber, W., Menta, D., Vernon, R., 1967 The R & D factor in international investment of United States industries. *J. Pol. Economy*, 75(1) Feb.
4. Harvey, E., 1968 Technology and the structure of organizations. *Amer. Sociol. Rev.* **33** (2) April
5. Hickson, D.J., *et al.* 1969 Operations technology and organization structure: An empirical reappraisal. *Admin Sci. Quart.* **14** (3): 378–97
6. Knight, K.E., 1963 A study of technological innovation–The evolution of digital computers. Unpublished Ph.D. dissertation, Carnegie Institute of Technology
7. Sherwin, C.W., *et al.*, 1966 First Interim Report on Project Hindsight, US Department of Defense, Washington DC, June 30
8. Stobaugh, R.B., 1972 How investment abroad creates jobs at home. *Harvard Bus. Rev.* Sept–Oct. pp. 118–26
9. US Bureau of Labor Statistics, 1970 Patterns of US economic growth. *BLS Bulletin* 1672
10. US Bureau of Labor Statistics, 1972 Indexes of output per man hour: Selected industries. *BLS Bulletin* 1758
11. Vernon, R., 1966 International investment and international trade in the product cycle. *Quart. J. Econ.* May
12. Williams, B.R., 1967 *Technology Investment and Growth*. Chapman and Hall, London
13. Woodward, J., 1958 *Management and Technology*. Tavistock Institute, London

3.4

Diffusion of Major Technological Innovations: Towards a Broader Analytical Framework

Bela Gold, William S. Peirce and Gerhard Rosegger

Decision-making Processes

Inasmuch as technological diffusion is rooted in decisions by the managements of individual firms, a brief review of the stages leading up to and following such decisions may help to clarify the processes that may be involved. As shown in figure 1, decision-making channels have continuous inputs representing pressures from product markets, factor markets, technological developments and internal operating as well as organizational problems. Proposals for coping with such pressures – including technological innovations – may be generated by external suppliers of facilities and services as well as by various internal groups. The bearing of such proposals on each of the channels of pressure is appraised by staff specialists and assembled into technical recommendations to higher levels of management. Whenever these fall short of yielding inescapable conclusions, because of data inadequacies or substantial uncertainties or implications not covered by prevailing policies, successive exchanges may take place involving responses to higher management's inquiries. Only then is a decision made: to initiate internal studies including research and development projects; or to reject action for the present; or to choose among available proposals, specifying both the level of commitment and the period within which resulting acquisitions or construction should become ready for use. After resources are allocated, a considerable period may elapse before production results are experienced (including technological findings, product yields and needed operational adjustments), followed by

Source: from B. Gold, *Technological Change: Economics, Management and Environment*, Pergamon, 1975.

another extended period before an adequate array of market responses is secured. Both of these delayed sets of results constitute additional inputs to the various channels of pressure as well as to the technical assessments prepared as the basis for the next round of decisions.

There are several reasons for introducing this broader framework. It emphasizes that decisions about technological innovations fit into general decision processes instead of being essentially unique, that technological pressures are continuous instead of arising only when technological innovations are proposed and that technological pressures affect a wide range of decisions in addition to those involving new facilities, often making it very difficult indeed to separate out the differential effects of technological innovations. Secondly, the focus on decision cycles stresses that many major decisions do not take the climactic form of once-and-for-all choices among alternatives, but rather represent temporary commitments subject to successive extensions or modifications on the basis of intervening results and other information.[3] Hence, it may be quite misleading to infer the bases for original decisions from eventual profits and other operating results. This view also serves as a reminder that diffusion rates are affected not only by firms which have hitherto delayed adoption, but also through progressively larger commitments by early adopters experiencing favourable results from exploratory initial applications. Thirdly, this flow chart highlights the problems of reaching beyond available authoritative data and recognized computational procedures to account for top level management decisions ([2]pp. 106–8) by suggesting the role of organizational objectives and managerial values both in reshaping technical judgements and in supplementing them to arrive at major decisions.[3] Finally, the chart underlines the likelihood that periods of one to ten years may elapse before the facilities embodying a major technological innovation are in place and functioning, with further delays to be expected before market responses are adequately plumbed. This raises questions about the significance of findings concerning utilization rates and other conditions at the time when innovations become operative – for wide differences in the periods required to implement decisions involving the adoption of innovations means that conditions at the time when these come on stream need have no consistent relationship to those in effect when the decisions were made, and it is presumably the latter which are of primary analytical concern.

The Decision Model

Decision models concerned with the adoption of major technological innovations seem to have succumbed to the temptations both of overgenerality and of overrationality. To say that major innovational decisions are based on profitability expectations adjusted for the estimated probabilities of adverse outcomes may be unobjectionable, and may even be correct in some sense; but it is certainly unenlightening. As an 'explanation' of past decisions, it offers nothing more than a tautology: i.e. if an innovation was

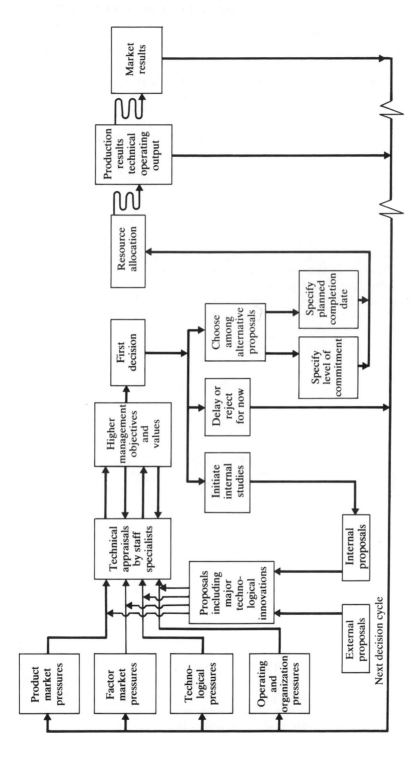

Figure 1   Decision process for major technological innovations

adopted, management must have expected it to be sufficiently profitable; and if not, not. And as an empirical 'test' of such expectations, it is hardly more helpful to demonstrate that the innovations which survive and achieve eventually wide diffusion are those whose utilization has been associated with profitability. Serious analysis surely requires digging deeper.

The diversity of diffusion patterns presented earlier suggests that firms appraising newly available major technological innovations usually emerge with widely differing estimates of the desirability of immediate adoption. In seeking to account for such disparate conclusions, it may be useful to examine possible sources of difference in the predecision environment, in the characteristics of the innovation and in the bases for making evaluations relative to each criterion as well as for combining the results into a decision.

Perhaps the most important elements of the predecision environment are the urgency of the needs leading to a consideration of major innovations, the nature of these needs, the availability of alternatives to technological innovations for coping with such needs and the existence of any marked preferences by management among such alternatives. Firms may obviously differ widely in the degree of felt urgencies, defined as the period within which management considers it essential that measures for dealing with specified pressures become operative. Such urgencies might include pressure to expand into newly burgeoning markets; or to offset serious losses in, or impending threats to, its competitive position; or a need to ease rapidly developing imbalances in its production flows – such as was experienced by the steel industry as a result of the shift to basic oxygen furnaces, and which is intensifying consideration of continuous casting. It may even be useful to distinguish between 'discretionary' decisions which allow management a choice between action and delay, and those permitting a choice only among alternative proposals, as suggested in figure 1. Those facing the greatest urgencies might then be expected: (1) to act sooner than others with comparable evaluations of the prospective benefits and burdens of available innovations; (2) to favour innovations promising more immediate gains; and (3) to be willing to accept greater risks relative to expected gains.

Differences in the specific needs constituting the pressures to action are also likely to engender dissimilar environments within which alternative innovations are evaluated by individual managements. Even if the list of such pressing needs is limited to a few broad categories – increasing or replacing capacity, increasing the scale of some or all operations, improving product quality, lowering requirements of inputs in short supply, decreasing production costs via reduced inputs relative to output levels, decreasing such costs through lower factor prices and increasing sales – it is clear that the potential contributions of technological as compared to marketing and other innovations are likely to be more dominant in respect to the first five than the last two; and it is also clear that individual innovations would likewise differ substantially in their usefulness in promoting such disparate ends. However, the more general the terms in which corporate urgencies are defined, e.g. as a need to increase profitability, the wider the range of alternatives to technological innovations which invite consideration. And inasmuch as firms may have deeply rooted preference orders as among alternative approaches to major problems – reflecting differences in managerial training, experience and past sources of company success – they may well differ both in their readiness to consider technological solutions at all, and in the relative levels of net advantage required of technological as compared with marketing or other preferred innovational alternatives.[3]

While the aspects of technological innovations which engender different managerial reactions are not yet clear, they may include the source and nature of the innovation as well as the kinds of prospective benefits and risks involved. The first of these concerns whether the innovation has been developed by the firm's own technical staff, or is being offered by an external source on an exclusive basis, or is available from a supplier seeking to promote widespread use. The second might include such matters as whether the technology involved is alien or kindred to the company's current fields of specialization and whether resulting products would flow through existing channels to established customers or would involve breaking new ground in these respects. For example, companies with products long rooted in mechanical engineering and staffed by such specialists tended to resist the early incursion of electronic approaches to effectuating similar design, manufacturing and production control objectives. Managerial responses to available alternatives among technological innovations are also obviously dependent on the distinctive arrays of benefits and burdens offered by each relative to company rankings of their respective urgencies and preferred forms of burdens. Merely defining each prospective innovation's relative contributions to each of the seven broad categories of need suggested above would at least partially explain divergent evaluations by companies seeking to cope with dissimilar patterns of urgencies. For example, even innovations offering lower costs might be of limited interest to firms confronted by urgent demands for higher quality products. At a deeper level of detail, innovations with a limited scale or product range would obviously be rated as less attractive by firms whose needs would consequently remain largely unmet. Moreover, innovations likely to yield marketable benefits only five to ten years hence are likely to be rated as relatively less desirable among firms facing critical shorter term pressures.

Among the deterrent factors associated with prospective innovations, it might suffice for current summary purposes to list four potential burdens (displacement of undepreciated assets, need for new investment funds, changeover costs including attendant labour and organizational turmoil, and the possibility of higher unit costs accompanying desired product improvements or other gains) and three potential risks (that the innovation will not function acceptably, that product or cost improvements will fall short of expectations and that the innovation under consideration is likely to be displaced by more advantageous innovations before recovery of the attendant investment can reasonably be expected). Inasmuch as available innovations are likely to warrant different ratings in respect to each of these deterrents and inasmuch as different firms might attach unequal emphasis to these several criteria, another important source of disparate evaluations of given innovations by prospective adopters has been identified.

Attention may now be focused on the decision-making model for which the predecision factors and the evaluations of the benefits and burdens of alternative innovations which have been discussed serve as inputs. Almost all managerial decision models assume that each criterion affecting decisions is somehow translated into cost and revenue equivalents and that these are then combined to yield profitability estimates as the ultimate determinant of the decisions made. Pervasive as they are, however, these assumptions are not without alternatives worth some consideration.

In respect to the overriding role of profitability, one of the more extreme dissidents was Keynes, who argued that many major business decisions are taken 'as a result of animal spirits – of a spontaneous urge to action rather than inaction' ([5]pp. 161–2). Others have emphasized the need to keep moving with the stream of change, even though its eventual outcomes are unclear and even though the short-range outcomes of individual projects do not seem unduly attractive. Some even argue that innovations are necessary to attract and hold good personnel and to encourage innovative efforts. An intriguing hybrid between 'profit-directed' and 'other-directed' views suggests that major technological decisions tend to be dominated by the judgements of technologists, who tend to *evaluate* the physical improvements of prospective innovations and to *assume* parallel economic benefits via *ceteris paribus* assumptions commonly associated with the work of economists. While the forms of such assessments and their shortcomings have been discussed elsewhere ([2]pp. 106 ff.), it should be recognized that such technical judgements tend to be regarded as convincing not only because of their apparent objectivity and precision, but also because too little research has yet been done to indicate the magnitude of the errors which may be involved in assuming parallel economic benefits. Moreover, the fact that most commitments to major innovations unfold progressively over several decision cycles may constitute a significant safeguard in that the dominance of technological judgements concerning gains in physical input–output relationships during early decisions may be gradually offset in later decisions as firmer cost and market data become available.

The logic of rigorous rational models implies a need to identify the major categories of benefits, burdens and risks considered in making a decision, to explain how their respective magnitudes are estimated and how the results are combined to yield a conclusion. If it should turn out that the available factual content of such efforts is often fairly low in respect to newly available innovations and that the estimates used to supplement them are recognized as subject to wide margins of error, the very focus of analysis may have to be shifted, however reluctantly, to more influential albeit less measurable determinants of such major policy decisions.[3] And it is quite possible, of course, that final managerial outcomes are based not on any formally weighted combination of favourable and unfavourable factors, but on the extent of 'concrescence' (to use Max Black's term in accounting for how 'scientific' a given methodology is)[1] among those pointing in a given direction – much like our current practice in interpreting the 'net' implications of 'leading' and 'lagging' business cycle indicators.

One means of attempting to circumvent these difficulties has been to use *ex post* performance data to infer the primary determinants of the original adoption decisions. But this approach raises several problems. The first, which was mentioned earlier, is that eventual results need not have much relationship to the expectations underlying the initial decision, which may have been modified repeatedly during successive stages of extending the original commitment to encompass more, larger and even improved facilities. A second problem is that estimates of the differential effects of given innovations on profits are frequently subject to serious error, because most innovations serve as parts of larger complexes of facilities whose results cannot be disentangled from one another because firms which are especially successful with major innovations are likely to have managements which are above average over a wider range of performance and because profits represent the interaction of technological improvements with a variety of adjustments in product and factor markets as well. 'The overwhelming majority of business investment projects are of this segmental or component type – replacements, improvements, expansions or some combination thereof. They become a part – usually a small part – of an existing operation. Since it is impossible in most cases to compute their separate revenue generation and operating cost incurments after they are in service, it is even more impossible to predict these magnitudes before the projects are acquired' ([7] p. 52). The third difficulty is that the differential benefits of innovations tend to be limited during the early period of their full functioning because of attendant efforts to learn how to operate them effectively[4] and how to adapt to market responses; and they also tend to erode with time under competitive conditions, as has been discussed at some length elsewhere.[2] This raises the questions of when evaluations should be made and, even more interesting, what the benefit patterns may be expected to look like through time. A fourth problem is that studies of the outcomes associated with eventually successful innovations can hardly provide effective models for deciding whether or not to adopt proposed new innovations, when the probabilities of failure may far overshadow those of successes. And finally, the data offered concerning the profit expectations 15–60 years ago, when recently studied innovations were first adopted,[6, 7, 8] represent recollections, inferences, estimates and guesses, most of which seem open to substantial margins of error.

One might well add that in order to account for empirical diffusion patterns, it would seem necessary to explain not only the decisions of 'pioneer' adopters, but also the changing bases for decisions confronting individual firms during later stages of diffusion as well as the characteristics of the changing mix of firms dominating adoptions during such successive stages. In particular, laggard managements would not only be confronted by changing estimates of technological and market uncertainties, but they might also face change in materials prices, wage rates and even in product prices and product specifications generated by market responses to the spreading innovation.[2] Thus, further adoptions might come to be motivated increasingly by threats to competitive position and even to survival rather than by the attractions of added benefits. As for the changing mix of adopters between early, intermediate and later stages of diffusion, there is little basis as yet for even formulating interesting hypotheses comparing diffusion rates among different classifications of firms inasmuch as the critical distinguishing variables are not yet clear.

References

1. Black, M., 1949 The definition of scientific method. In: Stauffer, W. (ed.) *Science and Civilization*. University of Wisconsin Press, Wisconsin
2. Gold, B., 1964 Economic effects of technological innovations. *Management and Science* Sept.
3. Gold, B., 1969 The framework of decision for major technological innovation. In: Baier, K., Rescher, N. (eds.) *Values and the Future*. The Free Press, New York
4. Hirsch, W.Z., 1956 Firm progress ratios. *Econometrica* April
5. Keynes, J.M., 1936 *The General Theory of Employment, Interest and Money*. Macmillan, London
6. Mansfield, E., 1968 *Industrial Research and Technological Innovation*. Norton, New York
7. Terborgh, G., 1958 *Business Investment Policy*. Machinery and Allied Products Institute, Washington, DC
8. *33/Magazine of Metals Producing*, June 1969

PART 4
Corporate Strategies for New Technology

4.0

Introduction

This part broadly looks at strategies for improving the implementation of new technology. Several of the previous readings have indicated the importance of relating decisions on technological change to the environment in which the productive unit is located. In reading 4.1 Braham considers the role of Marks and Spencer in the garment industry. This is a sector which has been the subject of considerable attention in terms of a shift to countries with lower labour costs. Given the scale of the relocation of production that has taken place and the differential of labour costs – even in relation to the UK – it is perhaps to be expected that the shift will continue and that the industry will continue to decline in significance. Braham shows that this is not necessarily the case. In part, the high productivity levels that can be achieved with the new technology that is continually becoming available provide a means of countering the advantages of producers in developing countries and of capitalizing on their locational disadvantages. But the study emphasizes that this is not simply a matter of applying new technologies but of relating these to market opportunities. Beside acting as a 'boundary scanner' for new technology for many of its smaller suppliers, M & S also provides very direct pressures for technological change to be market led. The example is one that should be instructive, particularly for those who might be accused of being over production oriented.

This provides a useful background for consideration of Small's proposals for revival programmes in manufacturing firms (reading 4.2). While directed at the detail of change, the reading follows along the lines of much that has been suggested by earlier authors. In part, Small is concerned to re-establish the key role of engineers in the initiation, design and achievement of change, their role having been generally subordinated to that of other groups such as accountants. His main concern is that technological change is necessarily being set within a broad re-examination of the goals of the production unit. This is likely to reveal possibilities for technological change (using the term in Macdonald's sense) on a much broader scale than may have been envisaged, some of which will not be centred around machines. Drawing on his experience as a prominent engineering consultant, Small shows that changes which are really a better way of organizing established technical systems will often produce large savings, sufficient to finance subsequent investment in new equipment. Yet he is somewhat critical of evolutionary change which, as earlier authors suggested, is typical of much technological development. Small regards much of the inadequacy of existing production systems as being a consequence of successive change decisions which were unrelated. Ultimately however, he appears to accept that this will remain the pattern but demands that it take place in a coordinated way such as through phased improvement plans.

The factors that may contribute to relative technological stagnation and to the associated loss of markets are taken up by Hayes and Abernathy in reading 4.3. Although they write within an American context, many of their conclusions are very relevant for a British audience for comparison reveals some striking similarities – allowing for obvious differences of scale and context – between manufacturing in the two countries. These similarities are not solely those of performance – declining investment, slow growth in productivity etc. – but of managerial attitudes and systems. Thus Hayes and Abernathy set out their diagnosis of the nature and causes of American business performance in familiar terms. For example, if the emergence of generalist management is the problem they suggest, it is a problem in both countries – although not, as yet, in Continental Europe or Japan. Similarly, their concerns about the detachment of much strategic thinking and policy from technological decision making, and about the pressure for short-term gains at the expense of long-term development, are very much in place in the UK setting. Factors like these have directed decision making away from technical issues when, in their view, 'industrial success depends, "to an unprecedented degree" ... upon competition in the marketplace on technological grounds – that is, to compete over the long run by offering superior products'.

This last point (and others raised by Hayes and Abernathy) is underlined by Sciberras' study of international competitiveness in the television industry (reading 4.4). Japanese firms have been extremely successful in penetrating both the European and the American markets. It is shown that this is ultimately due to Japanese strategy on the one hand, and the inadequate response of their competitors on the other. The Japanese correctly saw that they could compete, not by following the strategy of marketing low cost and fairly basic sets, but by marketing on the basis of product quality and reliability. Sciberras relates their success to a number of factors. Of particular interest is the importance attached by the Japanese to process R & D, to process innovation and to the conditions which encourage process innovation. He describes the integration of product design – particularly component reduction – with process changes that were concerned much less with the reduction of production costs than with the improvement of product quality. This was a direction effectively closed to European and North American firms since they generally applied limited and short-term financial criteria to the evaluation of technological change. The acceptance by Japanese companies of intangible factors in the justification of investment and of longer pay-back periods enabled a more rapid use of automated methods.

Without some modification of attitudes to technological innovation, (particularly among the nontechnical managers involved in decision making), and without changes to formal appraisal criteria, the experience of subsequent developments in production processes in manufacturing may well follow that of the television industry. In reading 4.5 Towill considers some of the issues that have to be dealt with in the selection and use of robots. These are often seen as 'a complete working tool ... requiring no special integration ... which will work immediately upon installation ... will require no debugging or engineering effort to keep operating at maximum throughput'. Towill deals with some of the technical issues of selection, assessment and utilization of robots. The article also indicates that if there is to be successful general application of robots a change in the underlying approach to firm-level innovation is essential. Among other things this requires that robots not be seen in crude terms as direct substitutes for people but that the many advantages of human operators be rediscovered and utilized. Additionally, while their effects may appear to be limited they present considerable problems of integration within manufacturing systems even when used on a limited scale.

Towill's article in common with the previous two, raises the question of relationships between people – managers particularly – who come from different disciplinary backgrounds and who now work in different functional areas of the firm, which may also be separated by geographical as well as other factors. Where technological change is sought, planned or effected there is generally a need for such interdisciplinary working. The means by which this can be achieved has received a considerable amount of attention in the project management literature and elsewhere. One of the favoured directions is through the development of matrix systems of organization. In reading 4.6 Stucki describes how this was achieved in a drug company which depended upon continuous innovation. While the example concerns product rather than process change it does demonstrate how a very high degree of disciplinary and functional differentiation, and of organization environment contact can be overcome. In common with much work on the innovation process, Stucki emphasizes the importance of a product/project champion or advocate who will sustain the commitment to and enthusiasm for the project even during periods of gloom and adversity.

However, as Stucki also points out, problems of this sort cannot be changed by organizational developments alone. In reading 4.7 Blumberg and Gerwin, reporting on a comparative study of the United States, Great Britain and the Federal Republic of Germany, suggest that the complexity of some new technologies is such that its control is beyond the capacities of most companies – a reflection of the points made by Godet. While much attention has been paid to technical development of advanced production systems, insufficient efforts have been directed towards the problems of coping experienced by managers, specialists and shop-floor workers.

In their view, it is at this level that many of the problems lie in the path of implementation. In the case of CIM systems, many of the established systems of appraisal and control are simply not valid or applicable and require replacement. Thus the development of a new infrastructure of technical support is needed. In some ways, their conclusions reiterate some of the points made by Small about planned or stepped progression towards the adoption of integrated production systems, allowing complexity to develop in step with learning and experience.

4.1

Marks and Spencer: A Technological Approach to Retailing

Peter Braham

Introduction

This case study is concerned with the installation of new technology in the garment industry; its point of departure is, however, the *selling* of a product rather than its manufacture. It seeks to explore the influence that a major retailing chain – Marks and Spencer – has had on the production methods of its suppliers.

It seems safe to suggest that Marks and Spencer ought to know what products the public want: it is the largest retailer in Britain; each week some 14 million customers shop at its 260 stores; and it receives £1 in every £7 spent on clothing in the UK. The relevance of this dominant market position to production methods and to the application of new technology may be less obvious. In part, this might be because, though the British have been called a nation of shopkeepers, the importance of retailing within the economic system has traditionally been downgraded, and its relationship to other parts of the system has been overlooked (De Somogyi 1967, p. 59). In addition, retailing is often visualized as a low technology enterprise.

These views are not shared by those who have studied M & S. Rees, for instance, discerned in M & S's operations a new conception of the function of a large-scale retailer: it was not only an interpreter to industry of consumer demand, but also 'a guide to the most rapid and fruitful applications of the advances of science and technology' (1973, p. 122).

Marks and Spencer uses the 'leverage' derived from its size as a customer to gain access to the best garment makers at all stages of production. It does so not simply to make demands on manufacturers about, say, quality and delivery, but also, as Rees suggests, as a means of resolving technological problems.

Thus M & S technologists may be involved in the appraisal of machinery in order to advise their suppliers of its existence or in order to specify the standard that can be obtained in a finished product. Sometimes this may even involve studying the latest technology used in producing and collecting a raw material such as cotton, in order to be able to agree with a manufacturer what grade of cotton can be obtained and who can supply it.

The critical elements of the relationship between M & S and its suppliers are, however, more elusive than these examples would suggest. We may begin with the statistic that M & S buys about 20 per cent of the output of the UK clothing industry. Marks and Spencer's policy of 'buying British' is well-known. Thus 90 per cent of the garments that it sells are made in the UK. It should, though, be added that 50 per cent of the *fabrics* from which these garments are made are imported. Ostensibly, this is because M & S must go abroad to purchase certain types of fabric and various ranges of design which are unobtainable in Britain. But even with this qualification, M & S's policy is in marked contrast to those of its principal high street competitors, all of whom import finished goods heavily. If we discount the possibility that this policy is based merely on sentiment, an immediate question is raised: how and why is it maintained, particularly in the face of what is taken to be a substantial and irreversible shift in the location of the centres of clothing and textile production to the less developed countries (LDCs) of Asia?

An Ailing Industry?

It has been widely assumed that, despite the prop provided by M & S, the UK textile and clothing industries are in steady, perhaps terminal decline. But reports of the death of these industries have been greatly exaggerated. Certainly contraction has occurred, but this has been primarily in the number of workers and the number of firms, rather than in the level of output.

This trend is not confined to Britain. For example, according to Frobel *et al.* (1980), clothing and textiles have long been viewed as crisis sectors within an otherwise prosperous West German economy. They argue, however, that while workers lost their jobs and plants closed, the companies continued to thrive by relocating in low-wage countries. Throughout the 1960s and 1970s the industrialized countries (ICs) as a whole commanded a declining share of world trade in clothing, accompanied by a continuous loss of jobs (though at a fluctuating rate). Before 1973 the *annual* loss of jobs was about 1 per cent of total work force in most ICs. Since then it has hovered at around 3 per cent per annum, though rising to 5 per cent in 1974–75 (Hoffman and Rush 1984).

In Britain the main aim of Government policy towards the textile and clothing industries has been to improve competitiveness by providing financial incentives to encourage companies to regroup and re-equip. In general, the take-up of such incentives may have been uneven, although in some cases they appear to have been the chief reason behind the decision to invest in new equipment (Fevre, n.d. p. 91). By the mid-1970s productivity in the clothing industry had improved faster than in any other area of manufacturing, but capital expenditure per employee was still the lowest of all manufacturing industries: £2,928 compared with any average of £6,089 (Business Statistics Office 1976, cited in Coyle 1984, p. 12).

The textile and clothing industries remain relatively labour intensive in all the ICs. In 1980 the share of clothing in the manufacturing output of ICs was 2.8 per cent, but clothing accounted for 5.7 per cent of manufacturing jobs. In textiles the figures were 4.4 per cent and 7.2 per cent (respectively) (*The Economist* 1984b, p. 71). In the ICs these industries have traditionally relied on cheap sources of labour. The experience of the postwar period suggests that reductions in unit labour costs – whether achieved by locating cheaper sources of labour such as immigrant workers, or by improvements in productivity – will still be insufficient to allow the ICs to produce garments as cheaply as LDCs. Indeed, LDC clothing exports have expanded dramatically, particularly since 1970. For example, between 1970 and 1978, they increased at an average annual rate of 30 per cent, rising from $1,300 million to $9,280 million, at which point they accounted for 37 per cent of world trade (Keesing and Wolf 1980, p. 34, quoted in Hoffman and Roche).

What amounts to a partial relocation of textile and clothing plants in LDCs is taken to demonstrate a significant change in the terms of comparative advantage in favour of LDCs. It is widely assumed that in selected industries, such as those in which research and development play a minor role and where production depends on plentiful, but unskilled labour – and here clothing and textiles are seen as examples *par excellence* – 'in these types of production the developing countries can successfully compete with the industrialized countries because of their low wage level ....' (Heimenz and Schatz 1979, p. 11). The principal means of trying to reduce this disadvantage has been the limitation of imports from low-cost countries. The first restrictions, imposed in 1961, were directed at cotton imports. In 1973 these were extended by the Multi-Fibre Arrangement (MFA) to include manmade fibres, both textiles and garments.

In Coyle's opinion, as far as the UK is concerned the MFA (renewed in 1977 and again in 1981) has been quite ineffective, serving 'only to maintain an ailing industry rather than facilitate its modernization' (Coyle 1984, p. 13). From another view, the threat posed by low-cost imports has provided the major impetus behind the rapid and extensive changes in organization and technology that have occurred in recent years.

One of the chief impediments to the modernization of the UK clothing industry lies in its structure and in the outlook which this structure engenders. The great majority of the firms are still small-scale family-owned enterprises. These encompass the sweat shop at one extreme and the modern factory at the other. At either extreme, however, the idea of spending many thousands of pounds on programmable sewing equipment or hundreds of thousands of pounds on CAD equipment may be beyond their resources and, perhaps more significantly, beyond their conceptual framework.

There are also many large-scale enterprises, often employing many hundreds of workers and using advanced equipment. The conventional argument is that in the face of competition from LDCs, their viability ultimately depends on the application of microelectronics to the highly repetitive tasks for garment making. However, this is rather a narrow perspective. It would be more accurate to say that the success of such firms rests on introducing *flexible automated systems*, which allow the production of the range of styles and sizes required at prices which match (or undercut) those of equivalent goods from LDCs. But it would be a mistake to see the installation of new technology solely as a bastion against imports from LDCs. Low-cost imports are not about to overwhelm the European market. In 1980, for example, the amount by which clothing and textile imports into Europe exceeded exports from Europe was equivalent to less than 3 per cent of total production, though without the MFA, the excess would, no doubt, have been greater (*The Economist* 1984b, p. 71). Moreover, a substantial percentage of UK clothing and textile imports, in terms of value, came from other high labour cost countries, notably the USA, Japan, the Federal Republic of Germany, Italy and Scandinavia. Though it might be thought that the UK clothing industry would benefit from having the lowest labour costs in the EEC, it was less advanced and less efficient than its counterparts in other ICs. In particular, it had neglected to make sufficient provision for research and development (Walter 1984, p. 8). Even those companies which had been prepared to modernize have often been content to wait for manufacturers to turn up with new machines, instead of trying to influence the design and capabilities of new technology.

Buying British

In the past, particularly before the First World War, M & S obtained much of its merchandise from abroad. But, as Israel Sieff's memoirs make clear, it became apparent that there were significant commercial advantages in buying from British suppliers (Sieff 1970). It may be argued that this change of perspective came about because M & S, by reason of its concern for quality, was obliged to place a higher priority on forming close contacts with its suppliers, rather than on buying in the cheapest market. This is true, but superficial. First, the commercial advantages in dealing with British garment manufacturers have been sustained and developed. For example, when M & S places an order (or a commitment to produce), no payment is made until it decides to call the merchandise from the supplier. In this way M & S maintains a minimum warehousing capacity, while it is the manufacturer that finances M & S's stock for an indefinite period. This system is accepted – and as far as is practical, costed for – by UK suppliers. In similar fashion, certain of M & S's Scandinavian suppliers are required by M & S to operate their own warehouses in the UK. It is, nevertheless, difficult to envisage this *modus operandi* being applied to LDC manufacturers. Second, the links that M & S established with its British suppliers would not have been maintained indefinitely, had they not simultaneously provided an environment which encouraged the search for new techniques of production.

This may be illustrated by describing what happened when M & S decided to sell men's suits. They discovered that only one UK supplier could produce goods of suitable quality. In contrast, Scandinavian, Italian or West German manufacturers employed advanced technology, offered superior design facilities and could produce garments on the scale envisaged by M & S, who aimed to sell 20 million suits a year within three to four years of entering the market.

A tour of inspection of Swedish factories was arranged to enable technologists from M & S and I.J. Dewhirst (one of M & S's oldest suppliers) to study the technology of suit-making. The study confirmed that British manufacturers had much to learn in terms of both garment design and production processes. It also revealed the parallels between the manufacture of shirts – especially in cutting and fusing techniques – which formed the great bulk of Dewhirst's production, and the manufacture of suits, in which Dewhirst had no experience.

Subsequently, 'the technical personnel of both companies jointly designed and built a brand new factory in the north-east of England, using the most up-to-date machinery .... The technology was new to everyone in Britain, and manufacturing was carefully planned to enable both the Dewhirst and Marks and Spencer personnel to get used to its capability and requirements' (Tse 1985, p. 80). By 1981, ten years after M & S's initial decision to sell suits, more than 60 per cent of its suits were made in the UK, and the proportion is still increasing. A vital factor in M & S's rapid penetration of the suit market was the ability to fit many more men with a limited range of sizes by selling suit jackets and trousers separately. This required computer-based quality control of fabric shade of an unprecedented order. The colour assessment instruments used for this task, developed jointly by M & S and Instrumental Colour Systems (ICS), gained a Queen's Award for Technological Achievement in 1984, the first time a retailer received this award.

At first glance, M & S's policy of buying British might be construed as a prop to a declining industry. The above example, and other instances, reveal it to be inseparable from one of M & S's basic principles, enunciated in a speech delivered by Lord Sieff in 1966, 'to encourage suppliers to use the most modern and efficient techniques of production, dictated by the latest discoveries in science and technology' (quoted in De Somogyi 1967, p. 55). As such it emerges as a sensible strategy in face of a fragmented and relatively inefficient industry.

Most large suppliers of garments to M & S, such as Dewhirst, Nottingham Manufacturing, S.R. Gent and Corah, have managed to ride out the depressed market conditions of the early 1980s. They have been able to increase both turnover and profits substantially, while many other manufacturers were making workers redundant and closing factories (Moreton 1982).

It might therefore be thought that the key to survival and prosperity in the clothing industry was simple: become a Marks and Spencer supplier. A more logical conclusion might well have been to recognize the importance of investing in new technology, as did firms such as Dewhirst. There is some evidence to suggest that this is becoming more widely appreciated in the industry. For example, British manufacturers are buying 10 per cent of the world's automated fabric cutters.

But this strategy is not appropriate in every case. The initial cost of the automatic cutter necessitates considerable further expenditure in the reorganization required to permit a centralized cutting operation on a 24 hour basis. Only in this way will the installation of the cutter be financially justifiable. It is significant that since 1980, M & S has made a noticeable shift towards larger suppliers with more advanced equipment (Moreton 1982, Kay 1983). In particular cases, this may indicate a failure by smaller suppliers to appreciate the impact of new technology. In general, however, it suggests that the present cost of such technology and the scale of manufacture required to make it pay, persuaded M & S that it could no longer sustain all its existing suppliers. In turn, this may presage a wider restructuring of the industry.

Relationship with Suppliers

Like any large retailer, M & S seeks to increase the flow of goods through its stores. In this way, the price paid by the customer can be contained; its suppliers can operate at or near full capacity; and its own profits remain buoyant.

Unlike other large retailers, M & S has also pursued another, though complementary, major goal. By assuring its suppliers that they can expect long production runs, it has given them the confidence to invest in the latest technology and to adopt the most efficient production techniques. In this sense, the history of M & S as a modern concern dates back to the mid-1920s. Before this, as was the custom, M & S purchased its goods from a wholesaler. The art of retailing consisted largely in choosing from the wholesaler's stock. In turn, this reflected what the manufacturer had decided to produce (Rees 1973, pp. 97–98). In fact, Simon Marks and Israel Sieff (who between them, ran M & S for some 40 years) believed that the business of M & S was not retailing as such, but social revolution. They wished to create a mass distribution system to supply goods of the highest quality at prices the working class and the lower middle class could afford to pay (cf. Rees 1973 and Tse 1985).

According to Drucker, to succeed in this task demanded, first, a basic strategy objective: namely the decision to change from being a successful variety chain to being a 'speciality' marketer, concentrating on clothing. Second, M & S had to formulate what Drucker calls its innovation objectives. Because the type and quality of textiles and clothing which M & S required simply did not exist, it had to transform the kind of quality control which would have been implemented by any large retailer, into something closely resembling what would, much later, be termed total quality control. Though nominally a retailer, M & S set up laboratories which 'developed new fabrics, new dyestuffs, new processes, new blends, and so on. It developed designs and fashions. Finally it went out and looked for the right manufacturer' (Drucker 1974, pp. 96–7).

In going directly to the manufacturer, with the promise of large orders, M & S intended not only to influence what was made, but also how it was made. By clearly defining standards of quality, M & S effectively set technical standards. In its dealings with suppliers, M & S has always placed more emphasis on constructing long-term relationships than it has on signing short-term contracts. Indeed, of its more than 800 suppliers, 150 have manufactured goods for M & S for more than 25 years and 50 for more than 40 years. Great effort is put into securing a common approach not just towards maintaining quality and improving production methods, but also to achieving something which is more difficult to define, but what might be called a general operating philosophy.

At one level, this can be seen in the drawing up of detailed specifications for each type of garment and in the provision of more or less free access for M & S personnel at all stages of production. At another level, it may prompt M & S management to concern themselves with wider matters which they, nevertheless, consider might adversely affect production or be otherwise undesirable. Occasionally this may be resented, as it was when in the 1930s M & S instituted an inquiry into wages in their suppliers' factories (Sieff 1970, p. 186).

Inevitably, this degree of involvement takes up an enormous amount of management time, not only on the M & S side, but also on the part of its suppliers. It also illustrates the point that what M & S purchase from their suppliers is not a product, but a production capacity.

A Technological Approach to Retailing

Marks and Spencer has been referred to sometimes as a 'manufacturer without factories'. In part, this reflects what was called within the company, the 'technological approach' to retailing. This approach required first a recognition of both the practical and the research effort necessary for product development. Second, and of equal importance, it implied a readiness to underwrite a share of the cost of such work – whether incurred in the laboratory or in the factory and (a significant point) which was often beyond the means of the individual manufacturer (Rees 1973, p. 197). It is worth adding, perhaps, that a distinction should be drawn here between two areas of technology, namely (a) product development and (b) manufacturing and process development. There is no doubt that M & S has made a major investment in product development, but its role in the development of new processes is much less one of direct investment, than of encouragement and dissemination of best practice.

Marks and Spencer has been closely concerned with scientific and technological development pertaining to textiles and clothing almost from the moment of its first direct dealings with garment manufacturers in the mid-1920s. If M & S saw as its primary purpose the improvement of the quality of its merchandise, one of the principal means by which this goal could be obtained was through the provision of technical services: 'We came to regard ourselves as a kind of technical laboratory. We felt it was one of our functions to provide our suppliers with expert technical information about new materials and processes which the advance of technology was making available. We saw ourselves as, in a limited way, production engineers, industrial chemists, laboratory technicians. We learned to exercise an active influence on production generally and on the textile industry in particular' (Sieff 1970, pp. 146–47).

In 1936 M & S set up a Merchandise Development Department which, able to draw on the work of the company's textile laboratory, began to investigate many of the problems encountered in the mass production of textiles and garments, problems that ranged from the properties of various raw materials to questions of colour fastness and shrinkage. In the textile industry, the pace of scientific discovery and technological advance has been high, particularly since the 1930s. For instance, by the mid-1960s something like 50 per cent of M & S's textile sales were made from manmade fibres, which only ten years previously had not been available in any appreciable quantity.

At each stage of the production of these fibres (including that of the original fibre) M & S technologists endeavoured to represent the interests of the consumer. As a retailer, the company was well-placed to remind manufacturers that, important though it was to preserve the properties of the fibre as it was spun, knitted, dyed and finished, it was just as important that garments that were produced appealed to potential customers in terms of colour, style and texture. To reconcile these objectives, M & S technologists collaborated with various manufacturers in order to develop new machinery, processes and production methods (Tse 1985, pp. 31–32).

In order that this technological approach to retailing could be realized in practical terms, it was not only essential that M & S technologists be consulted when a merchandiser detected a 'technical problem', but also that they should play an active part in the process of commercial decision taking. To this end each division within a product group (such as men's wear) contains both business managers and technical managers, the latter including specialists in fabrics, garments and machinery in their ranks.

Over 350 technologists are employed at M & S's head office, a much greater number than is employed by any other retailer in the UK. The areas in which they can offer assistance to M & S suppliers include the choice of raw materials, processes of textile and clothing production and problems of factory layout and equipment.

In some instances, M & S technologists have suggested improvements to existing machinery. In other cases, their work has produced significant advances. For example, M & S technologists have contributed to the development of the ICS colour computer by deriving their own algorithms for use in colour matching of garments. The system comprises two parts: a spectrophotometer which measures colour and a computer which translates these readings into numerical values and can define a shade uniquely

(*Financial Times*, 12 October 1984, p. 30). This enables an entire garment to be scanned for colour consistency both more quickly and more accurately than can be done by personal judgement. The system has been piloted by Fermark, a supplier of underwear to M & S. They found that it reduced production lead times by 25 per cent, so repaying its £20,000 cost in 15 months. It is likely that the system will be installed by many of M & S's garment suppliers. This development has a wider advantage, in so far as M & S sells an increasing range of colour coordinates, for which more reliable methods of colour matching and of testing the reactions of toning colours to different light conditions will prove particularly useful.

New Technology in Garment Manufacture

There are three main areas in which new technology can be applied to the manufacture of garments. These are the use of:

(1) microelectronics and computers to improve existing processes;
(2) robotics and related techniques to handle fabric, direct the operations of machines and control the selection and movement of goods; and
(3) alternative processes to replace existing methods of cutting, fusing and so on (Walter 1984, p. 3).

Space precludes detailed examination of each of these areas. But some of the chief advances in the first two categories will be described, in part to disabuse those who visualize technology in the garment industry as consisting of little beyond a few sewing machines.

Microelectronics and computers

(a) The application of CAD and computer-controlled machinery to the once largely separate and highly skilled, pre-assembly stages of design, grading, pattern making and cutting, represents the most significant use of microelectronics in the industry. It should be explained that the various constituent parts of a garment do not increase or decrease *pro rata* over a range of sizes. Prior to the introduction of CAD, grading was a lengthy process which might take several days to complete. Now it can be done in an hour or so. The graded parts are then portrayed on a VDU (figure 1 gives an example), at which point the optimum pattern layout can be calculated and displayed.

In most manufacturing industries, computerization of the design phase is the vital precursor towards the automation of production. The information gathered in this phase can be used to guide the operation of microelectronically controlled machinery at all subsequent stages. Thus far, the single most important 'downstream' development has been the manufacture of CNC cutting machines (see below).

By comparison with assembly and finishing, pre-assembly operations are now highly capital-intensive. Nevertheless, a fundamental problem remains: how can three-dimensional drawings of garments be transferred to the computer screen? At first sight this problem seems no different from that encountered by other industries (e.g. steel manufacture), where it has been overcome with relative ease. But in these industries the raw material is rigid (or rapidly becomes so), whereas garment manufacturers must cope with the way different fabrics drape, stretch and fold. These variations, in turn, require alterations in, for example, patterns, fitting of darts and character of seams. CAD equipment for the garment industry must, therefore, be developed to measure degrees of drape and so on, a task which is being left largely to the equipment manufacturers.

A US manufacturer, Gerber Garment Technology has produced a number of computerized pattern drawing systems, one of which is known as the AM5. This machine can take a pattern and grade it over a number of sizes, facilitate optimum use of any width fabric and automatically mark out the pattern prior to cutting.

(b) Together with its associated CNC cutter, the lowest priced version of this system costs about $600,000 to install and there is an annual maintenance charge which exceeds 5 per cent of capital cost. The cutter employs a self-sharpening knife, and by using a vacuum, to clamp the fabric, it can be cut in multiple-ply. At the moment, Gerber, protected by its patents, enjoys a virtual monopoly of CNC cutting equipment. (Laser cutting equipment does not rely on the patented Gerber vacuum clamp system.) Other manufacturers, such as Investronica (Spain), have produced or are producing cheaper alternatives, but their availability will be restricted unless licensed by Gerber or until the Gerber patents expire in 1988. Though presently extremely expensive, the installation of CNC cutters can be justified (provided, of course, the throughput is sufficient) on the grounds that the cost of the fabric represents 50 per cent of the cost of the exfactory garment. Thus a saving of as little as 1 per cent of fabric utilized is significant.

(c) In this context, there is a very much cheaper computer-aided machine that can help to save fabric at the pre-assembly stage, by improving the detection of flaws, so that the garment maker pays the textile supplier only for what fabric can be used. This machine is the Butex II Fabric Inspection System, one of the very few examples of British-made new technology for garment manufacture.

(d) In addition, Methods Workshop (now owned by Scicon, the computer subsidiary of British Petroleum) has developed a computer system known as General Sewing Data (GSD). GSD's computer memory contains precise information about the multiple operations a machinist must perform in order to complete a particular sewing task. This data resource facilitates calculations about costing and timing of an entire manufacturing process. M & S suppliers account for more than half the subscribers to GSD (Walter 1984, p. 7)

First computer run: width 1120 mm; length 8.14m; efficiency 79.5%

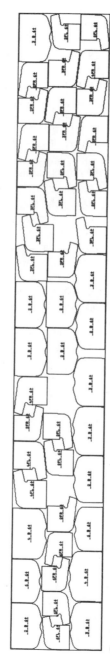

Last computer run: width 1120 mm; length 7.85m; efficiency 82.4%

Figure 1   Printout of VDU pattern grading showing how the computer increases the percentage of material utilized (reproduced by kind permission of Courtaulds Clothing Ltd).

Handling techniques

(a) The softness and floppiness of fabric also presents severe problems for the designers of advanced sewing machines and automatic handling equipment in assembly, finishing and packing. In Walter's graphic description, 'the order of the cut-stack is replaced by the disorder of the bundle system' (1984, p. 8).
It is desirable when sewing the constituent parts of what are, after all, three-dimensional garments to hold them flat and rigid. Nevertheless, progress can be made even where this is out of the question. For example, both Courtaulds and Abbey Hosiery have installed the Rimoldi Gemini 'twin-machine', which allows the operative to sew both seams of a pair of men's briefs simultaneously. This Italian-made machine costs about £10,500, considerably more than twice as much as would a 'single' equivalent. But its throughput is 150 dozen garments per day compared with 80 dozen on an ordinary machine.

(b) In fact, 'needle time' occupies only a relatively small part (20–30 per cent) of the assembly stage. The majority of time is taken up by handling the material prior to sewing, or in transferring it between work stations. In this latter sphere the largest advance has been the development of automated garment transfer systems such as the Swedish Eton system. This transports parts of garments to individual sewing stations by means of an overhead conveyor. Each part is fitted to a clamp which, in turn, is hooked to the conveyor. The clamp can be identified by the system and accordingly delivered to the appropriate machinist in the correct sequence. Corah installed Eton in its main Leicester factory, linking 160 sewing stations at a cost of £1,400 per station. The system is said to pay for itself in 9–12 months through greater efficiency of operation. For example, Courtaulds, who have installed Eton in their Meridian factory, claim that it has reduced throughput (i.e. from loading parts onto the system to boxing the finished garment) from 2½ weeks to 4½ hours. Developments in the computer capabilities of such systems will make it possible to reveal the route that a finished garment has taken; can locate who was responsible for a certain fault or, on the basis of automatic monitoring, select the 'best' machinist for a given task.

(c) Research into, and development of, robotic handling systems in garment manufacture is also underway, though it is perhaps in its infancy. For example, Corah is using a number of dedicated advanced handling systems, each costing £4,000, made by Sahl, an Austrian manufacturer. This machine is capable of detecting when a machinist has finished sewing a piece of fabric; its arm then picks up the item and places it on a bundle which is then removed mechanically for its next operation (Marsh 1984). Corah and M & S are co-sponsoring a research group at Hull University that used a Puma robot and added a prototype arm which, by using air jets that cause the top layers of fabric to flap, can pick individual pieces of cloth from a bundle and position them for a machinist to sew (*The Economist*, 1984a, p. 94).

M & S as a 'Catalyst'

Marks and Spencer has made an important contribution to the introduction of new technology in a number of ways. In practice, these contributions overlap to some degree. Moreover, given that M & S is a 'manufacturer without factories', each of them can be regarded as a form of what is called 'pull-through technology'. That is, M & S's pre-eminence as a retailer has enabled it to set technical standards for much of the clothing industry and thereby allowed it to influence the design of machinery. Nevertheless, these contributions can be distinguished as follows.

First, as has been argued, by its original demand for goods of a type and quality which were then unavailable, M & S eventually set new technical standards. In order to conform to these standards, manufacturers were often obliged to install new machinery, as well as to modify existing techniques of production. Thereafter, M & S has maintained and advanced these standards by issuing to its suppliers detailed specifications which set out the quality, quantity and type of material to be used; the sizes and dimensions of garments including, for example, the number of stitches per inch; and even the manufacturing processes to be applied and the machinery to be utilized. There are, however, occasions when the supplier may think that the imposition of new specifications has a negative impact. This is more likely to be resented if the specifications in question are seen as, strictly speaking, extraneous to the product itself or to product quality, but concern, say, packaging. A case in point might be where M & S is intent on streamlining its retail operations without being sufficiently sensitive to the effect this might have on investment in new technology already undertaken by its suppliers. For instance, a machine that was purchased to perform a certain task more efficiently than before, may have its cost-effectiveness undermined retrospectively by the issuing of new specifications which add to the complexity of the task and so necessitate expensive modifications. Second, M & S technologists have been involved in new processes, new machines and new systems. Third, M & S has acted as a catalyst in the use and development of advanced technology. It is to the third of these activities that we now turn.

Marks and Spencer's general aim of persuading manufacturers to use the most advanced forms of production takes various forms. For example, suppliers are provided, free of charge, with specialist advice in such areas as production engineering and technology; they are encouraged to establish their own laboratories and specialist services; and, because several suppliers may be producing from the same raw materials to the same or similar specification, M & S encourage manufacturers to share their experience of new fabrics and new machinery (De Somogyi, 1967, p. 56). Through close contact with its many suppliers, M & S disseminates knowledge about the best available manufacturing processes from one manufacturer to another rather like a bee pollinating flowers. This positive role, as a guide to efficient methods of production (made more important still by M & S's prominence in the marketplace), nevertheless contains a problematic aspect. This kind of interchange about new technology may benefit all manufacturers to a certain degree, but it is of much greater value to some than to others. The often substantial cost of investing in R & D, of installing and evaluating new machinery and so on, undertaken by the more dynamic of M & S's suppliers may, in effect, go to subsidize their weaker, less adventurous brethren. This is resented by many manufacturers and, indeed, several have gone so far as to close their doors to M & S, though still remaining major suppliers.

Clive Walter, a senior technologist in M & S, has set out a number of ways in which his company's general aim of encouraging the use of advanced forms of production could be furthered (to which are added a number of comments, in parenthesis). These are:

(1) developing contacts with manufacturers of clothing machinery, whether in Europe, Japan or the USA;

(2) visiting major exhibitions of machinery (M & S attends all important garment and fabric machinery fairs);

(3) surveying developments in other industries, such as engineering, that may be relevant to garment manufacture;

(4) locating centres, whether companies or research institutions, that have developed appropriate expertise. Together with these centres and with M & S suppliers, identifying potential development projects. [M & S probably sees inside more manufacturers of either garments or garment machinery than any other concern. For example, knowing that in the industry generally, difficulties in maintaining new equipment have often proved a disincentive to installing new machines, when they discovered that *Arrow*, a giant US shirt manufacturer, had established an advanced training school to familiarize its mechanics with new ranges of machinery, M & S alerted its own manufacturers, with the result that one of the largest has set up its own training school. However, other manufacturers may be reluctant to follow suit. They may think it more cost-effective to make use of the comprehensive training facilities provided within easy reach in the UK by all the machine suppliers;

(5) undertaking or commissioning studies about the reliability of manufacturing processes in production conditions and relaying the findings to equipment manufacturers;

(6) organizing technical seminars. (M & S is most effective in organizing useful courses and seminars. For example, M & S has arranged courses to promote awareness amongst its suppliers of CAD in garment production);

(7) arranging training courses to develop the expertise of both its own technologists and the technical personnel of its suppliers. (As far as its suppliers are concerned, M & S has for the last nine years organized a two-week course for junior managers designed to increase awareness of new technology. They have also run courses for chief mechanics, which include some management training); and

(8) keeping in touch with DTI schemes to finance and support investigations into the application of new technology to industry. (Unfortunate-

ly, it has proved extremely difficult for manufacturers in the garment industry to obtain DTI funding. At the time of writing, the DTI has withdrawn its support from the Innovation Grant Scheme) (Walter 1982, p. 6).

M & S's Suppliers and the Implementation of New Technology

M & S suppliers have been in the forefront of the effort to apply new technology to the production of garments in the UK. For example of the 441 Gerber CAD grading and lay planning systems installed worldwide, 39 are in the UK and of these, 28 are operated by M & S suppliers. (The respective figures for the installation of Gerber computer-controlled automatic cutters are 291:21:14 (Walter 1984, p. 7).) The way in which M & S suppliers approach new technology may be illustrated with reference to specific manufacturers.

We may distinguish two types of M & S garment suppliers: those that can best be described as 'M & S dedicated', such as Abbey Hosiery, and large companies (often part of conglomerates) with a more diversified pattern of outlets, such as Courtaulds Clothing Ltd (referred to subsequently as Courtaulds).

Courtaulds

Courtaulds is the largest clothing manufacturer in Europe. It employs 23,000 people; it operates at more than 100 factory sites in the UK, as well as six sites overseas; it makes more than four million garments per week; and its production facilities include fabric knitting and dyeing and every aspect of garment sewing, making-up and finishing. In the year ending 31 March 1984, Courtaulds Contract Clothing had a turnover of £216 million, equivalent to 60 per cent of the turnover of Courtaulds Consumer Product Division, which in turn accounted for 16 per cent of the turnover of the entire Courtaulds group. A considerable proportion of Courtaulds' contract clothing output is destined for M & S. Indeed, many of the 100 Courtaulds factories produce solely for M & S. In recent years, Courtaulds have taken over a number of garment manufacturers that were already supplying M & S. Presumably, these acquisitions were influenced by the market security promised by their existing links with M & S, as well as by efficient organization of production and high quality of output widely associated with M & S suppliers.

Courtaulds have endeavoured to keep abreast of technical developments in the industry in several ways. They have installed a wide range of advanced machinery: for example, automatic fabric laying equipment, computerized lay planning and pattern grading equipment; mechanized handling systems; programmable sewing machines; and automatic cartoning and bagging machines.

Courtaulds Technical Development Centre has the task of ensuring that the company possesses the most advanced machinery. It also has a major responsibility for research into, and development of, new technology. For example, it is pursuing long-term research into the development of flexible automated manufacturing systems. In conjunction with M & S, Vantona Viyella and S.R. Gent (both major M & S suppliers), Courtaulds has sponsored a research project at the Cranfield Institute of Technology into the 'transfer of technology' for automating the production of shirts and shirt-like garments. The co-sponsors have a guarantee of 12-month's access to the project findings before they are made available on the open market. It will be interesting to see what complications arise given (a) that the project is part-funded by the DTI and (b) the need to involve an equipment manufacturer who will almost inevitably be non-British and will expect to secure some commercial advantage through being involved.

Finally, Courtaulds seek to ensure that the standard and the efficiency of its production methods are maintained. It does this partly by implementing thorough quality control and partly by means of quality circles which have been set up at many of their sites in order to identify and resolve problems at an initial or early stage of production. It is worth mentioning that such was the enthusiasm for the idea of quality circles at one of Courtaulds' Meridian (Nottingham) factories, that throughout a protracted strike, but with the blessing of their Union (the National Hosiery and Knitwear Workers' Union) the members of the quality circle continued to meet at a nearby pub. (M & S technologists have strenuously promoted the concept of quality circles to their suppliers. By late 1982, 14 of them had established an active quality circle programme (Tse 1985, p. 110).)

Abbey Hosiery

Abbey Hosiery was established in 1919 by the present owners' grandfather. Then it employed 40 people and had an annual turnover of £20 K. In 1975 when its turnover stood at £600 K, about 40 per cent of its output went to M & S. In 1984 its turnover had increased to £13 million and the proportion of its output sold to M & S had also risen substantially, to 95 per cent. It now operates five sites, all in Nuneaton and employs nearly 1000 people. Its main product is men's briefs, though it produces thermal underwear, baby wear, boy's pyjamas and an increasing range of men's leisure wear as well.

Abbey Hosiery (AH) has responded to the challenge presented by the development of new technology with considerable enthusiasm. It has installed CAD and a centralized 'high tech' cutting operation, but perhaps its most enterprising undertaking has been to set up a 'high tech' men's brief production unit (HTU) in a 'greenfield site'. This was designed with the intention of building up a centre of expertise to produce a new garment (i.e. in terms of its construction) utilizing new technology. The HTU was organized as a separate company selling its output to AH which, in turn, sells it solely to M & S. The project was planned and initiated by the AH board of directors, but its detailed supervision was left to a small team which included the manager of the HTU, the chief mechanic and the company's project engineer.

The HTU employs about 120 machinists who mainly operate Italian-made Rimoldi machines, each equipped with a computer. The machines can be programmed to carry out the specialized operations (of which there are 40) within the total manufacturing process. Although the machines had proved themselves in other types of garment manufacture abroad, they had not previously been used to produce men's briefs. The project team, therefore, began by videoing every operation, so that on the frequent visits to Rimoldi's Italian factory, the modifications that they required could be discussed in detail. In addition, their mechanics were trained at the Rimoldi factory to the stage where any of them could strip the machines and deal with the electronics. Finally, though the HTU achieved impressive results in terms of both output and quality, the search for more advanced machines continued.

Marks and Spencer took a keen interest in this project for two reasons. They wished to examine ways of reducing the cost of producing underwear in the expectation that competition in this area – particularly from LDCs – would greatly intensify in the coming years. They also regarded it as a 'helping hand' project in so far as, by employing only female school leavers as machinists, it would cement links between both AH and M & S and the local community. In fact, AH's decision to employ a 'green' work force was primarily intended to avoid recruiting workers who were attuned to existing technology and work practices associated with it. However, problems were encountered in terms of work rate, absenteeism and so on. The balance of recruitment was eventually altered such that one-third of the work force is now described as 'more experienced', and is expected to set the pace of, and the attitude to, work for the school leavers who now form only two-thirds of the total. Perhaps M & S's most vital contribution to the project was to place a very large order one year in advance. In this way, not only could AH make the necessary investment with reasonable confidence of a return on their expenditure, but they had a substantial period in which to iron out technological and manufacturing problems before having to start mass production.

Conclusion

At least a section of the UK garment industry, having weathered the recession of the middle and late 1970s and emerged from it with a more rationalized structure, is in a comparatively healthy state. Without a detailed comparison of the respective fortunes of M & S suppliers and non-M & S suppliers, it is difficult to quantify M & S's part in this revival. Nonetheless, it seems that many of the aims of M & S's concerted policy of encouraging its suppliers to invest in new technology have been attained. This is not simply a question of the degree of automation achieved but, more fundamentally, of the appreciation on the part of manufacturers of the need to introduce advanced machines, to keep abreast of innovations and to undertake their own research and development. It might be suggested that M & S's suppliers can now largely cope with the demands of new technology if not unaided, at least without the same degree of aid previously forthcoming from M & S.

Indeed, the balance of comparative advantage between production in ICs and LDCs may be beginning to change in favour of the former group, largely – though not entirely – because of the impact of new technology. Thus it is now being claimed that automation in certain key areas will permit UK manufacturers to undercut LDC imports, e.g. in the production of shirts. In addition, the requirement for shorter lead times and the need to react swiftly to Electronic Point of Sale (EPOS) data collected by the retailer, may benefit technologically advanced UK manufacturers.

In the main the above developments have been of benefit to medium-sized manufacturers (e.g. AH) and large manufacturers (e.g. Courtaulds) who have had the resources to invest in new technology. There are signs, however, that the benefits of automation may become more widely available as intermediate technology is developed that is better-suited to the smaller clothing company, which lacks the turnover to support more sophisticated and productive machinery. For example, Gerber has developed the Gradamatic-5. This is a comparatively low-cost pattern grading and pattern cutting system, which can be upgraded to an AM5 should the purchaser wish to install a lay planning system. (*Apparel International* January 1984, p. 8).

Encouraging as these developments are, one important issue remains. If an industrialized country is defined as one that not only manufactures goods, but also manufactures the machines that are used to produce these goods, then the UK, if judged solely in terms of the garment industry, may be described as partially, not fully, industrialized. From one view, the virtual absence of UK manufacturers of machinery for the garment industry is not a serious handicap. Thus Courtaulds and AH maintain close links with various foreign suppliers of equipment (as indeed do M & S) and can influence the design and capability of the machines which they purchase. From another view, a parallel can be drawn between the problems which arise for retailers when relying on foreign suppliers of garments (several large UK retailers have 'burnt their fingers' on foreign suppliers, having encountered problems with modifying orders, in terms of the quality of garments etc.) and those which might arise for manufacturers who must depend on overseas suppliers of equipment (for example, in terms of frequency of access to discuss modifications). Moreover, in so far as many machines are manufactured abroad, a calculation that it is worth investing in expensive new machinery may have to be revised if the £ Sterling falls against other currencies.

The author would like to acknowledge the help of the following individuals in the preparation of this article: Patrick Arnold-Baker, Technical Executive, Marks and Spencer; John Catherall, Garment Technologist, Technical Development Centre, Courtaulds Clothing Ltd; Mike Drury, Chief Executive, Technical Development Centre, Courtaulds Clothing Ltd;

David Shaw, Consultant to Central Textile Technology, Marks and Spencer and to Abbey Hosiery; Clive Walter, Senior Technologist, Marks and Spencer.

References

1. *Apparel International*, 1984 The 'affordable' automated grading/pattern cutting system, p. 8
2. Business Statistics Office, 1976 *Business Monitor: Census of Production*, HMSO
3. Coyle, A., 1984 *Redundant Women*, The Women's Press (chapter 1: The clothing trade)
4. De Somogyi, J., 1967 St Michael's marketing philosophy. *Br. J. Marketing* Autumn, pp. 54–59
5. Drucker, P., 1974 The power and purposes of objectives: The Marks & Spencer story. *Management: Tasks, Responsibilities*. Heinemann, London
6. *The Economist*, 1984a Robots take the sweat out of the rag trade, September 15th
7. *The Economist*, 1984b The MFA is too costly a Joke, December 22nd
8. Fevre, R., n.d. *The Labour Process in Bradford*. EEC/DES Transition to Work Project, Bradford College
9. Frobel, F. *et al.*, 1980 *The New International Division of Labour*. Cambridge UP
10. Heimenz, U., Schatz, K., 1979 *Trade in Place of Migration Geneva*. ILO
11. Hoffman, K., Rush, H., 1984 *From Needles and Pins to Microelectronics – the Impact of Technical Change in the Garment Industry*. Science Policy Research Unit University of Sussex
12. Kay, W., 1983 How they sold their soul to St Michael. *The Sunday Times* 19 June
13. Keesing, D., Wolf, M., 1980 *Textile Quotas Against Developing Countries*. Trade Policy Research Centre, London
14. Marsh, P., 1984 Underwear by robots. *The Financial Times*, 3 May
15. Moreton, A., 1982 Marks & Spencer suppliers: How close links pay off. *The Financial Times* 28 May
16. Rees, G., 1973 *St Michael: A History of Marks & Spencer*. Pan
17. Salmans, S. 1980 Mixed fortunes at Marks & Spencer. *Management Today* November
18. Sieff, I., 1970 *Memoirs*. Weidenfeld and Nicolson
19. Tse, K., 1985 *Marks & Spencer – Anatomy of Britain's Most Efficiently Managed Company*. Pergamon Press
20. Walter, C., 1982 *The Application of Advanced Technologies to Garment Manufacture*. Marks & Spencer, September
21. Walter, C., 1984 Fashion response technology, *Apparel International*, January

4.2

Wealth Generation – Our Essential Task

B.W. Small

Introduction

Wealth generation is our essential task. If we do not accept this responsibility our profession should and will decline.

This reading is about how we engineers can help generate wealth. After all it is what the bulk of us do in our daily work that decides our national well-being rather than what government or corporate management can do. They can create a sympathetic involvement but the major responsibility is with us.

Hopefully, there are some ideas in this reading of specific help to practising engineers wanting to take action, especially those with management responsibilities for our engineering industry and those who have a hand in deciding how we make our products including those who design them.

We are all only too clearly aware from the media that our engineering industry is on the decline.[1, 2, 3] Reversing the trend of national decline is not only possible, it is also an absolutely practical proposition. After all, there are a number of engineering businesses in this country which are excellent in their economic performance, their marketing, their product engineering and in their manufacturing. If we all were as good as the best 10 per cent we would be dominating world trade.

More than ever before manufacturing performance is determining success or failure in world markets. Effective manufacturing has become as important a competitive tool as marketing and product innovation. We need to improve our manufacturing effectiveness if our engineering industry is to survive and prosper.

I suggest a revival programme with five ingredients. We need to:

Source: from *Institution of Mechanical Engineers, Proceedings 1983*, 197 (58).

(1) Accept that too many of our factories are obsolete. Many of them, if designed anew for today's purpose, would not remotely resemble existing operations.
(2) Stop being tolerant of the 15–30 per cent excess cost in our products.
(3) Become more commercially aggressive in our application of the new technologies including CAD/CAM and FMS.
(4) Substantially improve our national skills and capacity to plan and implement more competitive operations.
(5) Encourage every engineering business unit, which does not have at least one major radical improvement programme under way, to start one now.

While most revival programmes require investment, the imperative ingredient is people, especially the creative engineer.

There is a misconception in this country that only by the introduction of new products, based on entirely new technology, will we improve our share of world markets and our national wealth. I do not believe this. Many of the applications for our engineered products have been in existence for decades and will continue to exist for as long as we live. Figure 1 provides typical examples. The factories exist. Collectively they represent a huge investment and commitment. They certainly cannot be changed overnight. It is critical, however, that we clearly understand their strengths and weaknesses if we are to accelerate our efforts at improving them.

Engines	Power stations
reciprocating	Electrical motors
rotary	Production equipment
Automobiles	Pumps
Off-highway vehicles	Valves
Gear boxes	Compressors
Consumer durables	Railway equipment
Cranes	Aeroplanes
Office equipment	Trucks

Figure 1   Many engineered products have existed and will exist for decades

Too Many of our Factories are Obsolete

Our path to manufacturing obsolescence, as elsewhere in the developed world, has been quiet and uneventful. Product changes have occurred slowly over decades with factories slowly absorbing the changes. Obsolete manufacturing concepts have sometimes been passed from generation to generation, step by step, so that even a modern-looking shop can be obsolete by criterion of cost-effectiveness and competitiveness.

Symptoms of creeping obsolescence are not difficult to spot:

(1) low profit margins despite price increases;
(2) stretched deliveries;
(3) high or difficult to control work-in-progress;
(4) increasing total labour content per unit.

Creeping obsolescence is also seen in these situations:

(5) a plant arrangement (by process or product) which has not been challenged despite a number of expansions and modifications;
(6) a historic pattern of bottleneck-to-bottleneck solutions to capacity problems;
(7) idle equipment in an otherwise busy plant;
(8) high indirect labour;
(9) increased scrap, rework and warranty costs.

Four clear origins of obsolescence stand out: the first three are related to the long-term effects of market changes on manufacturing; the fourth relates to equipment, methods and facilities. The three origins that stem from market changes are:

(1) Volume has changed while manufacturing concepts have not changed.
(2) Product mix has become more complex.
(3) Product design has changed significantly.

Volume has changed

Many plants still operate under manufacturing concepts initiated when product volumes were much different from today's. The only way these old concepts may be cost-effective now is by chance. A given set of manufacturing concepts generally are cost-effective only for a range of volumes.

Expanding incrementally to keep pace with business growth is not necessarily wrong. Rather, at some point it becomes more economical to switch basic manufacturing concepts. Incremental contraction has a similar set of problems, if more painful. Exactly when this point occurs, when radical action is needed, is not always easy to discern. But once reached, unit costs begin to climb and responsiveness to market requirements declines.

Product mix more complex

The trend is for product mixes to proliferate, and rightly so. Most of today's markets – from heavy capital goods to consumer items – demand more variety than 20 years ago.

Production managements have naturally responded to gradual product changes and additions by piecemeal modification to the manufacturing system, adding a new machine here and there, developing new fixtures, installing more conveyors or extending the factory buildings. After a time, the plant ends up making products for which it was not designed. At the time, each change probably seemed to be a good idea but the cumulative result of many changes is a conceptually weak plant which is no longer cost-effective.

Product design has changed

Few of today's products bear much resemblance to their predecessors of 10 or 20 years ago. Functional improvements, market stimulation for change, government legislation, material substitutions and new applications all have their effects on product design, and consequently on obsolescence in manufacturing.

Creative manufacturers can often find a way to accommodate any given product change at acceptable cost. But too few managements then go on to question the full effect of these piecemeal changes and ask 'Can we make the product at less cost some other way? What about new processes, new methods? Part grouping? More or less automation? More or less specialized equipment? Different scheduling practices? Different plant arrangements? Different manufacturing organization?' The alternatives are many.

An obsolete plant can also impede product improvements. A product may be driven off the market because the manufacturer can no longer fully respond to what his customers really want. He becomes vulnerable to his more flexible competition even though he may perceive his customers' needs and knows how they should be fulfilled.

A subtle and obvious truth

A universal truth underlying obsolescence, both subtle and obvious, is this:

> *If most manufacturing systems were designed from scratch for today's products, volumes, and mixes plus today's technology, the new shop would not remotely resemble the present operation.*

Manufacturing costs would be reduced and we would remove many of the following typical non-value-adding activities:

(1) fork trucks hauling materials everywhere, damaging and losing parts and pallets full of work-in-progress;
(2) progress chasers and half the supervisory staff trying to find parts for the next operation;
(3) operators going for tools while their machines sit idle;
(4) aisles full of people dodging the fork trucks;

(5) complex production control systems trying to keep track of thousands of transactions each day;

(6) paperwork – the preparing, handling and reading of it;

(7) inspectors trying to separate bad parts from good throughout the process;

(8) late deliveries;

(9) delivery lead time measured in months for products with a routed manufacturing time measured in a few hours;

(10) multiple layers of supervision because controlling such an environment is, in fact, a big job and

(11) above all the cost of poor quality, scrap and rework.

Not only have we allowed too many of our plants to become technologically obsolete in equipment and facilities, we have also allowed them to become out of phase with the products we are making today. It is the combination of all these factors which has contributed to our declining competitiveness, especially to the cost of our products.

Stop Being Tolerant of Excess Cost in our Products

We lost sight of our cost problem – it changed some time ago.

Too many of our engineers, production managers and facilities planners have inappropriately focused their attention on direct labour costs. This cost element has been given so much attention for so long that today it is only a small part of the cost of sales. It is almost not worth worrying about any more. This seemingly shocking statement gains credibility if we ponder the irrevocable fact that direct labour in most engineering plants now ranges between only 5 to 15 per cent of total manufacturing cost. At the same time, plant overhead costs often reach 40 per cent, with material making up the other 50 per cent (figure 2).

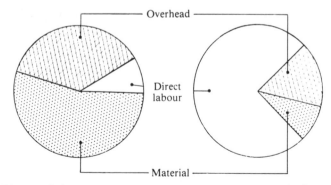

The way it is   The way we attack them

Figure 2   We lost sight of the problem

But, guided by traditional training, our talent continues to pound away at what has become a fraction of the cost differential between best and worst world prices and costs. Something else *must* be done.

This traditional misconception of today's cost problem makes it hard, if not impossible, to justify new manufacturing technology.

The well-published notion that the Japanese, for example, have lower costs and prices because they have employed manufacturing technology better is tragically misleading to top managers. Certainly some countries use robotics in greater quantities than we do. Some have also done remarkable things in the fields of flexible manufacturing systems (FMS) and automation. But the reason they did it was to attack unit costs in total.

Too often our formal requests for major capital expenditures are still concentrated on direct labour. It is almost as if we do not believe that the costs of overhead are reducible. For example, inventory turns of two or three per year in a plant producing products with a routed time of only a few hours is accepted. The cost of inventories today can actually exceed the cost of direct labour.

Once we begin to focus on and attack the large cost issues, we will be able increasingly to support the use of more new manufacturing technology. It can be cost justified if we seek the justification in overheads and in materials.

By allowing our factories to evolve incrementally and sometimes haphazardly, for reasons I have described, we were forced to permit our overhead costs to grow. We needed more overhead people to manage and control activities which were becoming more complex, because of obsolescence. Overhead people costs are now bigger than direct labour costs and, remember, capital investment is rarely a major part of total unit cost.

We have left nobody responsible for the largest manufacturing cost element – materials. The cost of materials seldom becomes as low as 40 per cent, is usually about 50 per cent and can reach 70 per cent of manufacturing cost. Most companies have substantial technical departments, accountants and payroll departments scrupulously tracking the small per cent of direct labour cost. Is comparable effort being put into the material cost portion that is four to six times as large? I believe not.

Our typical organization structures fragment control over material costs:

(1) product engineering being concerned with 'will it work?';
(2) production engineering concerned with 'can we make it?';
(3) purchasing negotiating prices with only a little help from engineers.

It is hard to imagine that we can compete successfully in today's tough international markets while leaving responsibility for about half of our manufacturing cost diluted between different functions. For all practical purposes this simply means there is *no* responsibility.

Our problem is that we are 'starting' with the waste in place and now have to extract it to compete. How are we to achieve this?

Attack material costs

Attacking material costs is not usually a glamorous task. However, the opportunities to reduce cost can be virtually limitless. There are two methods of attack, neither involve fundamental design – that is another subject.

Criticize, part by part, manufacturing operation by operation, those specific items and details that create material cost. The process is iterative. A new shape can suggest a different raw material which may suggest a better manufacturing process or vice versa. Evaluation of the material specifications can reduce the variety and reduce the purchasing task.

Compare the product with competitive products, in detail. Some of the differences will be minor. Nevertheless, the differences should be costed and totalled. The totals are often impressive.

Spearhead the attack with the creation of an independent group dedicated to this single task. Make the team multidisciplined. Mix designers with manufacturing people, especially if you sense there has been a brick wall between the two groups in the past. Use people who are not committed to the present situation. Encourage them to conduct 'should cost' manufacturing studies on bought-in components and assemblies to see if you can help suppliers reduce their costs.

If there is the opportunity to combine the attack on material costs in parallel with the attack on overhead costs, the results will be even better.

Attack overhead costs

Somehow we engineers have too often abdicated this task to the accountant. Their approach to reducing overhead costs is by necessity a negative one. They are forced to reduce the resources to match economic necessity *without* challenging the engineering fundamentals involved. They cannot challenge why a multitude of indirect tasks are apparently necessary to design and make our products. For example:

(1) Why products are inspected in a way which forces men and machines to be idle while inspectors perform their task – the cost being charged to overhead.
(2) Why the shop is organized so that it requires progress chasers to force products through – as an overhead cost.
(3) Why drawings are prepared so that the information has to be constantly regenerated into new formats to suit the factory's purpose – yet another overhead cost.

Present excessive costs are not, generally, in the direct tasks necessary to transform materials into products. It is what is done between these physical tasks that causes the costs to escalate.

A partial list of overhead items we cannot afford are:

(1) work-in-progress;
(2) people: progress chasers, stores keepers, quality inspectors, industrial engineers, accounting clerks;
(3) materials being handled dozens of times;
(4) paperwork, paperwork, paperwork and more paperwork;
(5) scrap, rework;
(6) direct labour doing indirect work;
(7) excess layers of supervision.

These overhead items make us *less* responsive not *more* responsive to our customers. They are tasks we are forced to perform in our obsolete factories because our factories are *out* of control.

There will be some who would immediately conclude that the only significant means of removing these overhead costs is in the application of the new manufacturing technologies such as flexible manufacturing systems (FMS), computer-integrated manufacture (CIM), computer-aided design (CAD), robotics, management information systems (MIS) etc. Of course we need them and some have needed them for a decade. However, much can be done today to improve the appropriateness and effectiveness of our factories. We cannot and must not wait. We certainly must not use the eventual benefits of these new technologies as the excuse why we should wait, yet again, for a few more years before we make improvements. Perhaps a 30 per cent reduction will only be possible after application of FMS in one case, or CAD/CAM in another, however there is probably 15 per cent or more, waiting to be removed in 1983.

Let me emphasize this point with a topical illustration. A popular trend currently, is to subcontract the machining of components that previously had been manufactured in-house. The lower prices being obtained are generally because of the lower overheads typical of the smaller subcontracting firm. There are many situations when the decision to subcontract is desirable. However, if the decision is taken without rigorously challenging existing internal overhead cost and organization first, the firm may be doing the exact opposite of 'attacking overhead costs'.

Become More Aggressive in Applying New Technologies

A possible reason why the application of robotics, CAD/CAM, FMS and CIM has been relatively slow to develop momentum in the UK has been because we have not seen them as competitive tools. When we engineers become preoccupied with new technology for its own sake, we shouldn't be surprised if our plans have low credibility with our more profit-minded business colleagues. We need to increase our emphasis on the identification and communication of the wealth potential of these technologies both for now and in the future.

There are a number of major new manufacturing technologies available for business application. I shall concentrate on flexible manufacturing systems (FMS) and CAD/CAM leading towards CIM (computer-integrated manufacture).

CAD/CAM: A manager's tool?

Computer-aided design is a well-established and valuable aid to engineering departments, which minimizes drudgery and improves consistency of engineering information, dimensioning etc. It is also proving to be a valuable sales engineering tool. Computer-aided manufacture, though too rarely integrated with CAD, is also helping streamline and simplify input to CNC machine tools, improve utilization of plate in fabrication and similar activities. As a tool for engineers the case is, I believe, proven. They have not generally had a major impact on reducing unit cost, as yet.

But is CAD/CAM destined to be only an engineer's tool or is it also going to be a manager's tool for controlling his business and, in the process, make a major contribution to unit cost reduction? I believe the latter. The major economic benefits will not, in my opinion, be significantly realized until it does become a manager's tool. This will be particularly so for our batch producing industry making products that need at least some, or a lot of, specialization to match differing customer needs.

For this type of company, where each specification is somewhat different and the detail design is not finalized until some weeks after the order date, the interface between design engineering, drawing preparation and management information systems is critical.

The customized product maker is always seeking an early accurate definition of what has to be made from its engineering department to cut lead times and manage costs better. What happens in reality is that too many companies are 'sailing blind' for too long wanting and waiting for 100 per cent detailed data. We are using creativity putting things right that would not have been wrong if the precise data had been available earlier. The problem is that this abuse of scarce creativity adds to product cost without adding value.

Computer-based management information systems (MIS) are also well-established, perhaps even more advanced than CAD/CAM. These systems are often directly linked to costing systems. However they are alpha-numeric systems and they need a structured bill of material as a starting point. They do not 'think' shapes as does CAD/CAM. Rarely is this powerful capability intimately linked with the engineering information source upon which it relies. For example, a design change cannot be exploded into a new shop schedule immediately the change is approved.

A projection of the type of integrated system that will exist within the next decade as routine, at least from an international dimension, is shown as figure 3. I think the reader will agree we are no longer talking about CAD/CAM, we are beginning to describe computer-integrated manufacture.

So what are the economic benefits of introducing CIM? There will be diverse benefits including better performance and lead times to our customers. The impact on unit costs, especially overhead and material costs, will be in the following areas:

(1) reduced work-in-progress;
(2) reduction in the real cost of quality;

(3) less paperwork, less clerical effort in the factory;
(4) less fitting and more assembly;
(5) less material wastage;
(6) fewer progress chasers;
(7) fewer inspectors;
(8) less clerical activity by engineers and draughtsmen.

There are hard commercial benefits to be gained by radically increasing the level of control we have over the creation, timeliness and accuracy of the engineering information needed and the automatic conversion of this information into the alphanumeric data needed to run the business. It is worthwhile to pursue these economic benefits.

Flexible manufacturing systems – A new technology?

Most of us still believe FMS is a new technology to be treated with caution. However, it has only been our lack of ability to apply the principles that results in it still being termed a 'new technology'.

What is an FMS? A report published last year entitled 'The FMS report' declined to give a single definition.[4] The economic definition given alongside several useful technological definitions was:

A technology which will help achieve leaner factories with better response times, lower unit costs and higher quality under improved management and capital control.

This improvement is achieved by imposing 'absolute physical control' over components and products for the total time they are in the factory. Flexible manufacturing systems give parts no chance to do anything but follow a predescribed, fully controlled 'path' in the plant. Flexibility is provided to match needs of the market place while at the same time radically reducing the need for flexibility to cope with the level of disorder existing in our obsolete factories and the associated costs.

Most important of all, FMS gives us the opportunity, as with CIM, to attack unit costs with the emphasis, in this case, on the following:

(1) reduction in work-in-progress;
(2) higher utilization of equipment for more hours per day;
(3) lower direct labour;
(4) lower indirect labour: less labourers, chasers, store keepers, inspectors;
(5) simplified production control;
(6) some less space;
(7) less scrap and rectification;
(8) less tooling and fixturing.

While there are still only a limited number of full FMS installations operating in the world, enough experience has been accumulated, especially

152

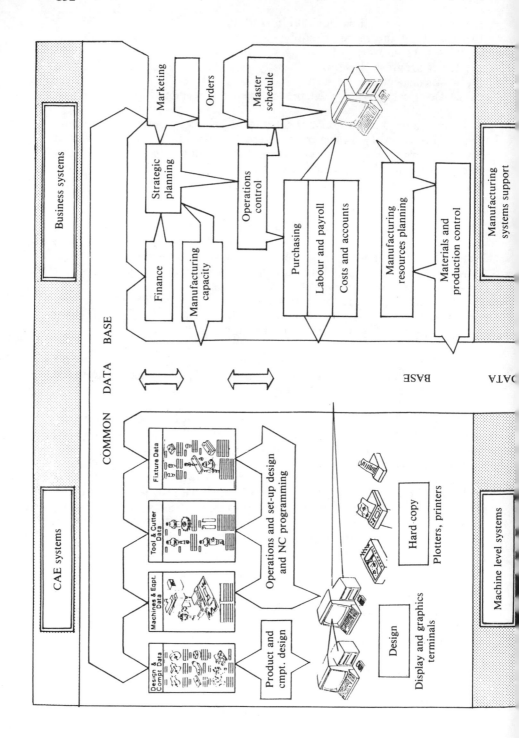

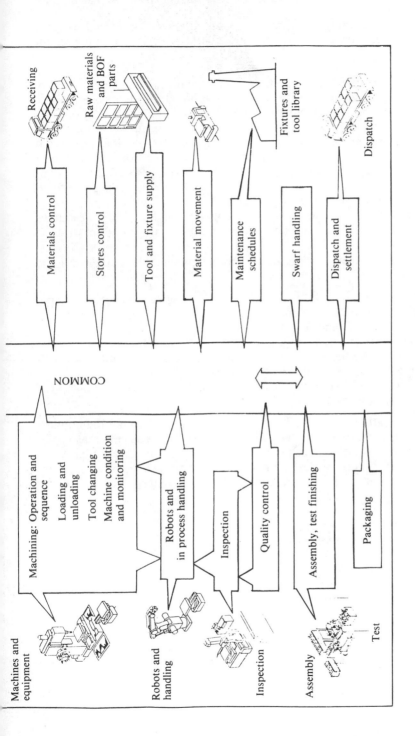

Figure 3   The main elements of systems in manufacturing

on flexible machining systems, to identify three key factors that differentiate between success and failure and the achievement of lower unit costs:

(1) Flexible manufacturing systems are not naturally flexible. Flexibility is selective and exists where it is specifically forced into the system by its planners. Each business needs to decide its priorities, in terms of degrees of flexibility, that match its own business goals: for example, to make to order rather than make to stock, to permit random design variations within a product range, to permit rapid launch of new products during the life of the system, to reduce unit cost etc. Introduction of FMS needs to be an integral part of a well-defined business and manufacturing strategy. Management must be fully involved.

(2) The strength of the production engineering effort and resources applied by the user, over and above the contribution by vendors to:

(a) fully assess the implications of the product range to be made within the system;

(b) plan functional requirements, write performance and interface specifications for each element;

(c) progress, inspect and test diligently;

(d) evaluate improvements in unit cost;

(e) understand the full implications of the software and its interface with existing management information systems;

(f) introduce the FMS into production and monitor the impact this has on all the people involved.

(3) Application of FMS principles is more important than the application of all the technology available. FMS is needed for business reasons, to improve response, to remove existing bad and costly practices. Eighty per cent of the risk is in the last 20 per cent of the move towards a totally integrated, unmanned system. FMS, with the right planning, can be phased.

Unit cost reduction – or improved competitiveness

Introducing FMS, CIM or one of the other major new technologies is painful and difficult. There are many pitfalls and problems, some still to be identified. Our creative engineering talent will be stretched. The capital requirements will be substantial. If the benefit was solely unit cost reduction I doubt that the difficulties involved could be justified. Bringing our factories fully under control means much more in terms of improved competitiveness. Benefits that the engineer is best qualified to perceive will follow – benefits that we must translate into business terms for our colleagues. Some of these benefits are:

(1) faster response to market changes: product mix, product volume, product design improvements;

(2) dramatic improvements in control over quality;
(3) more reliable and shorter deliveries;
(4) improved management control.

Decisions about manufacturing are becoming an increasingly important part of any company's business strategy. We are entering a new era.

Paying for technology

Funding capital investments is, and always will be, difficult. Temporary overfunding creates cash flow difficulties, liquidity ratio problems and the like which any corporate management team would prefer to be without. However, there are three weapons we have at our disposal:

(1) Produce unit cost reduction projections of 20–30 per cent and the other financial analysis techniques used [such as ROI (return on investment), etc.] will also give positive results.
(2) FMS and CIM both fundamentally reduce lead times, stock levels and work in progress levels. Arguments based on total capital employed (more capital investment and less work-in-progress) are likely to be convincing.
(3) Incremental small investments rarely provide attractive economics. Create a master plan for a five-year programme with step-by-step investment in parallel with unit cost reductions and again the application is more likely to succeed.

Improve Our Skills and Competitiveness

Those who decide to apply the ideas outlined in this reading will quickly meet the problem of deciding how best to use the scarce creative talent available. My recommendation is to concentrate these resources rather than dilute them on too broad a front, on too many projects. I recommend this for three reasons.

Get your strategy right first

Too little feasibility and planning work at the beginning results in more work during implementation, a late project and worst of all, only partial achievement of the projected improvements. This logic is so obvious and so simple it frequently gets ignored. Then lack of achievement creates more cynics and less opportunity to do it again.

Attack traditional thinking systematically

Breaking with traditions is difficult. It certainly takes a little longer.

'Custom and practice' developed over years causing the excess cost, whether in factory or product or both, takes time, creativity and persuasiveness to remove. Superficial changes will not remove 15–30 per cent of unit cost. It involves challenging the creators of the product in engineering, and what may be even more difficult, challenging the management structure and activities that make up the overhead portion of the cost. Without this critical, time consuming, challenging phase the results will simply not materialize.

Implement aggressively

Having established what is needed, the programme must not be allowed to erode. Our product-producing industry has much to learn from our contract engineers in this regard.

Programmes with sufficient content to *change* the wealth-generating capability of a business will not be effectively realized as a superimposed work load on an existing departmental structure. It will take too long, have less impact and most important of all 'custom and practice' will creep back into the programme and dilute the results.

It may only need a small team but a project orientated group is necessary. Give the team enough authority and they will measure themselves against their ability to achieve the targeted change on time and within cost.

In the UK we have an additional problem to overcome. From 1971 to 1975 I worked in the USA. On my return to the UK, I could not understand the over-caution and sometimes lack of professionalism in our engineers when handling improvement programmes. Yet these engineers often had better practical and academic experience than those doing similar work in the USA. The answer was simple. Many of the US manufacturing engineers I had met had been responsible for, or participated in, capital expenditures of three to twelve times that experienced by their British counterparts. I suspect the situation has not changed that much. I am afraid that improving this national skill will need to come from making change, not talking about it.

Encourage Improvement Programmes

Of course virtually every company has programmes for improvement. Many would not have survived the current recession if they had not. Improving performance in our products and from our factories every year is our constant challenge.

The hard evidence, however, is that we have to do more than this if our share of world markets is to improve. And to think we can change this situation overnight would be naive. What we can do, and must do urgently, is create 'pockets of excellence' – wealth-generating excellence.

We need to measure these 'pockets of excellence' by absolute criteria, or at least by best international standards and not parochially by measuring improvement with historic performance within the company itself. We have concentrated too much on this second criterion of measuring performance because this is the criterion our accountants prefer.

By this means we will achieve a stronger momentum, building on success and experience, that shows the best is achievable in practice; the norm, rather than accepting less. But these grand thoughts do not, in themselves, achieve change nor do they help us even get started.

But how do we get started?

Each company has different needs. However, I make the following general suggestions:

(1) Keep the project specific enough so that the time-scale is less than 18 months. The programme needs to be *seen* to be successful as well as be successful.

(2) Avoid single department studies. Major unit cost reductions usually mean that it is necessary to attack a major proportion of the cost elements.

(3) Establish tough business objectives first, engineering objectives second and technology objectives third, e.g.:

(a) reduce unit cost by 27 per cent, lead time by 60 per cent etc.;
(b) reduce weight of product by 22 per cent and machining content by 14 per cent, rationalize product range from seven to four, reduce inspection costs by 50 per cent etc.;
(c) introduce FMS principles, or robotics or CAD/CAM etc.;

Make them tougher than you believe can be achieved, not less.

(4) Assign a full-time core task force supported by part-time specialists as required. Do not underresource. Make sure that senior management and shop floor participate. A team creates its own momentum – individuals rarely do.

(5) Allocate a proportion of the task force's time to communicating with others in the company. Get senior management excited by involving them.

(6) Do not allow capital constraints to be imposed before the options and economic returns are fully considered.

And perhaps most important of all do not plan to start – get started now.

People create change

In the introduction, five actions were suggested as being the ingredients of a revival programme to improve our wealth-generating capability. Each of these actions calls for initiative by engineers. If we do not act, our engineering industry will continue to decline. In addition to our engineering skills, we must also be stimulators, antagonists, communicators and persuaders for change. These skills are probably more important at this time than our engineering ability. If our profession is to prosper, we must *lead* in this vital task of creating and implementing radical improvement in our engineering industry. Wealth generation is our essential task.

References

1. Carey, P., 1979 UK industry in the 1980s. *Proc. Instn Mech. Engrs.* **193**: 439–46
2. Barlow, W., 1982 The acceleration of change in British engineering. *Proc. Instn Mech. Engrs.* **196**: (40)
3. Osola, V.J., 1982 Innovative response to a changing world. *Proc. Instn Mech. Engrs.* **196**: 347–56
4. Ingersoll Engineers, 1982 *The FMS Report*. IFS, Bedford

4.3

Managing Our Way to Economic Decline

Robert H. Hayes and William J. Abernathy

During the past several years American business has experienced a marked deterioration of competitive vigour and a growing unease about its overall economic well-being. This decline in both health and confidence has been attributed by economists and business leaders to such factors as the rapacity of OPEC, deficiencies in government tax and monetary policies and the proliferation of regulation. We find these explanations inadequate.

They do not explain, for example, why the rate of productivity growth in America has declined both absolutely and relative to that in Europe and Japan. Nor do they explain why in many high-technology as well as mature industries America has lost its leadership position. Although a host of readily named forces – government regulation, inflation, monetary policy, tax laws, labour costs and constraints, fear of a capital shortage, the price of imported oil – have taken their toll on American business, pressures of this sort affect the economic climate abroad just as they do here.

A German executive, for example, will not be convinced by these explanations. Germany imports 95 per cent of its oil (we import 50 per cent), its government's share of gross domestic product is about 37 per cent (ours is about 30 per cent) and workers must be consulted on most major decisions. Yet Germany's rate of productivity growth has actually increased since 1970 and recently rose to more than four times ours. In France the situation is similar, yet today that country's productivity growth in manufacturing (despite current crises in steel and textiles) more than triples ours. No modern industrial nation is immune to the problems and pressures besetting US business. Why then do we find a disproportionate loss of competitive vigour by US companies?

Our experience suggests that, to an unprecedented degree, success in most industries today requires an organizational commitment to compete in the marketplace on technological grounds – that is, to compete over the

Source: from *Harvard Business Review* July/August 1980. ©President and Fellows of Harvard College.

long run by offering superior products. Yet, guided by what they took to be the newest and best principles of management, American managers have increasingly directed their attention elsewhere. These new principles, despite their sophistication and widespread usefulness, encourage a preference for (1) analytic detachment rather than the insight that comes from 'hands on' experience and (2) short-term cost reduction rather than long-term development of technological competitiveness. It is this new managerial gospel, we feel, that has played a major role in undermining the vigour of American industry.

American management, especially in the two decades after the Second World War, was universally admired for its strikingly effective performance. But times change. An approach shaped and refined during stable decades may be ill-suited to a world characterized by rapid and unpredictable change, scarce energy, global competition for markets and a constant need for innovation. This is the world of the 1980s and, probably, the rest of this century.

The time is long overdue for earnest, objective self-analysis. What exactly have American managers been doing wrong? What are the critical weaknesses in the ways that they have managed the technological performance of their companies? What is the matter with the long-unquestioned assumptions on which they have based their managerial policies and practices?

A Failure of Management

In the past, American managers earned worldwide respect for their carefully planned yet highly aggressive action across three different time frames:

(1) *Short term* – using existing assets as efficiently as possible.
(2) *Medium term* – replacing labour and other scarce resources with capital equipment.
(3) *Long term* – developing new products and processes that open new markets or restructure old ones.

The first of these time frames demanded toughness, determination and close attention to detail; the second, capital and the willingness to take sizeable financial risks; the third, imagination and a certain amount of technological daring.

Our managers still earn generally high marks for their skill in improving short-term efficiency, but their counterparts in Europe and Japan have started to question America's entrepreneurial imagination and willingness to make risky long-term competitive investments. As one such observer remarked to us: 'The US companies in my industry act like banks. All they are interested in is return on investment and getting their money back. Sometimes they act as though they are more interested in buying other companies than they are in selling products to customers'.

In fact, this curt diagnosis represents a growing body of opinion that openly charges American managers with competitive myopia: 'Somehow or other, American business is losing confidence in itself and especially confidence in its future. Instead of meeting the challenge of the changing world, American business today is making small, short-term adjustments by cutting costs and by turning to the government for temporary relief .... Success in trade is the result of patient and meticulous preparations, with a long period of market preparation before the rewards are available .... To undertake such commitments is hardly in the interest of a manager who is concerned with his or her next quarterly earnings reports.'[7]

More troubling still, American managers themselves often admit the charge with, at most, a rhetorical shrug of their shoulders. In established businesses, notes one senior vice president of research: 'We understand how to market, we know the technology, and production problems are not extreme. Why risk money on new businesses when good, profitable low-risk opportunities are on every side. Says another: 'It's much more difficult to come up with a synthetic meat product than a lemon–lime cake mix. But you work on the lemon–lime cake mix because you know exactly what that return is going to be. A synthetic steak is going to take a lot longer, require a much bigger investment, and the risk of failure will be greater.'[2]

These managers are not alone; they speak for many. Why, they ask, should they invest dollars that are hard to earn back when it is so easy – and so much less risky – to make money in other ways?[6] Why ignore a ready-made situation in cake mixes for the deferred and far less certain prospects in synthetic steaks? Why shoulder the competitive risks of making better, more innovative products?

In our judgement, the assumptions underlying these questions are prime evidence of a broad managerial failure – a failure of both vision and leadership – that over time has eroded both the inclination and the capacity of US companies to innovate.

Familiar Excuses

About the facts themselves there can be little dispute. Table 1 and figure 1 document our sorry decline. But the explanations and excuses commonly offered invite a good deal of comment.

It is important to recognize, first of all, that the problem is not new. It has been going on for at least 15 years. The rate of productivity growth in the private sector peaked in the mid-1960s. Nor is the problem confined to a few sectors of our economy; with a few exceptions, it permeates our entire economy. Expenditures on R & D by both business and government, as measured in constant (noninflated) dollars, also peaked in the mid-1960s – both in absolute terms and as a percentage of GNP. During the same period the expenditures on R & D by West Germany and Japan have been rising. More important, American spending on R & D as a percentage of sales in such critical research-intensive industries as machinery, professional and scientific instruments, chemicals and aircraft had

Table 1   *Growth in labour productivity since 1960 (United States and abroad)*

	Average annual percent change	
	Manufacturing 1960–78	All industries 1960–76
United States	2.8%	1.7%
United Kingdom	2.9	2.2
Canada	4.0	2.1
Germany	5.4	4.2
France	5.5	4.3
Italy	5.9	4.9
Belgium	6.9*	–
Netherlands	6.9*	–
Sweden	5.2	–
Japan	8.2	7.5

*1960–1977
Source: Council on Wage and Price Stability. July 1979 Report on Productivity. Executive Office of the President, Washington, DC.

dropped by the mid-1970s to about half its level in the early 1960s. These are the very industries on which we now depend for the bulk of our manufactured exports.

Investment in plant and equipment in the United States displays the same disturbing trends. As economist Burton G. Malkiel has pointed out: 'From 1948 to 1973 the [net book value of capital equipment] per unit of labor grew at an annual rate of almost 3%. Since 1973, however, lower rates of private investment have led to a decline in that growth rate to 1.75%. Moreover, the recent composition of investment [in 1978] has been skewed toward equipment and relatively short-term projects and away from structures and relatively long-lived investments. Thus our industrial plant has tended to age.'[6]

Other studies have shown that growth in the incremental capital equipment-to-labour ratio has fallen to about one-third of its value in the early 1960s. By contrast, between 1966 and 1976 capital investment as a percentage of GNP in France and West Germany was more than 20 per cent greater than that in the United States; in Japan the percentage was almost double ours.

To attribute this relative loss of technological vigour to such things as a shortage of capital in the United States is not justified. As Malkiel and others have shown, the return on equity of American business (out of which comes the capital necessary for investment) is about the same today as 20 years ago, *even after adjusting for inflation.* However, investment in both new equipment and R & D, as a percentage of GNP, was significantly higher 20 years ago than today.

The conclusion is painful but must be faced. Responsibility for this competitive listlessness belongs not just to a set of external conditions but

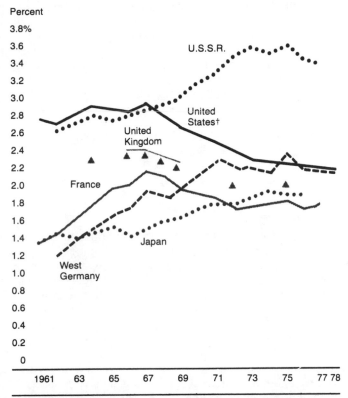

Gross expenditures for performance of R&D including associated capital expenditures.
†Detailed information on capital expenditures for R&D is not available for the United
States. Estimates for the period 1972-1977 show that their inclusion would have an impact
of less than one-tenth of 1% for each year.
Source: *Science Indicators – 1978* (National Science Foundation,Washington, D.C.), p. 6.
Note: The latest data may be preliminary or estimates.

*Figure 1   National expenditures for performance of R & D as a per cent of GNP by
country, 1961–78*

also to the attitudes, preoccupations and practices of American managers.
By their preference for servicing existing markets rather than creating new
ones and by their devotion to short-term returns and 'management by the
numbers', many of them have effectively foresworn long-term technological superiority as a competitive weapon. In consequence, they have
abdicated their strategic responsibilities.

The New Management Orthodoxy

We refuse to believe that this managerial failure is the result of a sudden
psychological shift among American managers toward a 'super-safe, no

risk' mind set. No profound sea change in the character of thousands of individuals could have occurred in so organized a fashion or have produced so consistent a pattern of behaviour. Instead we believe that during the past two decades American managers have increasingly relied on principles which prize analytical detachment and methodological elegance over insight, based on experience, into the subtleties and complexities of strategic decisions. As a result, maximum short-term financial returns have become the overriding criteria for many companies.

For purposes of discussion, we may divide this *new* management orthodoxy into three general categories: financial control, corporate portfolio management and market-driven behaviour.

Financial control

As more companies decentralize their organizational structures, they tend to fix on profit centres as the primary unit of managerial responsibility. This development necessitates, in turn, greater dependence on short-term financial measurements like return on investment (ROI) for evaluating the performance of individual managers and management groups. Increasing the structural distance between those entrusted with exploiting actual competitive opportunities and those who must judge the quality of their work virtually guarantees reliance on objectively quantifiable short-term criteria.

Although innovation, the lifeblood of any vital enterprise, is best encouraged by an environment that does not unduly penalize failure, the predictable result of relying too heavily on short-term financial measures – a sort of managerial remote control – is an environment in which no-one feels he or she can afford a failure or even a momentary dip in the bottom line.

Corporate portfolio management

This preoccupation with control draws support from modern theories of financial portfolio management. Originally developed to help balance the overall risk and return of stock and bond portfolios, these principles have been applied increasingly to the creation and management of corporate portfolios – that is, a cluster of companies and product lines assembled through various modes of diversification under a single corporate umbrella. When applied by a remote group of dispassionate experts primarily concerned with finance and control and lacking hands-on experience, the analytic formulas of portfolio theory push managers even further toward an extreme of caution in allocating resources.

'Especially in large organizations', reports one manager, 'we are observing an increase in management behaviour which I would regard as excessively cautious, even passive; certainly overanalytical; and, in general, characterized by a studied unwillingness to assume responsibility and even reasonable risk'.

Market-driven behaviour

In the past 20 years, American companies have perhaps learned too well a lesson they had long been inclined to ignore: businesses should be customer oriented rather than product oriented. Henry Ford's famous dictum that the public could have any colour automobile it wished as long as the colour was black has since given way to its philosophical opposite: 'We have got to stop marketing makeable products and learn to make marketable products.'

At last, however, the dangers of too much reliance on this philosophy are becoming apparent. As two Canadian researchers have put it: 'Inventors, scientists, engineers, and academics, in the normal pursuit of scientific knowledge, gave the world in recent times the laser, xerography, instant photography, and the transistor. In contrast, worshippers of the marketing concept have bestowed upon mankind such products as new-fangled potato chips, feminine hygiene deodorant, and the pet rock ....'[1]

The argument that no new product ought to be introduced without managers undertaking a market analysis is common sense. But the argument that consumer analyses and formal market surveys should dominate other considerations when allocating resources to product development is untenable. It may be useful to remember that the initial market estimate for computers in 1945 projected total world-wide sales of only ten units. Similarly, even the most carefully researched analysis of consumer preferences for gas-guzzling cars in an era of gasoline abundance offers little useful guidance to today's automobile manufacturers in making wise product investment decisions. Customers may know what their needs are, but they often define those needs in terms of existing products, processes, markets and prices.

Deferring to a market-driven strategy without paying attention to its limitations is, quite possibly, opting for customer satisfaction and lower risk in the short run at the expense of superior products in the future. Satisfied customers are critically important, of course, but not if the strategy for creating them is responsible as well for unnecessary product proliferation, inflated costs, unfocused diversification and a lagging commitment to new technology and new capital equipment.

Three Managerial Decisions

These are serious charges to make. But the unpleasant fact of the matter is that, however useful these new principles may have been initially, if carried too far they are bad for US business. Consider, for example, their effect on three major kinds of choices regularly faced by corporate managers: the decision between imitative and innovative product design, the decision to integrate backward and the decision to invest in process development.

Imitative vs. innovative product design

A market-driven strategy requires new product ideas to flow from detailed market analysis or, at least, to be extensively tested for consumer reaction before actual introduction. It is no secret that these requirements add significant delays and costs to the introduction of new products. It is less well-known that they also predispose managers toward developing products for existing markets and toward product designs of an imitative rather than an innovative nature. There is increasing evidence that market-driven strategies tend, over time, to dampen the general level of innovation in new product decisions.

Confronted with the choice between innovation and imitation, managers typically ask whether the marketplace shows any consistent preference for innovative products. If so, the additional funding they require may be economically justified; if not, those funds can more properly go to advertising, promoting or reducing the prices of less-advanced products. Though the temptation to allocate resources so as to strengthen performance in existing products and markets is often irresistible, recent studies by J. Hugh Davidson and others confirm the strong market attractiveness of innovative products.[3]

Nonetheless, managers having to decide between innovative and imitative product design face a difficult series of marketing-related trade-offs. Table 2 summarizes these trade-offs.

By its very nature, innovative design is, as Joseph Schumpeter observed a long time ago, initially destructive of capital – whether in the form of labour skills, management systems, technological processes or capital equipment. It tends to make obsolete existing investments in both marketing and manufacturing organizations. For the managers concerned it represents the choice of uncertainty (about economic returns, timing etc.) over relative predictability, exchanging the reasonable expectation of current income against the promise of high future value. It is the choice of the gambler, the person willing to risk much to gain even more.

Conditioned by a market-driven strategy and held closely to account by a 'results now' ROI-oriented control system, American managers have increasingly refused to take the chance on innovative product/market development. As one of them confesses: 'In the last year, on the basis of high capital risk, I turned down new products at a rate at least twice what I did a year ago. But in every case I tell my people to go back and bring me some new product ideas.'[2] In truth, they have learned caution so well that many are in danger of forgetting that market-driven, follow-the-leader companies usually end up following the rest of the pack as well.

Backward integration

Sometimes the problem for managers is not their reluctance to take action and make investments but that, when they do so, their action has the unintended result of reinforcing the status quo. In deciding to integrate backward because of apparent short-term rewards, managers often restrict their ability to strike out in innovative directions in the future.

Consider, for example, the case of a manufacturer who purchases a major component from an outside company. Static analysis of production economies may very well show that backward integration offers rather substantial costs benefits. Eliminating certain purchasing and marketing functions, centralizing overhead, pooling R & D efforts and resources, coordinating design and production of both product and component, reducing uncertainty over design changes, allowing for the use of more specialized equipment and labour skills – in all these ways and more, backward integration holds out to management the promise of significant short-term increases in ROI.

These efficiencies may be achieved by companies with commodity-like products. In such industries as ferrous and nonferrous metals or petroleum, backward integration toward raw materials and supplies tends to have a strong, positive effect on profits. However, the situation is markedly different for companies in more technologically active industries. Where there is considerable exposure to rapid technological advances, the promised value of backward integration becomes problematic. It may provide a quick, short-term boost to ROI figures in the next annual report, but it may also paralyse the long-term ability of a company to keep on top of technological change.

The real competitive threats to technologically active companies arise less from changes in ultimate consumer preference than from abrupt shifts in component technologies, raw materials or production processes. Hence those managers whose attention is too firmly directed toward the market-place and near-term profits may suddenly discover that their decision to make rather than buy important parts has locked their companies into an outdated technology.

Further, as supply channels and manufacturing operations become more systematized, the benefits from attempts to 'rationalize' production may well be accompanied by unanticipated side effects. For instance, a company may find itself shut off from the R & D efforts of various independent suppliers by becoming their competitor. Similarly, the commitment of time and resources needed to master technology back up the channel of supply may distract a company from doing its own job well. Such was the fate of Bowmar, the pocket calculator pioneer, whose attempt to integrate backward into semiconductor production so consumed management attention that final assembly of the calculators, its core business, did not get the required resources.

Long-term contracts and long-term relationships with suppliers can achieve many of the same cost benefits as backward integration without calling into question a company's ability to innovate or respond to innovation. European automobile manufacturers, for example, have typically chosen to rely on their suppliers in this way; American companies have followed the path of backward integration. The resulting trade-offs between production efficiencies and innovative flexibility should offer a stern warning to those American managers too easily beguiled by the lure of short-term ROI improvement. A case in point: the US auto industry's huge investment in automating the manufacture of cast-iron brake drums probably delayed by more than five years its transition to disc brakes.

Process development

In an era of management by numbers, many American managers – especially in mature industries – are reluctant to invest heavily in the development of new manufacturing processes. When asked to explain their reluctance, they tend to respond in fairly predictable ways. 'We can't afford to design new capital equipment for just our own manufacturing needs' is one frequent answer. So is: 'The capital equipment producers do a much better job, and they can amortize their development costs over sales to many companies.' Perhaps most common is: 'Let the others experiment in manufacturing; we can learn from their mistakes and do it better.'

Each of these comments rests on the assumption that essential advances in process technology can be appropriated more easily through equipment purchase than through in-house equipment design and development. Our extensive conversations with the managers of European (primarily German) technology-based companies have convinced us that this assumption is not as widely shared abroad as in the United States. Virtually across the board, the European managers impressed us with their strong commitment to increasing market share through internal development of advanced process technology – even when their suppliers were highly responsive to technological advances.

Table 2   Trade-offs between imitative and innovative design for an established product line

Imitative design	Innovative design
Market demand is relatively well known and predictable	Potentially large but unpredictable demand; the risk of a flop is also large
Market recognition and acceptance are rapid	Market acceptance may be slow initially, but the imitative response of competitors may also be slowed
Readily adaptable to existing market, sales and distribution policies	May require unique, tailored marketing distribution and sales policies to educate customers or because of special repair and warranty problems
Fits with existing market segmentation and product policies	Demand may cut across traditional marketing segments, disrupting divisional responsibilities and cannibalizing other products

By contrast, American managers tend to restrict investments in process development to only those items likely to reduce costs in the short run. Not all are happy with this. As one disgruntled executive told us: 'For too long US managers have been taught to set low priorities on mechanization projects, so that eventually divestment appears to be the best way out of manufacturing difficulties. Why?

'The drive for short-term success has prevented managers from looking thoroughly into the matter of special manufacturing equipment, which has to be invented, developed, tested, redesigned, reproduced, improved, and so on. That's a long process, which needs experienced, knowledgeable and dedicated people who stick to their jobs over a considerable period of time. Merely buying new equipment (even if it is possible) does not often give the company any advantage over competitors.'

We agree. Most American managers seem to forget that, even if they produce new products with their existing process technology (the same 'cookie cutter' everyone else can buy), their competitors will face a relatively short lead time for introducing similar products. And as Eric von Hipple's studies of industrial innovation show, the innovations on which new industrial equipment is based usually originate with the user of the equipment and not with the equipment producer.[5] In other words, companies can make products more profitable by investing in the development of their own process technology. Proprietary processes are every bit as formidable competitive weapons as proprietary products.

The American Managerial Ideal

Two very important questions remain to be asked: (1) Why should so many American managers have shifted so strongly to this new managerial orthodoxy? and (2) Why are they not more deeply bothered by the ill effects of those principles on the long-term technological competitiveness of their companies? To answer the first question, we must take a look at the changing career patterns of American managers during the past quarter century; to answer the second, we must understand the way in which they have come to regard their professional roles and responsibilities as managers.

The road to the top

During the past 25 years the American manager's road to the top has changed significantly. No longer does the typical career, threading sinuously up and through a corporation with stops in several functional areas, provide future top executives with intimate hands-on knowledge of the company's technologies, customers and suppliers.

Figure 2 summarizes the currently available data on the shift in functional background of newly appointed presidents of the 100 largest US corporations. The immediate significance of these figures is clear. Since the mid-1950s there has been a rather substantial increase in the percentage of new company presidents whose primary interests and expertise lie in the financial and legal areas and not in production. In the view of C. Jackson Grayson, President of the American Productivity Center, American management has for 20 years 'coasted off the great R & D gains made during World War II, and constantly rewarded executives from the marketing, financial, and legal sides of the business while it ignored the production men. Today [in business schools] courses in the production area are almost nonexistent.'[4]

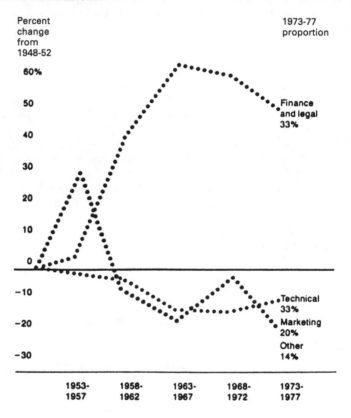

Figure 2   Changes in the professional origins of corporate presidents per cent changes from baseline years (1948–52) for 100 top US companies
Source: *Golightly and Co. International 1978*

In addition, companies are increasingly choosing to fill new top management posts from outside their own ranks. In the opinion of foreign observers, who are still accustomed to long-term careers in the same company or division, 'High-level American executives ... seem to come and go and switch around as if playing a game of musical chairs at an Alice in Wonderland tea party.'

Far more important, however, than any absolute change in numbers is the shift in the general sense of what an aspiring manager has to be 'smart about' to make it to the top. More important still is the broad change in attitude such trends both encourage and express. What has developed, in the business community as in academia, is a preoccupation with a false and shallow concept of the professional manager, a 'pseudoprofessional' really – an individual having no special expertise in any particular industry or technology who nevertheless can step into an unfamiliar company and run it successfully through strict application of financial controls, portfolio concepts and a market-driven strategy.

The gospel of pseudoprofessionalism

In recent years, this idealization of pseudoprofessionalism has taken on something of the quality of a corporate religion. Its first doctrine, appropriately enough, is that neither industry experience nor hands-on technological expertise counts for very much. At one level, of course, this doctrine helps to salve the conscience of those who lack them. At another, more disturbing level it encourages the faithful to make decisions about technological matters simply as if they were adjuncts to finance or marketing decisions. We do not believe that the technological issues facing managers today can be meaningfully addressed without taking into account marketing or financial considerations; on the other hand, neither can they be resolved with the same methodologies applied to these other fields.

Complex modern technology has its own inner logic and developmental imperatives. To treat it as if it were something else – no matter how comfortable one is with that other kind of data – is to base a competitive business on a two-legged stool, which must, no matter how excellent the balancing act, inevitably fall to the ground.

More disturbing still, true believers keep the faith on a day-to-day basis by insisting that as issues rise up the managerial hierarchy for decision they be progressively distilled into easily quantifiable terms. One European manager, in recounting to us his experiences in a joint venture with an American company, recalled with exasperation that 'US managers want everything to be simple. But sometimes business situations are not simple, and they cannot be divided up or looked at in such a way that they become simple. They are messy, and one must try to understand all the facets. This appears to be alien to the American mentality'.

The purpose of good organizational design, of course, is to divide responsibilities in such a way that individuals have relatively easy tasks to perform. But then these differentiated responsibilities must be pulled together by sophisticated, broadly gauged integrators at the top of the managerial pyramid. If these individuals are interested in but one or two aspects of the total competitive picture, if their training includes a very narrow exposure to the range of functional specialties, if – worst of all – they are devoted simplifiers themselves, who will do the necessary integration? Who will attempt to resolve complicated issues rather than try to uncomplicate them artificially? At the strategic level there are no such things as pure production problems, pure financial problems or pure marketing problems.

Merger mania

When executive suites are dominated by people with financial and legal skills, it is not surprising that top management should increasingly allocate time and energy to such concerns as cash management and the whole process of corporate acquisitions and mergers. This is indeed what has happened. In 1978 alone there were some 80 mergers involving companies with assets in excess of $100 million each; in 1979 there were almost 100. This represents roughly $20 billion in transfers of large companies from one owner to another – two-thirds of the total amount spent on R & D by American industry.

In 1978 *Business Week* ran a cover story on cash management in which it stated that 'the 400 largest US companies together have more than $60 billion in cash – almost triple the amount they had at the beginning of the 1970s'. The article also described the increasing attention devoted to – and the sophisticated and exotic techniques used for – managing this cash hoard.

There are perfectly good reasons for this flurry of activity. It is entirely natural for financially (or legally) trained managers to concentrate on essentially financial (or legal) activities. It is also natural for managers who subscribe to the portfolio 'law of large numbers' to seek to reduce total corporate risk by parcelling it out among a sufficiently large number of separate product lines, businesses or technologies. Under certain conditions it may very well make good economic sense to buy rather than build new plants or modernize existing ones. Mergers are obviously an exciting game; they tend to produce fairly quick and decisive results, and they offer the kind of public recognition that helps careers along. Who can doubt the appeal of the titles awarded by the financial community; being called a 'gunslinger', 'white knight' or 'raider' can quicken anyone's blood.

Unfortunately, the general American penchant for separating and simplifying has tended to encourage a diversification away from core technologies and markets to a much greater degree than is true in Europe or Japan. US managers appear to have an inordinate faith in the portfolio law of large numbers – that is, by amassing enough product lines, technologies and businesses, one will be cushioned against the random setbacks that occur in life. This might be true for portfolios of stocks and bonds, where there is considerable evidence that setbacks *are* random. Businesses, however, are subject not only to random setbacks such as strikes and shortages but also to carefully orchestrated attacks by competitors, who focus all their resources and energies on one set of activities.

Worse, the great bulk of this merger activity appears to have been absolutely wasted in terms of generating economic benefits for stockholders. Acquisition experts do not necessarily make good managers. Nor can they increase the value of their shares by merging two companies any better than their shareholders could do individually by buying shares of the acquired company on the open market (at a price usually below that required for a takeover attempt).

There appears to be a growing recognition of this fact. A number of US companies are now divesting themselves of previously acquired companies; others (for example, W.R. Grace) are proposing to break themselves up into relatively independent entities. The establishment of a strong competitive position through in-house technological superiority is by nature a long, arduous and often unglamorous task. But it is what keeps a business vigorous and competitive.

The European Example

Gaining competitive success through technological superiority is a skill much valued by the seasoned European (and Japanese) managers with whom we talked. Although we were able to locate few hard statistics on their actual practice, our extensive investigations of more than 20 companies convinced us that European managers do indeed tend to differ significantly from their American counterparts. In fact, we found that many of them were able to articulate these differences quite clearly.

In the first place, European managers think themselves more pointedly concerned with how to survive over the long run under intensely competitive conditions. Few markets, of course, generate price competition as fierce as in the United States, but European companies face the remorseless necessity of exporting to other national markets or perishing.

The figures here are startling: manufactured product exports represent more than 35 per cent of total manufacturing sales in France and Germany and nearly 60 per cent in the Benelux countries, as against not quite 10 per cent in the United States. In these export markets, moreover, European products must hold their own against 'world class' competitors, lower priced products from developing countries and American products selling at attractive devalued dollar prices. To survive this competitive squeeze, European managers feel they must place central emphasis on producing technologically superior products.

Further, the kinds of pressures from European labour unions and national governments virtually force them to take a consistently long-term view in decision making. German managers, for example, must negotiate major decisions at the plant level with worker-dominated works councils; in turn, these decisions are subject to review by supervisory boards (roughly equivalent to American boards of directors), half of whose membership is worker elected. Together with strict national legislation, the pervasive influence of labour unions makes it extremely difficult to change employment levels or production locations. Not surprisingly, labour costs in Northern Europe have more than doubled in the past decade and are now the highest in the world.

To be successful in this environment of strictly constrained options,

European managers feel they must employ a decision-making apparatus that grinds very fine – and very deliberately. They must simply outthink and outmanage their competitors. Now, American managers also have their strategic options hedged about by all kinds of restrictions. But those restrictions have not yet made them as conscious as their European counterparts of the long-term implications of their day-to-day decisions.

As a result, the Europeans see themselves as investing more heavily in cutting-edge technology than the Americans. More often than not, this investment is made to create new product opportunities in advance of consumer demand and not merely in response to market-driven strategy. In case after case we found the Europeans striving to develop the products and process capabilities with which to lead markets and not simply responding to the current demands of the marketplace. Moreover, in doing this they seem less inclined to integrate backward and more likely to seek maximum leverage from stable, long-term relationships with suppliers.

Having never lost sight of the need to be technologically competitive over the long run, European and Japanese managers are extremely careful to make the necessary arrangements and investments today. And their daily concern with the rather basic issue of long-term survival adds perspective to such matters as short-term ROI or rate of growth. The time line by which they manage is long, and it has made them painstakingly attentive to the means for keeping their companies technologically competitive. Of course they pay attention to the numbers. Their profit margins are usually lower than ours, their debt ratios higher. Every tenth of a per cent is critical to them. But they are also aware that tomorrow will be no better unless they constantly try to develop new processes, enter new markets, and offer superior – even unique – products. As one senior German executive phrased it recently, 'We look at rates of return, too, but only after we ask "Is it a good product?"'[2]

Creating Economic Value

Americans travelling in Europe and Asia soon learn they must often deal with criticism of our country. Being forced to respond to such criticism can be healthy, for it requires rethinking some basic issues of principle and practice.

We have much to be proud about and little to be ashamed of relative to most other countries. But sometimes the criticism of others is uncomfortably close to the mark. The comments of our overseas competitors on American business practices contain enough truth to require our thoughtful consideration. What is behind the decline in competitiveness of US business? Why do US companies have such apparent difficulties competing with foreign producers of established products, many of which originated in the United States?

For example, Japanese televisions dominate some market segments, even though many US producers now enjoy the same low labour cost advantages of offshore production. The German machine tool and automotive producers continue their inroads into US domestic markets, even though their labour rates are now higher than those in the United States and the famed German worker in German factories is almost as likely to be Turkish or Italian as German.

The responsibility for these problems may rest in part on government policies that either overconstrain or undersupport US producers. But if our foreign critics are correct, the long-term solution to America's problems may not be correctable simply by changing our government's tax laws, monetary policies and regulatory practices. It will also require some fundamental changes in management attitudes and practices.

It would be an oversimplification to assert that the only reason for the decline in competitiveness of US companies is that our managers devote too much attention and energy to using existing resources more efficiently. It would also oversimplify the issue, although possibly to a lesser extent, to say that it is due purely and simply to their tendency to neglect technology as a competitive weapon.

Companies cannot become more innovative simply by increasing R & D investments or by conducting more basic research. Each of the decisions we have described directly affects several functional areas of management, and major conflicts can only be reconciled at senior executive levels. The benefits favouring the more innovative, aggressive option in each case depend more on intangible factors than do their efficiency-oriented alternatives.

Senior managers who are less informed about their industry and its confederation of parts suppliers, equipment suppliers, workers and customers or who have less time to consider the long-term implications of their interactions are likely to exhibit a noninnovative bias in their choices. Tight financial controls with a short-term emphasis will also bias choices toward the less innovative, less technologically aggressive alternatives.

The key to long-term success – even survival – in business is what it has always been: to invest, to innovate, to lead, to create value where none existed before. Such determination, such striving to excel, requires leaders – not *just* controllers, market analysts and portfolio managers. In our preoccupation with the braking systems and exterior trim, we may have neglected the drive trains of our corporations.

References

1. Bennett, R., Cooper, R., 1979 Beyond the marketing concept. *Business Horizons* June, p. 76
2. *Business Weekly*, 16 February, 1976 p. 57
3. Davidson, J.H., 1976 Why most new consumer brands fail. *Harv, Bus. Rev.*, March–April, p. 117
4. *Dun's Review*, July 1978, p. 39

5. von Hippel, E., 1975 *The Dominant Role of Users in the Scientific Instrument Innovation Process*. MIT Sloan School of Management Working Paper 75–764, January
6. Malkiel, B.G., 1979 Productivity – The problem behind the headlines. *Harvard Business Review*, May–June, p. 81
7. Suzuki, R., 1979 Worldwide expansion of US Exports – A Japanese View. *Sloan Management Review*, Spring, p. 1

4.4

Technical Innovation and International Competitiveness in the Television Industry

E. Sciberras

Introduction

Colour television sales account for a third of the consumer electronics industry in the US and Japan and for around half the industry in Europe.[1]
Innovation has affected the television sector through:

(1) *Changes in the product.* These have included development of small (1″ to 5″) and large screen (Sony has 37″) and the development of colour, remote control and Hi-Fi sound. More 'exotic' developments have included flat screens, multiscreen sets, small screens within the main screen and sets with 'fax' capability able to produce paper copy of the screen image.

(2) *Changes in components.* These are often difficult to distinguish from changes in the product. Colour or screen size could be regarded as set changes or as changes in the tube component. Technical changes in componentry have included developments in the cathode ray tube (CRT) such as the development of in-line guns, convergence CRTs, the wide-angle deflection CRT as well as changes in circuitry. Solid-state components, particularly integrated circuits and microprocessors, both reduce component numbers and permit programmable control of video recorder sets and home information systems.

(3) *The development of related products and changes in the role of the set.* Video games, video cassette recorders and disc players widen the area of application of television sets and increase users' control over programmes. Users can record programmes on an alternative channel while viewing another; they can recall past programmes if they are not interested in the programmes being broadcast. Technical change has made it possible for television sets to be used to provide home information as well as

Source: from *Omega*, 10(6), 1982.

entertainment. In addition to passive information transmission such as Teletext and Oracle, interactive systems such as the PRESTEL service can link television with the telephone and with cable-based information transmission, as in the Japanese Hi-Ovis system. The home computer plugged through the television is another aspect of the widening of the area of application of television from entertainment to home information system.

(4) *Changes in the process of manufacture.* Television manufacture involves the process of assembly of components, soldering and the testing of subassemblies and finished products. It also involves the sourcing and testing of components as well as the handling of subassemblies and the packaging of the completed sets. Innovation has affected each part of this process, the relationship between them and also other activities of the firms such as marketing and product policy. Significant changes in the manufacturing process are as follows:

(1) Since the early 1970s, the number of components in a set has been reduced from over 1,400 to around 400. This has resulted from the design of circuits using integrated semiconductor components rather than discrete devices.

(2) Between 1970 and 1978, the number of integrated circuits [i.c.s] in a typical 20″ television set increased from two to nine, while the number of discrete transistors and diodes fell from 130 to around 70. This has reduced the labour time necessary for assembly, and therefore the share of direct labour costs in the value of sets.

(3) Greater capital intensity of production has been made possible by better design and by the dramatic fall in the number of components. Before the number and diversity of components could be reduced, the relative inflexibility of machines in identifying and handling diverse components made automatic insertion uneconomic.

(4) The high cost involved in disruptive downtime due to faulty parts or products in automated production, and the increasing importance of reliability in international competitiveness has enhanced the need for testing. The process of testing components, subassemblies and finished products has also been affected by technical change. Computer-based equipment has been used increasingly to test assembled boards and subassemblies before they are assembled into final products.

(5) Automatic insertion increases the need to ensure that components of high reliability rather than just low cost are used. Thus, the purchase of components requires greater scrutiny and consideration. Design of final product has increasingly to include consideration of these component constraints.

These changes increase the need for planned integration of all the various stages of production. This is necessary to ensure the regular supply of appropriate components and subassemblies and the arrangement of flows, including testing, to avoid duplication of transportation, bottlenecks and downtime of expensive machinery.

These component, product and process innovations do not operate independently. Full exploitation of developments in one area necessitates developments in others. The wide angle CRT required more sophisticated solid-state components for power and picture definition. Developments in solid-state components are necessary for product changes such as remote control. The changed role of the set also is dependent upon component developments such as decoder chips and microprocessor units (MPUs) for Teletext and Viewdata. Changes in the manufacturing process are dependent upon integration to reduce the number and variety of components to a level 'handleable' by automated machinery.

The interdependence of changes has dramatically increased the complexity and importance of manufacturing process design. The principal benefit achieved has been high quality and reliability. The Japanese have recognized this in the development and manufacture of their sets for some time.

Quality and Reliability in Competitive Strategy

There are three principal aspects of quality for consumer products such as television sets: product differentiation – from appearance and style to features including Hi-Fi and remote control; quality of the function – picture sharpness, definition and sound; and reliability – freedom from breakdown. There are significant differences in the role of quality and reliability in competitive strategy between US and European firms on the one hand and Japanese firms on the other.

Product differentiation

US and European firms have emphasized product differentiation, features and price rather than reliability in the past. Marketing opportunities have been the major factor in the introduction of new products by all the leading US and European firms. The role of technical factors in new product introduction varied between the 'big league' and the 'little league' of the industry. Generally, the big league firms were extremely conscious of competitors' new products and prices. Competition relied on low prices to protect market share, particularly in 'bread and butter' products such as large screen sets. Leading firms also offered full ranges of sets either by manufacture or, where their own costs of manufacture were too high, by purchase from other manufacturers. The leaders competed on price and service and generally avoided 'minor' technical changes until customer acceptance was proven. The concern to remain price competitive induced avoidance of the risks of cost increase with 'minor' features. Leading firms were generally reluctant to innovate before markets were proven to exist; although, exceptionally, they did introduce major innovations such as video discs and in-line and wide-angle deflection CRTs. The little league firms tended to introduce 'minor' technical changes such as automatic tuning (VIR) and comb filters, first. This was part of a strategy of up-market 'quality' competition to differentiate products from leaders and to command a price premium. Technology-induced new product introduction was more significant among these firms.

US firms assumed that consumers preferred to buy low price sets rather than sets of higher quality and this led firms to supply relatively cheap but standard sets without special features. Poor reliability was compensated for by dense networks of service facilities. Firms concentrated on improving maintenance facilities in preference to incurring higher set costs by improving quality and introducing special features. Such strategy began to change only after Japanese competition had captured almost half the market in the US, proving that customers would pay higher prices for better quality. US firms are still more concerned about low price in new product introductions than are Japanese competitors. Japanese firms' prices are significantly higher than, or at least equal to, those of the US leaders in the most popular product line (see table 1).

Table 1   Retail price of 19″ colour television sets in the US market (1977)

Brand	Country	Price ($)
Sony	Japan	580
JVC (Matsushita)	Japan	450
Zenith	US	450
Panasonic (Matsushita)	Japan	420
Toshiba	Japan	400
RCA	US	400

Source: Television Digest, 4 July 1977

European firms' strategies vary. Price and service rather than features or product reliability have played the major role in the UK market, as in the US. UK firms face consumer preference influenced by a combination of a rental market and lower living standards. Rental businesses require standard products to reduce service costs and have tended to push the market away from complex sets. Advanced technology received less emphasis in UK firms' product ranges than in those of their European and Japanese competitors.

Rental companies have wished to promote what they have seen as their strongest asset – efficient repair and maintenance service. They have not emphasized the special characteristics of the sets. This has delayed the UK big league firms' introduction of features such as remote control, sensor tuning and programmable TV which had been exploited by competitors in Europe sometimes more than five years earlier. Advanced sets being introduced by Japanese new entrants have recently begun to change this. The attractiveness of rental arrangements compared to direct purchase declined as the reliability of sets available for purchase dramatically improved. Rental companies found that they could no longer compete on service alone. The fall in the rental share of the market has compelled them to consider using features as a form of competition. UK rental companies have only recently widened their product base to include video recorders and portable colour sets.

In Germany and Scandinavia, the strategies of the firms have been different. Although set reliability has not been impressive until very recently (see table 2), high quality in the form of features and offered at high prices is an important aspect of competitive strategy. This emphasis on quality in competition may contribute to the relatively successful German export performance when compared to the UK and the US.

Table 2   Proportion of sets with no repairs in the last 12 months (1978 and 1980)

Country	Sample size		Sample percentage	
	1978	1980	1978	1980
UK	1373	1129	48	65
Other European	169	95	38	81
Japanese	338	481	91	96

Source: Which? January 1978, January 1980

Unlike the major US and UK firms, the German and Japanese large firms have been quick to introduce new products. However, the Japanese were the first to exploit small-screen TV and video recorder products which are growing to major shares of the video market in the US and Europe. Where difficulties of a technical or commercial nature have delayed entry, Japanese firms formed joint agreements in order to enter new markets quickly. In 1980, Sony entered into a technology exchange agreement with Philips, to access the laser based video disc technology of its European rival. Faced with the control of almost 50 per cent of the US TV retail network by RCA and its licensee, Zenith, Matsushita and its subsidiary, JVC, entered into a joint venture with the industry number three, GE, to market their video disc in the US.[4] But Japanese strategy towards innovation, particularly their emphasis on product reliability and on innovation in production engineering, differs markedly from that of all their US and European competitors. The leading Japanese firms introduce products as a planned combination of marketing, technical and competitive factors. Three of the leading firms are multidivisional and the close cooperation between the various divisions means that the video division is constantly aware of product opportunities made available through new developments in the component divisions of their firms. This integration between set and component development means that even minor technical changes will be adopted if they result in improved quality. In the view of the Japanese firms, customers are prepared to pay for higher quality and reliability and this is reflected in their 'up market' pricing strategy in Europe and the US.

Process innovation

The adoption of automated techniques by European firms is essentially a competitive reaction to the Japanese. But European firms have been less successful than the Japanese. Their products had not initially been designed for automatic insertion (AI). The extent of automatic insertion varied from 30 to 60 per cent with average insertion around 40–45 per cent. Most European firms only automatically assemble passive components. Transistors and i.c.s are still assembled manually. Only two – one a leading UK firm – had clear plans to increase automatic insertion significantly and to automate the assembly of transistors and i.c.s.

Automatic testing is adopted by most European firms who had moved towards automatic insertion and some firms also adopted automated soldering. Traditional testing, soldering, handling and packaging techniques accounted for the largest share of manufacturing processes for the European firms, however. All the firms adopting automatic insertion now have introduced new chassis specifically designed around the requirements for automatic insertion. Generally, European firms now recognize the need for higher quality components for automatic insertion. They have experienced difficulties in obtaining supplies of such components without resorting to Japanese sources.

US firms do not incorporate a significant extent of automation. Automatic insertion only accounts for a small share of assembly. One firm made a significant attempt to replace manual assembly lines in the early 1970s. This proved to be noncompetitive.

Paradoxically, the success of Japanese competition did not initially serve to convince the US firms of the need for automated insertion to improve quality, but made large-scale capital investment increasingly unattractive in the face of declining profitability. While acknowledging that quality and reliability are important in competitive success, US firms have continued to compete with lower labour costs of assembly through offshore location in the Far East and Mexico. Products are being designed around smaller numbers of chassis to ease service costs and attempts are being made to improve component reliability. Some recent findings by a major US component user show that for some standard integrated circuits reliability of Japanese devices is between six to nine times better than for comparable US devices.[3] According to Hewlett-Packard's figures, faulty US-made devices had been detected at rates between 0.11 and 0.19 per cent and Japanese failure rates varied between 0.010 and 0.019 per cent.[5]

All the Japanese firms adopt automatic insertion for component assembly. This is very extensive – accounting for over 75 per cent of components inserted on printed circuit boards (PCBs). i.c.s are inserted by semiautomatic machines but all discrete components are inserted automatically in some firms. Japanese firms began automatic insertion using the machines as isolated production units. Rising costs, particularly yen revaluations, compelled greater attention to more cost-effective use of automation. From the mid-1970s a greater share of R & D was directed towards this.

The effort concerned not only automation of insertion and testing but total production engineering. The result was an awareness of 'space engineering' – the design of total production flow to maximize the return to the total factory area. This includes the movement and handling of components and subassemblies and final products through to packaging. All Japanese firms take this total approach to automation. Insertion and testing are seen only as a part in this approach to innovation in manufacturing processes.

Firms using automatic processes both in Europe and Japan stress the quality-improving features. Direct cost advantages, relative to manual insertion, particularly in the Far East, are not proven. Comparison of rates of insertion indicates that automatic insertion may only be three to four times faster than 'best practice' manual. Generally, however, automatic insertion is claimed to be around eight to ten times faster than manual insertion.

Estimates of the cost savings per set arising from the use of automatic insertion are not dramatic. These have varied from between 20p per set to 40p per set. The difference between manual and automated insertion costs per component is 0.08p. Assuming plant volumes of 650,000 sets per year (about the largest European volumes), and 500 semiconductor components per set, savings are estimated at £260,000 per year. With the cost of the automated equipment to handle this volume of production at £1 million, payback would take just four years. But investment payback is expected in around two years in the European and US industry. So, for them, automated insertion techniques are not economically justifiable by direct labour cost savings alone. Japanese firms adopt longer time horizons in calculating profitable return on investment. They, therefore, are prepared to undertake long-term capital investment which would be rejected by their US and European competitors.

Because US, UK and most European firms have been more concerned in the past with cost rather than quality, automated insertion and testing have not been attractive to them. Japanese firms are different. Automatic insertion of radial and integrated components raises the share of components inserted. Rates of insertion therefore can be higher, lowering costs of insertion per component. These components are not inserted automatically by the US and European firms and this may contribute to the lower success with automatic insertion.

Most important, the higher volumes of production in Japan yield greater scale economies. Japanese plants are on average twice the size of the largest West German plants and around six times larger than the largest UK plants (see table 3). The learning curve cost advantages of this larger volume have been realized by the Japanese for some time. By 1970, their larger CTV volume offered a cost advantage of around 20–30 per cent over their US competitors (see figure 1). This preceded the revolution in automated assembly which was first successfully adopted by the Japanese. The cost advantage over the US since 1970 may have widened considerably as a result. The UK is especially disadvantaged in this respect relative to its Japanese competitors. Not only is plant volume generally much lower, but the disadvantages are exacerbated by the greater number of chassis produced within each plant. On average, Japanese firms produce four times as many sets per single chassis as UK firms.

The greater willingness to invest and the more effective adoption of process innovation is clearly reflected in the higher levels of manufacturing productivity and reliability achieved by the Japanese firms compared to their European competitors (see table 4). Although plant overheads are higher, as a share of total costs they are not unduly high for the Europeans. The most significant differences between the Europeans and the Japanese competitors are in labour costs. Labour costs are a function of labour time for manufacture of a set and the hourly labour costs. West Germany does not have significantly higher costs per hour and UK labour costs per hour are lower than Japan. The cause of higher European labour costs per set is poor manufacturing productivity *not* expensive labour, the result of poor performance in design, reduction and quality of components, and production automation. These factors interrelate with each other, and are difficult to assess individually. For example, design improvements which reduce the number of boards in a set from around a dozen to one or two can be made possible by reduction in component numbers.

Until recently, Japanese sets had only two-thirds of the components of European sets on average. While the Europeans have attempted to narrow this difference, the Japanese have gone even further in the use of Medium Scale Integration (MSI) circuits and with developments in CRTs which continue to reduce component content, or offer improved reliability and enhanced features without increasing component count significantly. These factors have reduced labour time in assembly and have also served to reduce the number of components necessary for interfacing. Such design improvement also makes possible more efficient and low cost testing and maintenance. This improves set reliability sharply. An analysis of the source of set failures indicates that between 20 and 40 per cent of breakdowns are due to development and design weakness, a further 40–65 per cent are due to quality of components and 15–20 per cent are due to workmanship.

The Japanese were most effective in dealing with each of these sources of failure; but the Europeans paid less attention to testing and product reliability suffered. In the face of Japanese competition, they made major efforts to improve reliability. The breakdown rate of European sets halved between 1978 and 1980 but remained significantly above that of the Japanese. Among European competitors, the UK firms' performance is the worst, and improved less over the same period. US performance has been no better than UK. US sets were found to fail at at least five times the rate of Japanese sets.[8] This is only now beginning to change in the face of Japanese competitive success.

The Causes of Competitive Success

Three sets of factors explain the differences in competitive performance between the Japanese 'new entrants' and the established European and US

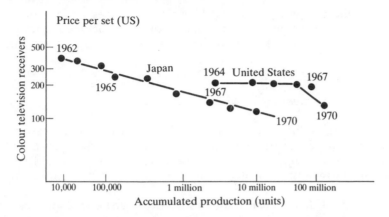

Figure 1   Estimated learning curve characteristics of the producers of colour television receivers
Source: *Cohen 1972*

Table 3   Average CTV plant volume (1977)

Country	Average CTV plant volume (thousand units)
UK	96 240
West Germany	350 650
Japan	500 1560

Source: Williams, 1979, p. 13

Table 4   Comparative production costs and direct labour costs per colour TV set

Comparative production costs	Japan	UK	West Germany
Direct labour costs (£)	5.7	10.6	15.1
Material costs (£)	100.0	126.0	119.0
Plant overheads (£)	11.0	20.0	17.0
Total production costs (£)	116.7	156.6	151.1
Direct labour costs			
Average manhours	1.9	6.1	3.9
Employment costs (£/hr)	3.0	1.74	3.85
Total labour costs (£)	5.7	10.6	15.0

Source: Williams, 1979, p. 13

firms in the television industry. These are attitudes to investment and research and development, the status of manufacturing and policies towards the reward, training and deployment of engineers and management.

Attitude to investment and research and development

There are major differences between US, European and Japanese firms in attitudes towards investment and firm profitability. Compared with the Japanese, US, European and UK firms in particular have short time horizons for investment write-off. In a recent study[9] it was found that:

> In most of Japan's successful industries, Japanese companies have invested more heavily and consistently than their Western competitors. Japanese companies typically evaluate investments differently from many Western companies. The latter usually isolate a project, grade it on a discounted cash flow (DCF) rate of return basis, and closely observe its effect on the current earnings of the individual profit centre. This favours conservative, early return projects and penalises more risky, long term programmes which may attempt a fundamental alteration of the business. ...
> In Japan's most successful industries ... companies did have a high debt equity ratio in the growth years of the 1960s. ... In the 1970s, as cost advantage was secured against both marginal domestic competition and Western producers, loans were paid off and debt levels fell. ...
> Their reward for investing more heavily than their competitors over the long period has been a low cost position and a more durable cash flow. The Japanese willingness to make investments which have a long payback period and their often superior production economies are related.

This applies to the television industry. The longer term view towards profit goals and capital write-off by Japanese firms means that they are prepared to undertake investment which would be regarded as unjustified by their US and European competitors. This has contributed to Japanese willingness to invest far more heavily than their competitors in the television industry in automated production and testing equipment.

A recent survey of the Japanese electronics industries argued that R & D and plant and equipment investments by American enterprises are concentrated in the avant-garde fields of military- and aerospace-related industry.[7] As a result, the streamlining and labour saving required to maintain competition in standard products such as consumer appliances and products have not materialized.

> In Japan, on the other hand, these measures have been taken with gusto and the Japanese have managed to open up a lead. ... Research fund expenditures are closely related to the production structure. ... The role

played by avant-garde technology in the electrical equipment industry is relatively low, and the emphasis tends to be to the development of application techniques.

The more appropriate strategy towards R & D and the better innovative performance of the Japanese relative to their competitors – both American and European – in the television industry confirms this.

There are wide variations in the extent and nature of firms' research and development efforts. Some Japanese consumer electronics firms allocate as much as 6.75 per cent of turnover to R & D. US and European, and particularly UK competitors spend proportionately much less. Japanese firms spend a higher share of turnover on R & D on consumer electronic activities, possibly because they are involved to a greater extent in other areas which tend to be research intensive. But the Japanese firm which spent the highest proportion on R & D was the one with the highest concentration on consumer electronics.

The most significant differences in R & D effort between US, European and Japanese firms are in the nature and extent of R & D in *process* relative to *product* development. Japanese firms place a far greater emphasis on production process development in their total development effort than their competitors. Generally, around one-third of total R & D effort is on process development. This extends to planning the design and purchase of components, automation of handling and packaging and 'space engineering' – efforts in the design of production layout. Such R & D is virtually nonexistent in many US and European firms. These differences derive from the relatively low status attributed to manufacturing in US and European firms. The status of manufacturing is much higher in Japanese firms.

The status of manufacturing

In US and European firms, design engineering is regarded essentially as a product development function. Technical effort and resources are directed to this product activity. Manufacturing engineering is generally much lower in status and manufacturing engineering departments are often the last to be consulted in product decision making. Manufacturing engineering is regarded as essentially a process of reducing the cost of a subassembly or chassis.

The training of product development and marketing managers in Japanese firms generally includes training in manufacture. Most managers are aware of the manufacturing requirements and manufacturing potential and limitations of the firm. Manufacturing management is intimately involved from the initial stages of new product development and the final product reflects significant manufacturing engineering input. Design of final products includes consultation with engineers in the components divisions and in the manufacturing equipment divisions. In US and European firms these divisions are operated on an 'arm's length' basis – or one 'prestigious' division such as Components dictates to the others. In Japanese firms the final product and production system is the result of closer *inter*action by all the design, product and production engineering departments concerned. Management's direct marketing and service experience of consumers' problems form an important input to product engineering in the design of sets and to manufacturing engineering in the production of sets. The result is products which are designed for more efficient manufacture but which also satisfy consumers' requirements for quality and reliability in use.

The competitive advantage of the Japanese firms stems not only from volume, larger R & D investment and capital availability free from the need for short-term returns or from diversified corporate structures, although these contribute substantially. The overall competitive advantage derives largely from the greater attention given to training of management. This training involves the greater consciousness of quality and manufacturing processes, and of the opportunities in the development of products. This is achieved by rotating management to the component, manufacturing equipment, service and manufacturing divisions within the firm, as part of their career experience.

Training, reward and deployment of manpower

Compared with Japanese firms, European and US firms have significantly different policies towards the training and reward of management and engineers. European and US firms are less aware of, or less successful in, training skilled manpower, rotation and skill deployment.

Graduates employed in European firms were found often to resist participation in long training programmes. Many trainees drop out of training. Engineers and technicians in European firms expect promotion and/or higher salaries automatically upon completion of training. Generally, the firms train graduates for specific functions in the firm. US firms do not adopt formal training schemes, formal career patterns or rotation for employees. Training tends to be 'on the job' and informal and varies from around six months to two years. The low status of manufacturing makes it difficult for US and European firms to hire and retain good engineers if they are allocated to this function. Japanese firms all adopt formal training for new recruits at graduate engineer and technician levels. This tends to be around two years. Training is a required feature of employment in Japanese firms and completion of training does not necessarily imply immediate promotion or higher salary as it tends to in US and European firms.

Attitudes to managerial rewards significantly influence firms' competitive strategy. A recent study[9] finds that:

> In many Western companies, managers' salaries are often tied to the short term performance goals of their individual profit centre. In addition, managers are moved frequently from one branch to another.

Projects which depress profitability even if they are expected to yield long term competitive advantage may thus be discouraged.

... Japanese companies do not usually employ the profit centre control system as often as Americans, and the compensation of management is less directly tied to this year's profit centre performance. Accordingly, the Japanese typically find it easier to initiate investments in projects with a long term payoff.

The use of the profit centre as the basis for managerial assessment encourages competition and mistrust between management in divisions of US and European companies. This is in the belief that such friction results in the optimum overall performance. But cooperation which may result in larger overall profits for the firm but lower profitability for any one division is inevitably discouraged.

Training of management in Japanese firms aims to prevent this. All new graduate employees undertake training which acquaints them with the whole range of activities of the firm before they are permitted to go on to training in their selected field. Thus, potential design or production engineers will spend up to six months gaining actual shopfloor manufacturing experience and a further six months retail or maintenance department experience. This means that new designers will be more aware of the manufacturing requirements for the products they design and of the nature of consumers' preference and of failures of the product in use. Often, commercial staff are given experience of manufacture to acquaint them better with the firm's products and with the problems of production requirements. They also learn to view managements in these divisions as colleagues and not merely competitors. Encouraging cooperation further, following training, Japanese engineers are rotated to a variety of divisions in the firm such as components and manufacturing equipment at three or four year intervals. This is not possible for all employees but firms manage to achieve around 30–40 per cent rotation.

Conclusion

Many small firms and small television activities of large firms, such as Telefunken and Nordmende in Germany, GEC, Rank and Decca in the UK, Tandberg in Norway, Luxor in Sweden and Warwick and Quasar in the US, have experienced financial crises, withdrawn from the industry or merged, often with Japanese firms.

Some of the larger European and US firms have responded to the Japanese threat much more positively, although the UK industry leader, Thorn, recently announced a link-up with Sharp, the only major Japanese firm still excluded from manufacturing in the UK, to 'share technology' in research, development and manufacturing.[6] Two European firms have redesigned sets, rationalized product lines and introduced automated insertion and testing on a large scale. Awareness of the importance of reliability has grown and dramatic improvements in reliability have been achieved by European firms in the last two years. Leading US firms have now acknowledged the importance of quality and revised their standards accordingly.

Imitating Japanese product designs and production lines or even resorting to high-quality Japanese components may yield dramatic short-term improvements in cost and quality. But the Japanese have a wide lead, greater experience and continue to develop new products and process technology. More fundamental changes are necessary. US and European firms need to give greater recognition to the importance of research and development in process engineering as well as products and to reconsider the role and status of manufacturing. The more selective hiring, intense training and appropriate deployment of engineers and management needs to be central to this reassessment.

If they are to be successful in competition, European and US firms must convince sources of funds of the need to adopt longer horizons for investment and to pursue profits in the long run. They must then adopt corporate objectives and strategies to achieve them. Crucial to this is the need to recognize that pervasive technical change results in increased need for integrated decision making to replace the crude short run 'profit centre' concept.

References

1. Chaniramonklasri, N., 1980 Technical change in the industrialised world. Unpublished MSc paper. University of Sussex
2. Cohen, J.B., (ed.) 1972 *Pacific Partnership: United States–Japan Trade: Prospects and Recommendations for the Seventies*. Lexington Books, Lexington, Massachusetts
3. CTC ESCAP, 1979 Transnational Corporations in the Consumer Electronics Industry of Developing ESCAP Countries. Working paper **5**, Bangkok
4. *Economist*, 14 June 1980. The plot thickens.
5. *Electronic Times*, 17 April 1980. Game, set and match goes to the Japanese i.c. men
6. *Financial Times*, 18 August 1980. Thorn's zest for alliances
7. Japan moves fast to tackle rapid international changes 1979 *J. Electron. Ind.* **38**: August
8. Juran, J.M., 1979 Japanese and western quality: A contrast *J. Electron. Ind.* March and April
9. Magaziner, I., Hout, T., 1980 *Japanese Industrial Policy* Policy Studies Institute, London
10. *Television Digest*, 4 July 1977
11. *Which?* January 1978 TV reliability
12. *Which?* January 1980 TV reliability and servicing
13. Williams, E., 1979 Time for colour TV makers to look again at restructuring. *Engineer*, 15 March, p. 13

4.5

A Production Engineering Approach to Robot Selection

D.R. Towill

Introduction

The use of robots in manufacturing industry is growing at an enormous rate. This growth can be gauged according to Jablonowski by the statistic that in 1978, a production engineer had 24 commercial robots from which to select candidates for a specific application, whereas in 1983 there were at least 280 models available.[6] Furthermore, the predicted USA market for commercial robots is $2 billion by 1990. It is therefore little wonder that production engineers are under increasing pressure from management to apply robots wherever possible. However, a number of recent papers suggest that many opportunities to increase productivity have been lost due to poor production engineering of robot installations.

In a scholarly study on robotics, programmable automation and international competitiveness, Gold reports the results of empirical research conducted on the effect of the introduction of new technology in a wide variety of USA industry, using a 25 year time span.[4] Two general conclusions have emerged;

(1) actual economic effects have almost invariably fallen far short of expectations,
(2) these exaggerated expectations have been due to their overconcentration on a limited sector of what is a complex, interacting situation, in which the interactions are as important as the individual elements in determining cost-effectiveness.

In engineering terms, Gold is arguing for an integrated systems approach to robot selection and use. This is a pleasing conclusion to observe, because it correctly emphasizes the role of the production engineer as systems designer responsible for the performance specification of each

Source: from *Omega*, 12(3), 1984.

interacting element in the manufacturing operation such that the global goal is achieved. Clearly, process planning is at the heart of the system's design. Only then can the proper choice between Detroit-like automation, robots or flexible modular machines be made. As Riley has pointed out, many manufacturing enterprises are managed by non-engineers, who regard the robot as a universal panacea which, by waving a five-axis wand, will cause all problems to disappear.[13] These points are confirmed in a UK-oriented paper by Marshall.[8]

This reading is aimed at providing the practising production engineer with background material which will help him or her to select the best tool for the job. In particular, some of the pitfalls of estimating robot workcycle times are highlighted. In turn this should lead to enlightened performance specification writing which will tie down the robot manufacturer into providing that best tool.

Production Engineering Guidelines for Robot Selection and Commissioning

An attraction to management is that the robot is seen as a complete working tool, requiring no special integration to be solved, which will work immediately on installation and which will then require no debugging or engineering effort to keep operating at maximum throughput. As we shall see later, the estimation of the maximum throughput which can be achieved by any given commercial robot is not a simple matter. It is very much task dependent, and requires a basic understanding of robot operation to predict.

The following practical guidelines offered to production engineers reacting to management pressure to introduce robots have been developed by Schehr based on his personal experience.[14]

(1) The robot is an integral part of a system. Equipment interfaced with the robot is usually much more complex and vaster in scope than the robot itself. (Sometimes the special support equipment can cost ten times more than the robot.)

(2) When bringing the robot on-line, the peripherals consume most of the debugging hours. Hence, when introducing robots, we must already have adequate support systems and technologies working within our plant.

(3) Technology advances are not just for cost reduction. Improved quality, better customer service, new product ranges and reduced process risks are equally relevant.

(4) All potential applications must be thoroughly analysed. There is no 'cook-book' solution, because robot selection is not a one-parameter situation. Robot installation has to be an orderly and professional, planned strategy.

(5) Getting started with robots means commitment, support and involvement. Judging technology requires an understanding of the details. All evaluation plans must be professionally supported.

(6) Payback calculations should take account of *all* immediate robot

Table 1   *Jobs that robots will do*
(A) *Estimated per cent of metalworking jobs which could be done by today's robots*
Source: Ayres and Miller 1981

Job	By simple robot	By advanced robot
1. Painting	44%	66%
2. Welding	27%	49%
3. Electroplating	20%	55%
4. Machine tool operation	20%	50%
5. Packaging	16%	41%
6. Inspection	13%	35%
7. Heat treating	10%	46%
8. Assembly	10%	30%

(B) *Estimated per cent of robot applications in Japan, in 1985*
Source: Aron 1981

Job	% of total value of robots
1. Assembly	22%
2. Machine tool operation	13%
3. Arc welding	11%
4. Inspection	10%
5. Spot welding	8%
6. Molding	3%
7. Other	28%
	100%

applications. Otherwise the answer may be less than optimal. A robot initially purchased for welding, could within months, be found painting, machine loading etc. as enlightened management seeks to make best use of the new technology.

(7)   Spare robots are hard to come by. Reliability must be high, and reversion to manual mode designed in.

It is difficult to see how this menu for the production engineer can be improved upon. The present author has had similar experiences when dealing with management wishing to introduce new technology. There has been little real commitment, support and involvement beyond the initial purchase of hardware or software, which was expected to perform miracles. Due to this lack of commitment, the new technology is underutilized, abused and ultimately becomes an object of scorn.

It is possible to infer target areas for future robot applications from the surveys undertaken by Aron and Ayres and Miller as summarized in table 1.[1,2] In this case it is reasonable to define an advanced robot as a device which has been enhanced via additional sensors and decision-making algorithms to permit the robot to (1) select between several objects, (2) sense proximity to target location, or (3) adapt performance according to local conditions.

Man and Robot

It is accepted that most robots installed to date have been seen as direct replacements for human operators, especially in boring or dangerous work. Unfortunately, according to McDaniel and Gong this has led to inappropriate comparisons betwen robots and human operators to the detriment of productivity.[9] No machinery has been so personified as the robot, by human-part labelling. The use of such descriptors as 'arms' and 'wrists' gives a humanoid impression, even when the comparison really ends with the limbs. McDaniel and Gong suggest that a change in language and emphasis is required. We need reminding that the robot purpose is to perform manufacturing tasks which humans are unwilling or unable to do. These are usually repetitive, dangerous or tedious functions, or have to be carried out in extreme conditions of heat, cold and noise. Psychologically McDaniel and Gong also feel acceptance of robots by the work force would be accelerated by subtle changes in job title such as 'robot supervisor' replacing 'robot attendant'. As we shall see shortly, such a change is more than justified when the role of the human operator in automated production is examined in more detail.

Figure 1 is a useful three-dimensional sketch by Schraft which highlights the effect of the three parameters: speed, duration and weight, on the decision as to whether the production engineer should opt for selecting man, man-with-aids, robot or special equipment, to perform a particular task.[15] What must not be lost sight of, is that within any manufacturing operation, the production engineer may need to integrate *all* of these *modus operandi* to best advantage. The correct role selection for the human operator is potentially so important that it is worthy of further examination at this point.

Many tables listing the relative advantages of men and machines have been published, such as Kamali *et al.*[7] The present author finds the simpler Fitts-type table 2 adequate for highlighting the relative roles of man and machine in manufacturing operations. It should be noted that the machine capabilities are changing as technology develops, whereas for our purpose human operator characteristics are unlikely to be significantly affected, except in some instances via properly designed aids, as figure 1 clearly shows. In fact, the 'advanced' robots referred to earlier are examples of improved machine design which help bridge the present very large gap between man and machine in the areas of perception and decision making.

To bring home the advantages of using the human operator, it is helpful to consider the conceptual block diagram of man as a systems element operating in a closed-loop manner. Figure 2 is an extension of the models reviewed in Towill.[16] Whilst certain loops and hence performance are clearly influenced by skill and experience, the major feedback loop can be

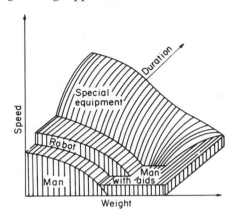

Figure 1   Suggested field of application of industrial robots

Table 2   Fitts-type list of relative advantages of men and machines

Property	Machine	Man
Speed of response	Can be very fast	= 1 second lag
Power	Can be large and consistent	2 HP ↔ 10 secs 0.2 HP ↔ Over working day
Consistency	Good at routine	Not reliable, subject to learning and fatigue
Memory	Good at literal reproduction	Better for principles and strategies
Reasoning	Good deductive Tedious to reprogramme	Good inductive Easy to reprogramme
Computation	Fast, accurate: poor at error correction	Slow, subject to error Good at error correction
Input selection	Poor pattern detection, inflexible	Good pattern detection Wide range of levels and types, e.g. eye deals with location, movement and colour affected by environment
Overload	Sudden breakdown	Graceful degradation
Intelligence	None, goal switching not possible	Can adapt to unexpected
Manipulative abilities	Specific	Great versatility and mobility

Men and machines are complementary; not easy to put numbers to advantages of men: as task becomes more uniquely defined, so does it become more suitable for machines

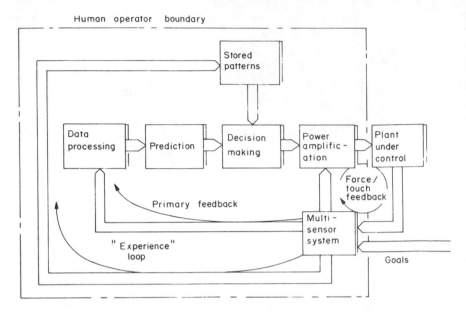

Figure 2   Block diagram of the human operator as a man–machine element

made operational in crude fashion under a wide variety of circumstances because of man's superb inbuilt sensor system. Through this system he can see, hear, touch, taste, smell and sense acceleration and angular velocity. It is therefore obvious that irrespective of the data processing, decision-making and power amplification capabilities of the human operator, we have a lot of sensors provided free. Frequently, this characteristic of the human operator is extremely expensive to replace in a fully automatic system.

Figure 2 is also helpful to the production engineer in another important way. If a complex, integrated man–machine system is conceived as the best solution to the problem, the block diagram encourages the system's designer to focus on providing the best aids. An activity chart with frequency of activity superimposed will soon determine whether the aids should take the form of 'intelligent' sensors, prediction algorithms, a stored menu of response patterns or just additional power amplification. Where long periods of inactivity are likely, but unavoidable, then effective alarms must also be provided. Concentration and job satisfaction may also be improved by involving the operator in the tactical decision-making procedure, and using the computer to monitor men, and not the other way round.

Finally, the block diagram of figure 2 clearly shows the human operator influencing performance via use of deductive reasoning, i.e. the 'human use of human beings' advocated so long ago by Weiner. The production engineer must utilize such skills fully in the design of manufacturing systems. It is bad practice to automate those aspects of the process which are economically and technologically viable on the false assumption that the human operator can pick up the rag-bag of tasks left and make the system operate efficiently; designer errors can be a major source of subsequent operating problems. As Bibby *et al.* have commented,[3] even highly automated systems need human beings for supervision, adjustment, maintenance, expansion and improvement of automated systems, for which both technical and human factors are important.

Robot Time Standards

In view of the 'humanoid' attitude towards robots observed by McDaniel and Gong[9] it is only to be expected that the concept of predetermined time standards such as MTM have been carried over to the prediction of robot performance. Here the main contributors have been from Purdue University, including Paul and Nof[12], Nof[10] and Nof and Lechtman.[11] Most of the research has been directed at the 'Stanford Arm' robot, using PAL as the robot programming language. Combined with the use of databases such as those formulated by Ioannou and Rathmill[5], RTM systems should provide a useful adjunct in the computer-aided design of manufacturing systems, provided that the proper division of tasks between man and machine have been taken into account. The general concept of the robot time and motion study (RTM) analyser[10] is shown in figure 3. RTM builds on MTM experience, although there is not a one-to-one equivalence, since some elements are beyond the capability of the robot. The production

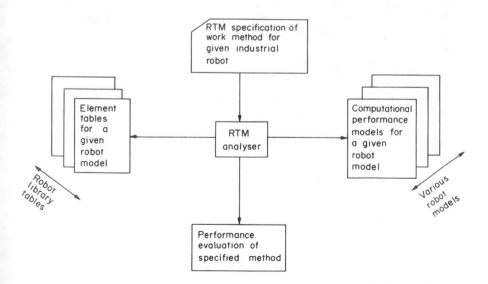

Figure 3   Concept of computer-based RTM analyser

engineer wishing to develop his own RTM system will need to refer to the three major references, and then almost certainly embark on a major exercise in deriving synthetic times appropriate to the robot under consideration. As Schehr has said, there is no cook-book solution for robot selection and implementation.[14]

Setting up an RTM covering a range of robots is no small task. However, there is an added difficulty. Contrary to popular belief, robots are not highly predictable devices. Task performance varies tremendously depending on the instantaneous position of the robot within the operating envelope, age of the robot, time of day etc. There is also considerable variation in performance between similar robots purchased from the same manufacturer. It is therefore important for the production engineer designing the manufacturing system to be aware of typical performance variations.

Robot Specifications

Schehr has emphasized the importance of the line balancing problem in robot selection.[14] The production engineer undertaking the preliminary design of a manufacturing system may therefore wish to obtain a 'ball park' estimate of robot throughput time so as to ensure that the product line is in reasonable balance before proceeding to undertake detailed studies. Without access to the RTM analyser, or to an identical robot on his own shop floor, how does he proceed to obtain such estimates? With difficulty, is the answer, until robot manufacturers provide dynamic performance data as an integral part of the machine performance specification made available to prospective purchasers.

For example, Jablonowski[6], following visits to various exhibitions in the USA, found that the following robot details were relatively easy to obtain for any given model;

 (1) Type (cylindrical etc.)
 (2) Load capacity (kg)
 (3) Number of axes
 (4) Drive (electric, pneumatic, hydraulic)
 (5) Horizontal reach (m)
 (6) Vertical reach (m)
 (7) Sweep around body (°)
 (8) Date introduced
 (9) Operator programming type (teach pendant etc.)
(10) Memory capacity (either positions or programmes)
(11) Repeatability (mm)
(12) Price (excluding debugging costs and end effector)
(13) List of possible applications
(14) Special features (DNC link, modularization, etc.)

Such information is very useful, but is noteworthy for the absence of any data which will allow the production engineer to estimate the rate at which the robot can deliver the goods. However, a little more information made available by the manufacturer can be made to go a disproportionately long way. For rapid growth, a great deal depends on robot manufacturers contributing vital facts for inclusion in the appropriate databases.

References

1. Aron, P., 1981 Robots revisited. Report No. 25, Daiva Securities, New York (summarized in *Robotics Today*, pp. 26–28)
2. Ayres, R., Miller, S., 1981 Preparing for the growth of industrial robots. Technical Report, Department of Engineering and Public Policy, Carnegie-Mellon University
3. Bibby, K.S., Marguilies, F., Rijnsdorp, J.E., Withers, R.M., 1975 Man's role in control systems. *Proc. 6th IFAC Congress*, Boston
4. Gold, B., 1982 Robotics, programmable automation and international competitiveness. *IEEE Trans.* **EM-29**(4): 125–45
5. Ioannou, A., Rathmill, K., 1982 Data base provides tool for robot selection. *Ind. Robot* **9**(3): 153–71
6. Jablonowski, J., 1982 Robots: Looking over the specifications. *Am. Mach.* **126**: 163–78
7. Kamali, J., Moodie, C.L., Salvendy, G., 1982 A framework for integrated assembly systems: humans, automation and robots. *Int. Jl Prodn Res* **20**(4): 431–48
8. Marshall, A., 1983 A poor man's guide to automating production. *Engrs. Digest* **44**(5): 10–16
9. McDaniel, E., Gong, G., 1982 The language of robotics: Use and abuse of personification. *IEEE Trans.* **PC-25**(4): 178–81
10. Nof, S.Y., 1982 Decision aids for planning industrial robot operations. *Proc. Annual AIIE Conf.*, pp. 46–55
11. Nof, S.Y., Lechtman, H., 1982 Now it's time for rate-fixing for robots. *Ind. Robot* **9**(2): 106–10
12. Paul, R.P., Nof, S.Y., 1979 Work methods measurement – A comparison between robot and human task performance. *Int. Jl Prodn Res.* **17**(3): 277–303
13. Riley, F., 1982 Why robots: What about the other option to manual assembling? *Prodn Engr.* **61**(6): 34–5
14. Schehr, L.H., 1982 Is robotics for you, now? *Annual AIIE Conf.*, pp 3–7
15. Schraft, R.D., 1981 Review of the use of industrial robotics by German industry. *Proc. 4th British Robot Ass. ANN, Conf.*, Brighton 12–21 May, pp. 29–36
16. Towill, D.R., 1980 Man–machine interaction in aerospace control systems. *Radio Electron. Engr.* **50**(9): 447–58

4.6

A Goal-oriented Pharmaceutical Research and Development Organization: An Eleven-year Experience

Jacob C. Stucki

The intended result of most research and development effort put forth by pharmaceutical companies is the approval for marketing of new drugs by government drug regulatory agencies. This R & D effort and the subsequent production and marketing efforts are always multidisciplinary and may even be interdisciplinary in character. R & D success requires interactive contributions of people with scientific backgrounds and experience in such fields as organic chemistry, pharmacology, toxicology, pathology, clinical chemistry, analytical chemistry, physical chemistry, biochemistry, pharmacy, medicine, clinical pharmacology, biostatistics, chemical engineering, computer engineering, microbiology, immunology, bioengineering and molecular biology. Full commercialization of a new product requires people with training and experience in production, marketing, sales, finance, law and business.

The drug discovery and development process is also pluralistic, requiring for success the involvement of organizations and people outside the corporation which undertakes the drug discovery and development effort. Drug development requires the active participation of physicians in clinics and hospitals outside the immediate control of the pharmaceutical company. Clinical evaluation of new drugs in patients of the type for whom such drugs are intended is, of course, the ultimate test of the benefit and the risk values which can be ascribed to these new drugs. The large number of patients required to demonstrate safety and efficacy and the diversity of approach to the application of drug therapy in illness can be found only through clinical studies with large numbers of private and institutional physicians. Typically, clinical studies supporting the successful drug regulatory agency approval of one new drug will require the services of as many as a hundred physicians.

Source: Paper presented to UMIST R & D Conference, 1981.

The pluralism of drug discovery is also evidenced in the acquisition of the basic knowledge upon which drug discovery is built. Such knowledge derives from the whole scientific community, not only from the scientists employed by the sponsoring pharmaceutical company. Not infrequently, drug substances that are developed by pharmaceutical companies are actually discovered and patented by scientists working in academic institutions. In these instances the drug is developed and sold under a license agreement. Such agreements also involve government agencies when the discovery effort was government funded or when the discovery was made in a government laboratory.

Time and cost are other dimensions which characterize the drug discovery and development process. Sarett estimated identifiable development costs between the time a new drug is selected for study and the time it is approved for marketing to average 1.2 million dollars in 1962 rising to 11.5 million dollars in 1972.[3] Schwartzman estimated that in 1973 the cost of discovering and developing each new drug was 24.4 million dollars if the cost of failures is prorated to the successful candidates.[4]

Typical development times reported by Clymer in 1969 range from 5 to 7 years following a 3- to 4-year discovery research effort.[1] Increasing regulatory pressures have greatly extended development times. Our own experience, which we believe is typical, includes a development time in excess of 17 years for one radically new therapeutic agent (minoxidil) which has recently been approved by several drug regulatory agencies.

Still another description of the process is the number of in-company individuals involved in each drug development project and the number of separate project activities associated with the development process between the time a decision is made to proceed to Phase III and the time the new drug is introduced into the marketplace. It is during this time period that PERT diagramming and scheduling, or similar techniques are employed to assist the project manager in controlling the project. Drug development projects we have undertaken each involve the active participation of well over 100 in-company people and, as previously mentioned, an additional number of outside physicians. PERT activities from the Phase III decision to marketing range from 75–150 or more, depending upon the level of detail preferred by the responsible project manager.

For some time it has been our opinion that drug discovery and development is of a complexity and nature such that it cannot be accomplished by functional departments that simply pass discovery and development information and problems along from one department to the next, but rather that the success of pharmaceutical research and development depends upon structured planning and people interactions and upon the utilization of project leaders, matrix organizations and matrix management.

Prior to 1968, Upjohn Pharmaceutical R & D was organized according to traditional disciplines into functional units. The drug discovery and development process was conducted by functional specialists assigned to project teams, mostly on a part-time basis. The system operated as a coordination matrix with project management shifting from one functional specialist to another as the stages of development of a candidate drug changed.

Steering committees at operating levels and at higher management levels provided oversight and an opportunity for reporting project status. The higher level steering committees made project decisions which moved candidates from one stage of development to the next. The very highest level of steering committee was chaired by the vice president for Research and Development.

Product goals had been established for the company through the cooperative effort of representatives of all functional departments in R & D and the marketing organization. These goals were translated into functional activities by the respective functional specialists and their managers. Because the functional units were relatively independent in the translation of goals to activities, the total organization experienced the problem of developing and maintaining concurrence on whether or not a given chemical compound was indeed a legitimate development candidate. Lack of consensus often led to delays in the creation and execution of development plans. Resolution of this type of conflict was often achievable only at the level of the R & D vice president. This was the lowest level of common leadership of the critical functional units which made, identified and proved therapeutic utility of the candidate drugs, viz., the chemistry, biology and medical units.

Project managers in the coordination matrix which existed before 1968 frequently were unable to negotiate solutions to truly serious project problems or to negotiate consensus for full commitment to potential development candidates. As a consequence, conflict resolution required an inordinate investment of top executive energy and time. Decisions were often inappropriately delayed or were made on what can be regarded to be an inappropriate basis, for example, on the basis of disciplinary or functional prerogative rather than scientific evidence.

In 1968 the functional organizational units of Organic Chemistry, Pharmacology, Biochemistry, Microbiology, Endocrinology, Immunology, Infectious Diseases, Virology, Clinical Pharmacology and Medical Development were abolished and a large part of their memberships were reorganized around programmatic commitments into seven goal-oriented Product Research Units each responsible for drug discovery and development in a separate disease or product area. Some of the members of the former functional units were separated from R & D and moved to Marketing to provide technical support for marketed products. The Product Research Units became responsible for the following product or disease areas:

(1) Central nervous system diseases
(2) Infectious diseases
(3) Cardiovascular diseases
(4) Hypersensitivity diseases
(5) Diabetes and atherosclerosis
(6) Cancer
(7) Fertility control products

Table 1   Two models of matrix organizations

	Leadership matrix	Coordination matrix

The model assumes people	–Tend to pursue their own goals:
	  – Professional
	  – Specialist
	– Need galvanizing into working to project goals
		– Are rational, objective
		– Act predictably on adequate information

Role of project leader is to:	– Motivate team to work to project goals	– Keep everyone informed about:
		  – Project status
		  – When their contributions will be needed

Consequences for project leader:	– Needs status, authority – Gets action by personal authority, influence, negotiating skills	– Is coordinator, with most complete information on status and future needs of project – Gets action by signalling deviations from plan
Consequences for functional manager	– Scope of authority, responsibility limited by project needs	– Must consider project needs in conjunction with needs of functional activities
Consequences for project team	– Must be cohesive group – Functional activities interfere with project work	– Meetings of nominated individuals – Project work interferes with functional activities

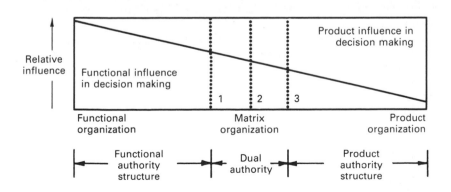

Figure 1   The range of organizational alternatives

Each Product Research Unit consisted of all the chemists, biologists and physicians required to make, identify and prove therapeutic utility of new drugs in their product or disease areas. Supportive Research Units such as Pathology and Toxicology, Pharmacy, Biostatistics, Analytical Chemistry etc., which are required to support all drug finding and development projects, continued to be organized functionally as they had been in the pre-1968 organization. These are units where resources are always in short supply and where project competition for resources is always manifested.

Individuals chosen to manage each of the new Product Research Units and to assume project management over the development projects in their disease areas were generally former functional managers who had had project management experience in the pre-1968 organization. The essential difference in the roles they played pre-1968 and post-1968 was that post-1968 the Product Research Unit managers assumed line management responsibility over a much larger part of the team required to complete the project task. Their positions post-1968 could be defined as those of relatively strong project managers in a leadership matrix. As previously noted, pre-1968 project managers were largely coordinators. Post-1968 project managers were in a somewhat better position to compete with each other for the scarce resources of the Supportive Research Units than were the pre-1968 coordinators. Of greatest importance was that post-1968 managers were better able to resolve conflicts surrounding candidates under development and conflicts over the identification of development candidates. They were also able to assure the rapid preparation and execution of a development plan.

The organizational actions which were taken and which are described above can be regarded as reflecting a significant shift in the decision-making process from functional dominance to product dominance and a change from a coordination matrix to a leadership matrix.

Not all pharmaceutical research and development needs can be served by units organized by disease or product area. Technical field activity dedicated to the identification and exploitation of product opportunities not covered by Product Research Units is lacking. The technology base in chemistry, biology and medicine must also be supported for long-term organizational health.

To serve these needs, Upjohn continued to organize a portion of its chemical, biological and medical effort along traditional disciplinary or functional lines. In the 1968 reorganization it created Experimental Chemistry, Experimental Biology and Experimental Medicine Units. The Product Research Units described above are largely independent of the Experimental Chemistry, Biology and Medical Units and vice versa. Exceptions occur when technology generated by either Product Research Units or by the Experimental Science Units becomes specifically applicable to the others. Experimental Science Units are also responsible for maintaining a discipline-oriented community for the continuing education and professional renewal of all chemists, biologists and physicians. Each Product Research Unit, each of the Experimental Science Units and each of the Supportive Research Units, e.g. Pathology and Toxicology, is

Table 2   R & D organizational types at Upjohn

Pre   – 1968
      – Functional units
         Discipline-oriented organization
         Functional organization
         Coordination matrix

Post – 1968
      – Product research units
         Goal-oriented organization
         Product organization
         Leadership matrix

      – Experimental sciences units
         Discipline-oriented organization
         Functional organization
         Coordination matrix

responsible for conducting basic research in its own area of interest. The Experimental Sciences Units can be regarded as constituting a coordination matrix not unlike that which existed in the entire R & D division prior to 1968. Synonyms for the two organization types which existed previously or now exist at Upjohn are shown in table 2.

Both a discipline-oriented organization or coordination matrix and a goal-oriented organization or leadership matrix have been in existence simultaneously for 11 years in the same company. Some internal comparisons are possible, but it must be stressed that by no means can our experience be considered to constitute a valid experiment in comparative organizational effectiveness. Rather, attention should be directed toward the interactions and influences each organization type has on the other, the synergism produced by their coexistence and their common top management and by the transfers of personnel, information and projects that occurred between the two organizations.

During our 11 years of post-1968 experience, all but one Product Research Unit was successful in bringing one or more new products to commercialization. Three of the units commercialized one product each, one commercialized two products and two commercialized four products each. The Experimental Sciences Units brought three products to the market. The total time required to develop new discoveries in the pharmaceutical industry is long and growing so only two innovations were brought through all stages of the process during this 11-year period. The remaining 14 products brought to commercialization during this period had already been identified as leads prior to 1968. All units currently have a full complement of candidates under development.

Prior discussion has focused on the role of the project manager in bringing projects to a successful conclusion. Our experience to date with both coordination and leadership matrices suggests that all pharmaceutical R & D effort directed toward the commercialization of a new product also requires a strong advocate who ensures continuing enthusiasm for the candidate drug. The advocate may or may not be the project manager. In our leadership matrix the advocate generally is the Product Research Unit Manager. In our Experimental Sciences organization, or coordination matrix, the advocate may be anywhere and is not infrequently some person high in the organization. Sometimes the advocate has been that person at the lowest level of management having line authority over all three critical elements in the development process, viz., chemistry, biology and medicine. It must be remembered that advocacy is not a work assignment but rather a state of mind or an adopted role. It appears that the organizational location of the advocacy role may be influenced by organization structure or matrix type.

Some of the problems inherent in functional organizations, viz., the development and maintenance of concurrence on whether or not a given compound is a development candidate initially, continued to exist with our Experimental Science Units. To deal with that problem and to enhance the adoption of advocacy roles at the most appropriate spots in the coordination matrix, a 'Troika' concept was adopted during the last half of this 11-year period. Experimental Chemistry, Biology and Medical Managers coadministered project team assignments and dealt with their supervisors as a triumvirate in matters of candidate selection and development. This approach was successful but it is absolutely dependent upon the existence of high mutual trust and well-developed conflict resolution skills.

The performances of individual scientists in the two types of post-1968 organization did not differ in any discernible fashion. Publication patterns were similar for individuals in both goal-oriented and functional units. At Upjohn, scientists have an opportunity to progress along a professional career path (analogous to academic rank). Promotions on the path are a result of individual performance as judged by a top level oversight committee. Progress of individuals along the career path and percentages of total population at any given career path level are similar for the goal-oriented and the functional units. In both types of organization, at Upjohn and elsewhere, confusion does exist over the sources of reward. Some scientists perceive rewards to be highest for goal achievement. Others believe rewards are highest for excellence in discipline. The oversight committee attempts to reward both kinds of contribution. Beliefs of this type appear to be more characteristic of individuals than of organizations.

A common criticism of goal-oriented organizations is that they lack ability to respond to new goals and opportunities that emerge over time. The simultaneous existence of a functional organization at Upjohn blunts this criticism. Our Experimental Sciences Units have responded well to new programme opportunities. Product Research Units have responded well to new opportunities that presented themselves within their broad programme areas. Another measure of flexibility, however, is the willingness of individuals to depart significantly from their current research activity or assignment. This kind of individual flexibility or inflexibility appears to be identical in both our goal-oriented and our functional organizations.

Our post-1968 goal-oriented departments appear to be less hierarchical in approach to technical decision making than our post-1968 discipline-oriented departments, but department type alone does not satisfactorily explain differences in approach. Some individuals are more comfortable with a career in a hierarchical, discipline-oriented department, perhaps because such departments are similar to academic departments experienced in graduate school. Such individuals will gravitate toward a discipline-oriented department when the opportunity to move arises. However, the original hope that individuals would frequently move between the Experimental Science Units and the Product Research Units as technical problems changed has not been completely realized. This is partly due to the long time required to complete any given project, primarily because of regulatory demands.

Our post-1968 leadership matrix organization was created to correct communication and decision-making defects that existed in our pre-1968 functional coordination matrix organization. Our experience with goal-oriented research management or leadership matrix management can be regarded as highly successful. Problems of consensus and commitment on candidate products largely disappeared, as did the problems of delay in preparation and execution of development plans in so far as they involved the members of the goal-oriented units. It must be concluded, however, that not all problems are amenable to solution by organizational shifts to leadership matrices.

It must also be concluded that the complexities of pharmaceutical R & D and the external demands placed upon it, the communication network it requires and the competition for resources inherent in the process preclude any organization other than a matrix. At issue is only the nature of the matrix and some of the process details. Implications for policy makers might be that under the general rubric of matrix, strategies are available that have profound and important effects on project success. A strategy that was successful for Upjohn involved deliberately shifting part of the organization from a coordination to a leadership matrix, the conferring of additional power on project leaders by making them line managers over key project resources, the maintenance of a competing and synergistic coordination matrix, a strong publication and basic research policy and a 'Troika' approach to candidate selection and development in the functional organization.

References

1. Clymer, H.A., 1968 The changing costs and risks of pharmaceutical innovations. In: Cooper, J.D., (ed.) *The Economics of Drug Innovation*. The American University Press, Washington

2. Galbraith, J.R., 1971 Matrix organization designs: How to combine functional and project forms. *Business Horizons* 14 (1):29–40
3. Sarett, L.H., 1974 FDA regulations and their influence on future R & D. *Research Management* 17 (2): 18–20
4. Schwartzman, D., 1976 *Innovation in the Pharmaceutical Industry*. The Johns Hopkins University Press, Baltimore and London

4.7

Coping with Advanced Manufacturing Technology

Melvin Blumberg and Donald Gerwin

Innovation in manufacturing processes may be regarded as one important strategy by which control [over the production process] is exerted.[5] For example, automation in mass and process production has contributed to the high degree of predictability associated with these methods of manufacturing. Abernathy showed how process improvements in the American automobile industry led to uncertainty reduction and high efficiency, although at the expense of new product development.[1] Batch production, the least automated of the three main types of manufacturing, has always been looked upon as the least controllable and predictable.

Recently, however, a major technological revolution has occurred in batch production with the advent of computer-aided manufacturing. One aim of this revolution is to introduce a degree of control comparable to that in mass and process industries. Consistent with this aim, the trend in computerization has been toward more complex, large scale, centralized, integrated and capital intensive production equipment. Computer-aided manufacturing began with the attachment of mini-computers to numerical control machine tools. It has progressed to the computer-integrated machining system which employs a central computer to control the operation of a battery of machining centres, information handling and possibly material handling as well. The vision for the future is the completely automated factory in which both machining and assembly are mechanized.

In this reading we seek to question the trend toward complexity and its associated characteristics in batch production technology by studying some organizational and human implications. Our thesis is that such technology, rather than contributing to the control of production processes, shows signs of being beyond the capabilities of most companies to control. Too much attention has been paid to technical development and not enough to the

Source: from International Institute for Management/Labour Market Policy, Berlin, 1981.

adjustments needed in organizations to accommodate the technology. This has produced a lack of fit between the demands made by the technology and the skills, attitudes, needs and values embodied in the social and technical structure of companies. The result is that the new technology raises both cognitive and motivational problems with which managers, staff specialists and workers have great difficulty in coping.

[As an example of CAM,] an individual palletized casting of the order of a one metre cube and weighing a metric ton is manually loaded onto a cart with the aid of cranes and fork lifts. Under control of an FMS computer the cart is transported to a work station where the pallet is transferred to the machine by a shuttle carriage. A DNC computer initiates direct numerical control transmission which causes the machine to process the part. After processing is completed the pallet is transferred back to a cart and routed to its next work station or returned to a loading station for removal of the finished part.

This process is carried on simultaneously for several dozen parts of various types which may be in the system at one time, routed in random order among the work stations. The work-group organization consists of one foreman, three operators who primarily monitor machining operations, three loaders, one fork lift driver for providing inputs and storing outputs, a repairman and a tool setter. The equipment is typically operated on a two-shift, five-day schedule, with an occasional skeleton crew on a third shift to provide critically needed castings.

Research Setting and Sample

Due to the current international interest in computer-aided manufacturing and in order to achieve a broad understanding of coping problems, firms were selected in three different countries: the United States, Great Britain and West Germany. The sample includes:

(1) a diversified American manufacturer with sales of $2 billion in 1980, which has a division producing tractors for which the major housings are machined on a flexible manufacturing system;

(2) a British producer of medium and large size electrical motors and generators having sales of $86,000,000 in the year ending March, 1980, which was in the process of installing a direct numerical control system to machine prime components;

(3) a British manufacturer of plain bearings for industrial engines, which has a rudimentary direct numerical control system for machining thin wall bearings; that is, a central computer controls six machines which operate independently of each other;

(4) a German aircraft manufacturer with sales of $420,000,000 in 1980, which was in the process of installing a flexible manufacturing system designed to machine over 200 different parts;

(5) a German producer of transmission systems for industrial and agricultural vehicles with sales of $150,000,000 in 1979, which is developing

its own flexible manufacturing system to machine gears and other rotary parts.

Data Collection

In order to obtain detailed insights in coping with problems it was decided to conduct in-depth case studies of the firms. In 1979, semi-structured interviews were held with managers and staff specialists representing various levels in manufacturing management, accounting, quality control, manufacturing engineering, maintenance, production preparation and data processing. Respondents were asked questions in three general areas: descriptive contextual information, the nature of the adoption process and implementation problems. Emphasis was placed on how the CIM affected the respondents' roles.

In addition to its impact on managers and staff specialists, complex manufacturing technology can also have a profound effect on first-line supervisors and direct workers. In Spring 1980 an intensive survey of foremen's and workers' reactions to an FMS was made in the American firm using a 177-item questionnaire. It was designed to tap several areas believed to be of major importance to employees on the basis of our knowledge of the literature and exploratory interviews in the company. These include job characteristics, individual growth, need for strength, job satisfaction, equitable rules, equitable pay, healthful working conditions, meaningful work, work-related stress, attitudes toward participation in work-related decisions, job switching, aspirations for the future and health and safety.

Results: Managers and Staff Specialists

The problems managers and staff specialists have in controlling CIMs are exemplified by difficulties in properly evaluating whether or not to adopt them, and in solving implementation problems arising out of their technical characteristics. The complexity and uniqueness of the new technology relative to the availability of skills and procedures for dealing with it are at the root of these difficulties. A recent study, which inquires into the relatively slow diffusion of advanced manufacturing equipment in batch production, illustrates the situation. The General Accounting Office of the United States surveyed almost 200 metal-working companies in 13 states and found that the use of automated methods was extremely limited.[2] For example, about 51 per cent of the firms did not use a computer, and only 17 per cent had one or more NC machine tools. When asked to cite the major barrier to adoption of advanced manufacturing methods, almost one in five companies mentioned the lack of widespread understanding of such technology. Difficulties stemming from the absence of common standards among the equipment and software of different manufacturers are a case in point. Small and medium firms in particular do not have the technical capability to interface components from diverse sources.

Evaluation Problems

Evidence is beginning to appear which indicates that companies have difficulty in rationally deciding whether or not to purchase advanced manufacturing equipment. They do not have the human skills and/or financial tools to perform meaningful analyses. Nabseth and Ray found that US and European companies had difficulties in assessing the anticipated profitability of NC machine tools and therefore decisions tended to be subjective.[6] A recent study of small- and medium-sized firms conducted by the Illinois Institute of Technology concluded that the majority of nonusers reject NC equipment because they are unable to properly evaluate and hence justify the investment.[7]

Our data indicate a similar situation for medium- and large-sized companies with respect to CIMs. The task force which investigates a CIM's feasibility is likely to discover that analytical financial tools such as discounted cash flow are of limited utility. Not enough information may be available to make sound estimates of future net returns. The situation will be even more problematical when, as was the case for two of the companies studied, a new product is being considered in conjunction with the equipment. One critical issue is how to quantify the benefits flowing from a CIM's flexibility. It is not possible to make financial comparisons between a transfer line which machines a few specified parts at low cost, and a CIM which machines the same parts at higher cost but can also produce other unspecified parts in the future. If decisions are to be made on objective grounds flexibility will not receive a proper weighting, and results will be biased against adoption. Consequently, companies which adopt CIMs do so more on the basis of intuition than forecasted net returns.

In at least three firms studied a special task force or individual had the main responsibility for decisions on a large capital improvement programmes in which a CIM was the main ingredient. In at least two firms the task forces prepared comprehensive financial analyses but only for their programme as a whole. The exact values of the numbers used could not be considered very reliable because of the uncertain future. However, the analyses served to meet corporate guidelines for rationality in capital budgeting and provided some indication of whether the minimum limits on acceptability were likely to be exceeded.

Within each firm sophisticated financial analyses were not made for individual equipment decisions including the CIM decision. In the American firm, for example, the decision to adopt a CIM was made by the head of manufacturing engineering and the head of the task force, both of whom favoured the new technology. Formal analysis was confined to an evaluation sheet which rated four machining systems on each of 26 dimensions including capital cost, labour requirements, flexibility to add parts, time to deliver and install and estimated amount of running time per week. The FMS came out far ahead of its closest competitor. In part the evaluation served to help crystallize the decision makers' thinking. However, the fact that their minds were already partially made up probably had some effect on the evaluation's outcome.

Implementation Problems

The most frequently mentioned implementation problems occurred in the quality control, accounting and maintenance areas. These are precisely the functions that manufacturing management counts on to control operations; that is, to set standards, compare standards to results and take corrective action if necessary. Consequently, the very basis for judging and improving the efficiency of operations is rendered problematical.

The new manufacturing systems also require raw castings with more precise specifications than are needed for conventional machines. One reason is that there are fewer opportunities for humans to make adjustments which can compensate for variations in the dimensions of raw materials. Originally, the two largest castings machined by the American firm's FMS were obtained from external suppliers. It was difficult to exert the necessary controls over raw casting specifications. The solution was to have the firm's foundry supply virtually all castings; a decision which required a considerable investment in upgrading facilities, and the transfer of the head of quality control to supervise operations.

Implementation of CIMs has profound impacts on accounting procedures. The principal issue involves determination of the cost standards against which actual performance is to be judged. First, companies discover that the standard cost of machining a part can no longer be expressed in terms of direct labour hours. It is no longer possible to unambiguously determine the amount of direct labour hours used in machining a particular part. Typically, a new cost system based on machining hours and the amount of time metal is being cut, must be employed. In at least the American firm this caused some unanticipated control problems for manufacturing managers. They had spent their entire careers dealing with standard costs based on direct labour hours and therefore found it difficult to understand the meaning and significance of the new concepts.

Second, three of the companies studied had problems in computing cost standards due to a lack of available data. None of the typical sources of data for calculating standard cost parameters, such as one's own shop or similar facilities elsewhere, were available due to the uniqueness of a CIM. It is necessary, at least initially, to base standards on intuitive estimates rather than existing data. The American company's experience has been that even after eight years of CIM operation a completely reliable performance history has still not been compiled. Total planned costs usually are a reliable benchmark, but the planned values of major cost components such as maintenance and rework are not very accurate. This makes it difficult to determine whether actual costs are within acceptable limits of control.

Due to these problems, manufacturing management is being evaluated by standards which they find difficult to understand and which are not completely reliable. The result is likely to be conflict between manufacturing and accounting concerning the assumptions on which the cost accounting system is based. Manufacturing will be at a distinct disadvantage, however, due to the effort which must be invested in understanding the intricacies of the accounting procedures.

CIMs pose complex maintenance problems, especially in the electronics and electrical areas. In one company studied, it was estimated that two hours of electronic and electrical maintenance are required per eight-hour shift. Firms forced to deal with these problems have two options, neither of which is very adequate. They can rely mainly upon vendors and contractors for repairs. Then the timing and quality of maintenance work is not under control. The head of maintenance in one of the English firms told of it taking three months before the vendor of one particular machine sent a serviceman. Further, the technology is so new and complex that even vendors are not very knowledgeable on how to deal with breakdowns. Three of the firms studied made this point. Part of the difficulty is that it is too time-consuming and expensive for the vendor to set up most CIMs in their own shops before delivery in order to test them out.

The other alternative is for a company to develop its own maintenance capabilities. However, it takes a long time to train repairmen, and then only a very few people are able to master the equipment's intricacies. These individuals are hard to retain because of the ample job opportunities. Once a CIM is installed it is difficult for maintenance people to learn about it because of the pressures from manufacturing to produce right away. Owing to the large capital investment, CIMs are run on second and even third shifts wherever possible, and planned maintenance may be reserved for weekends. It is hard to provide adequate maintenance coverage at these times because skilled technicians do not want to work then. A breakdown occurring at the beginning of second shift may not be handled until the next day.

The significance of CIM maintenance problems should not be underestimated. Breakdowns of computers, material handling systems and machine tools are a major contributor to lower machine utilization than anticipated, and in some cases lower utilization than less complex equipment. One firm, for example, gets about 15 to 20 per cent less uptime from its CIM than from its stand-alone equipment. And utilization is a key variable to manufacturing management because of its impact on meeting production schedules and customer deadlines.

Conclusions

Our findings raise a number of important issues concerning control in production systems containing CIMs. The main question to be asked is just who is in control. The answer appears to be that the new technology is partially out of control from the viewpoint of manufacturing management, staff specialists and workers. Complex manufacturing equipment introduces significant uncertainties into the production process which are not easy to cope with.

In general manufacturing managers do not have the expertise to second guess the judgements of technical experts when evaluating whether or not to purchase equipment. Consequently, they tend to employ mainly financial criteria in judging requests. They are likely to find however that financial analysis techniques are of limited value in evaluating CIMs.

During implementation of CIMs manufacturing managers may also find that their primary tools for controlling operations from accounting, quality control and maintenance are not very reliable. Neither may they be completely understandable. At the same time traditional, informal procedures for exerting control will no longer be valid. CIM foremen in particular are caught between having more performance pressures than their conventional counterparts, and more dependence upon staff and service personnel. Consequently, they are relatively high on work-related stress but not necessarily low on satisfaction.

With respect to staff and service personnel, engineers who prepare requests for the purchase of CIMs do so more on faith than on rational evaluations of costs and benefits. Control systems developed by accounting and quality control to govern the operation of CIMs cannot completely meet the demands of the new technology. They must also reflect what employees find comfortable to deal with, and the limits of control system technology. A CIM's complexity also affects its operating reliability causing intricate maintenance problems. And it is difficult to train and retain enough skilled maintenance workers to service the new technology around the clock.

Direct workers, at least in the company surveyed, suffered from loss of control. Respondents in all four nonsupervisory job classifications reported a relatively low opportunity to exercise discretion (autonomy). All nonsupervisory job classifications also believed they had too little participation in work-related decisions such as what is to be done and how it is to be accomplished. Workers in three of the four classifications perceived their jobs as being relatively low on permitting completion of an identifiable piece of work (task identity) and in feelings of accountability for results (responsibility). Workers, except for repairmen, also were dissatisfied with important aspects of their jobs and found their jobs to be stressful. Loaders and operators who perform the most routine work were the most affected.

Control problems arise out of a CIMs complexity, especially the tight interconnections among subsystems. Automating one aspect such as material handling leads to difficulties with another aspect such as quality control. To solve these problems it is often necessary to increase the amount of human intervention thus partially reversing the trend toward automation. Thus, difficulties in applying financial analysis techniques lead to human intuition playing a greater role, and problems in quality control lead to more human inspection. The same holds true for systems and procedures used to support CIMs. Production scheduling researchers at the *Institut für Produktionstechnik und Automatisierung* (IPA) in Stuttgart are developing production control models for flexible manufacturing systems. They originally had wanted to automate the entire procedure but now realise it must be interactive. It proved impossible for a model to effectively react to unanticipated changes such as machine breakdowns and work piece faults.

Since humans enter the production process at points where critical control problems exist they acquire power to significantly influence effectiveness. In the American firm for example, where direct workers make inspections while the machines are temporarily stopped, they can influence the product's quality through the accuracy of their checks, and the product's cost through the speed of their checks. This conclusion is supported by Doring and Salling who noted that the human element continues to be a major factor in the utilization of NC machine tools.[3] While the machine basically controls the manufacturing cycle, if the operator fails to service it promptly utilization suffers.

Where human intervention to handle control problems is not practical, massive changes in a CIMs immediate environment may be necessary in order to reduce undesirable variations. A good example is the American firm which had to improve the functioning of its foundry in order to control raw material specifications. In this regard Nabseth and Ray who studied NC diffusion among 142 firms in the United States and Europe, found that radical changes in the production process usually are connected with introducing NC.[6]

Is there an underlying explanation for the findings of our study? We believe it can be discerned using some basic principles of sociotechnical systems analysis,[9] and Schumacher's analysis of economic development.[8] The primary assumption is that the technical and social aspects of production systems are interrelated. Development does not depend solely upon technical equipment. It also requires an infrastructure consisting of people and their education, organization and discipline. New technology will be viable only if it can be supported by the existing infrastructure. If it depends upon a more sophisticated human support system than exists, it will not achieve integration and is likely to cause unanticipated problems. Thus, relatively unsophisticated technology in a compatible environment will be more effective than the latest technology in an unprepared environment.

We believe that there is a growing imbalance in sophistication between the technical and social aspects of production systems in industrialized nations. New manufacturing technology is being designed with little regard for the skills, attitudes and systems and procedures necessary to support it. Consequently, technical complexity is outstripping the capabilities of firms to deal with it. An overwhelming majority of companies do not have infrastructures that can properly evaluate whether to purchase a CIM, or that can control a CIM's operations once it is implemented. And this sophistication gap is likely to widen as single-minded pursuit of the automated factory continues.

Firms in the best position to take advantage of a CIM are those in a highly advanced state of NC development. The more highly developed is NC within a firm, the more its infrastructure will be able to cope with additional technical advances. Ettlie found for example that successful NC implementation is associated with a company having prior exposure to it.[4] Yet, experience with stand-alone NC equipment cannot completely prepare a firm to deal with the intricacies of an integrated system as was discovered by some of the companies interviewed. The novel problems are compounded by the fact that it is often impractical for a vendor to temporarily install a CIM in its testing facilities. This hampers the transfer of technical knowledge to the purchaser's infrastructure. Three firms interviewed indicated that vendors are not sufficiently knowledgeable in the new technology. Further, once a CIM is installed the huge capital investment creates pressures to use it immediately. Support personnel must learn about the system while it is producing on a daily basis. Two of the companies made this point.

For the vast majority of companies with infrastructures that cannot support CIM evaluation and implementation a logical alternative is stand-alone CNC machine tools. They require less control than a CIM because they are less vulnerable to unforeseen difficulties. They can be purchased in unitary increments which facilitates a gradual buildup of experience. By allowing parts programming at the machine they can provide operators with an added measure of control. Further, they can ultimately be hooked up into an integrated system as experience accumulates. This policy will still make for difficult problems but they should be more manageable than those associated with CIMs.

Finally, what are the implications of our findings for the automated factory concept? It is likely that the problems we have identified will grow as automation becomes more comprehensive. As machinery becomes more complex it also becomes less reliable. Hence, the ideal of the automated factory may be illusory. It is often depicted as a smoothly operating system free of human variability. Instead, it may well be a nightmare of uncontrollable problems requiring unanticipated human intervention at critical points and unexpected upheavals in the workflow structure. While the number of employees will undoubtedly be reduced, those remaining will enjoy greater power over the production process. It is time to give serious consideration to less complex manufacturing concepts which stress interaction between human and technical components.

Notes and References

Professor Gerwin wishes to acknowledge the generous financial support of the International Institute of Management, Berlin. Arndt Sorge made helpful comments on portions of the reading.

1. Abernathy, W.J., 1978 *The Productivity Dilemma*. The Johns Hopkins University Press, Baltimore
2. Comptroller General of the United States, 1976 *Report to the Congress: Manufacturing Technology – A Changing Challenge to Improved Productivity*. US General Accounting Office, Washington, DC
3. Doring, M.R., Salling, R.C., 1971 A case for wage incentives in the N/C age. *Manufacturing Engineering and Management* 66: 31–3
4. Ettlie, J., 1973 Technology transfer – from innovations to users. *Industrial Engineering* 16: 16–23
5. Gerwin, D., Tarondeau, J.C., 1981 *Uncertainty and the Innovation Process for Computer Integrated Manufacturing Systems: Four Case Studies*. School of Business Administration Working Paper, University of Wisconsin-Milwaukee (reading 6.7)
6. Nabseth, L., Ray, G.F., 1974 *The Diffusion of New Industrial Processes*. Cambridge University Press, London
7. Putnam, G.P., 1978 Why more NC isn't being used. *Machine and Tool Blue Book*. pp. 98–107
8. Schumacher, E., 1973 *Small is Beautiful*. Blond, London
9. Trist, E. L., Higgin, G.W., Murray, H., Pollack, A.B., 1963 *Organizational Choice*. Tavistock, London

PART 5
Financing Innovation

5.0

Introduction

The readings in part 5 try to chart a path between the formal requirements for financial investment appraisal and control, and the processes which seem to operate in practice.

Senker's study of the issues raised by the introduction of CAD and CAM systems (reading 5.1) raises similar problems to those raised by reading 4.7, i.e. the difficulties of controlling the introduction of complex new technology systems. It shows that they are evident even at a fairly simple level such as in the first use of CAD. He also shows how difficult it can be to apply the prevailing formal systems of appraisal because of the uncertainties inherent in radically new technologies and because many of the benefits, if not indeterminate, are intangible. The consequence, as he shows, is that approval is sometimes gained only through intricate man-oeuvering within the internal political system of the firm which may sometimes require elaborate subterfuge. This process necessarily continues after approval and equipment installation, otherwise managers seek to find means of justifying failure to realize some of the predicted tangible benefits. In this process, 'union problems' may provide a convenient and acceptable means of explanation.

In reading 5.2 Leenders and Henderson draw on some interesting examples of change with rather different types of technology to those considered in earlier readings like 5.1 and 4.7 – the continuous casting of steel. They show a striking contrast between the startup times – from project approval to production operations – predicted by the supplying company and anticipated by customers in their financial and other forecasts. In the installation of 30 largely similar plants, the startups ranged from eight months to six years, averaged 2½ years and compared with expectation of rather less. Their research found similar problems of cost escalation and delays to those we referred to above (p. 5). Reinforcing some of the points made by Blumberg and Gerwin they find that, the greater the degree of technological advance, and thus, of required learning, the greater the problems that will be encountered. This indicates the need for caution by potential early adopters of radically new technologies – although the probable high levels of 'debugging' and other problems can be offset by the advantages of entering before the competition. Finally,

Leenders and Henderson suggest a number of measures which may limit the difficulties, whether in larger or in small-scale projects.

The task of anticipating problems like those identified in reading 5.2 with a reasonable degree of accuracy is a central function in the broad process of capital budgeting, which is the concern of reading 5.3, an extract from J.L. Bower's *Managing the Resource Allocation Process*. Since Bower carried out the research for his book (in the late 1960s) there have of course been changes in the prevailing patterns of corporate organization, most relevantly in terms of decentralization. Thus his depiction of the budgeting process culminates in the type of climactic 'yes/no' decision criticized by Gold *et al.* (reading 3.4). The process of strategic level capital budgeting has also been beneficial because of improvements in the tools available, not least the decentralization of computing systems and, more important, developments in software. Nevertheless, such changes do not, in our view, undermine the essential validity of Bower's model of the process. He first considers the formalized budgeting systems which rest on a number of assumptions about organizational and operational rationalities and which largely depend upon the encapsulation of a project within a limited number of quantitative measures. By applying one or other methods of net present value calculation for a project, those responsible for corporate decision making are able to approve or reject a project from a selection or portfolio of proposals.

Bower found the formal, 'rational' model to contrast with the reality of practice, as does Senker in reading 5.1. Bower's research indicated a hierarchical, 'bottom-up' process of initiation and approval which is similar to that described for the US company in reading 6.7. The initiating level of proposals for capital spending tends to be that of the production unit management who will also be responsible for carrying projects through. The way in which managers at this level define a project is seen as being shaped more by factors such as: the ways in which the performance of individuals and of productive units are actually gauged; the way in which jobs at that level are defined; and, by the availability of information at that level. (The problematic nature of information at workplace level, particularly in relation to new technologies is underlined in other readings.) Proposals are subsequently transmitted upwards towards 'integrating' managers at a divisional or similar level where they are likely to be rejected or modified, amongst other things, according to personal assessments of career-related gains or costs. Consequently, when proposals are finally reviewed at the most senior level, those considered to be 'difficult' or risky are likely to have been weeded out so that those finally submitted are unlikely to be rejected, not least because they have been 'screened, revised and politically disinfected so that ... [they] ... now tell an attractive story in professional terms'. Bower argues that capital budgeting is not principally about discounting problems but about strategic problems in general management. Thus adjustment to systems of reward and recognition may well prove to be fruitful if the quality and quantity of proposals is to be changed.

However, this does not deny the significance of formal appraisal systems. These are reviewed in an American context by Hayes and Garvin in reading 5.4, an article that was in some ways a follow-up to reading 4.3. The authors are concerned with declining real levels of capital investment in the USA and review some of the possible reasons for this. They focus in particular upon the consequences which they see as following from the now widespread adoption of Discounted Cash Flow (DCF) systems of financial appraisal. Like Bower, they point to a selective framework of analysis which is linked to DCF and which ignores some important economic factors and consequences of investment decisions. Through an emphasis on short-term returns at the expense of long-term strategy, DCF systems can act as a barrier against investment in new capital stock. This is now a fairly widespread criticism, also expressed by Sciberras in reading 4.4, but Hayes and Garvin go beyond this to associate DCF with what they term a 'disinvestment spiral'.

Whatever the difficulties, those involved in new technology projects have to try and encompass both the uncertainties of the technology and of the environment. They also have to meet the need for decision makers to have an acceptable indication of the various elements of project performance (time, costs, expenditure/revenue relationships and so on). The following two readings try to get to grips with this dilemma at the levels of the individual project.

Morgan and Luck (reading 5.5) develop the concept of a total investment system. Given the dynamics of organizational environments and of technological change, this should be the subject of regular review and continuously adapted in the same ways as, say, a production control system. Investment systems are seen as comprising formal and informal elements (terms applied differently to Bower), as well as technical and social components. If successful, such a system should not only generate profits but facilitate adjustment to changes in external conditions. In achieving this, it should encourage not only a flow of new ideas but, in each case explicitly examine possible alternative courses of action.

Primrose, Creamer and Leonard look at the issues of appraisal in the case of CAD systems. They take up a number of issues including some of those raised by Senker in reading 5.1, in particular, the difficulties associated with so-called 'unquantifiable' factors. The magnitude of the evaluation problem is indicated by the number of variables they identify – 45 – which is not an exhaustive list and which relates to what is a relatively limited type of change. They argue that in practice, many 'unquantifiables' can be drawn into quantitative evaluation. This is by giving approximate but meaningful values to at least some of the variables which can then be incorporated in DCF evaluations. This task is assisted by the use of purpose-written software which also permits rapid calculation and evaluation of a series of options. Apart from providing a more comprehensive indication of potential benefits this also takes pressure off the crude work force numbers type of calculation illustrated by Senker. Additionally it may show up benefits or costs in departments other than that in which the new technology is located. However while this may overcome some of the problems associated with DCF, it does not deal with the issue of short time horizons.

Apart from systems of appraisal, experience of the adoption of new technology is also shaped by other elements of an organization's financial systems. Morgan and Lucks' emphasis on the need to continually review and develop the investment system extends to the financial system as a whole as is taken up by Starr and Biloski in reading 5.7. They examine the implications of the comprehensive change in production systems associated with moves to FMS (flexible manufacturing systems) which potentially may affect all batch systems of production – and thus the majority of production systems. The reading considers the general range of issues involved in FMS adoption including the implications for organizational size, the work force and approaches to marketing. But much of it is concerned with the implications of FMS adoption for conventional methods of financial appraisal and of cost control. Whereas batch systems tend to be labour intensive and variable costs form a large part of total costs, FMSs are capital intensive with high levels of fixed costs. This means that costing needs to be approached in very different ways, for example shifting from emphasis on labour inputs to machine-time inputs (see reading 6.7 for the difficulties of this).

But most problematic are the benefits from FMS, which are more likely to be quality related than cost saving and the additional costs, such as in marketing, which may arise outside the production system. Ultimately, the answers suggested by Starr and Biloski lie in a systemic analysis and management of technological development in which technical, financial, marketing and other issues are more readily perceived and are more readily acknowledged as being interactive.

5.1

Implications of CAD/CAM for Management

Peter Senker

Background

During the last five years, CAD techniques known as 'interactive graphics' have diffused rapidly in Britain. Interactive graphics CAD handles and produces pictures. It can be used wherever there is a need for a visual presentation of a design: it can be used to generate mathematical representations of designs – for example coordinate data; it can be used to 'draw' in two or three dimensions, to modify drawings and to compile drawings using descriptions of standard components stored in component databases.

The first important use of CAD interactive graphics was during the 1950s when refresh screens and light pens were used in the US SAGE early warning radar system. SAGE was funded by the military, and so the expense involved in using mainframe computers then necessary to drive refresh graphics was acceptable.

By 1965, McDonnell and Boeing were all experimenting with mainframe-based interactive graphics. Lockheed's CADAM system was developed under contract with the Department of Defence and IBM subsequently negotiated to market CADAM under licence. Space and military funding supported most new interactive graphics technology in the US until about 1970. But there was also important early development work in the US motor industry. By the mid-1960s, both Ford and General Motors had made progress with systems to help the motor industry fulfil its specialized requirement to produce and digitize mock-ups of new model bodies.

The role of the US military was indirect in promoting the development of CAD for microelectronics applications. The US military have consis-

Source: from *Omega*, 12(3), 1984.

tently supported programmes designed to develop smaller, more complex integrated circuits. The large-scale integration (LSI) circuitry which came into use in the late 1960s would have been almost impossible to design without CAD.

In the UK, CAD interactive graphics developments were several years behind the US. Nevertheless, by the end of the 1960s, Elliott Automation, Ferranti, ICL and Marconi were all offering systems or services based on their own interactive graphics hardware and Racal was offering a service based on Elliott Automation hardware. CAD activity at the end of the 1960s was all based on expensive mainframe computers.

In the period 1968–71, three innovations were made which paved the way to the more widespread diffusion of interactive graphics technology:

(1) relatively cheap minicomputers;
(2) the storage tube;
(3) structured programming and virtual memory techniques.

Minicomputers reduced the costs of the enormous computing power required to handle pictures. Storage tubes required substantially less computer power than refresh tubes and provided stable, flicker-free images. Virtual memory techniques allowed large programs to be stored and used in relatively small computers with the minimum of penalty in terms of computer response time. With the help of these techniques, minicomputers could be made to respond to CAD commands almost as fast as the much more expensive mainframes.

Original mainframe-based interactive graphics systems were developed to meet specialized application requirements – in aerospace, in motor vehicles, in microelectronics and in printed circuit board design. Early in the 1970s, intense competition arose as firms from various backgrounds – from aerospace, from electronics or set up by academics – began to develop and market 'general purpose' systems based on the new technologies. These new systems were designed as 'turnkey' systems – designed, in principle, so that customers could buy a system, plug it in, 'turn the key' and produce usable production drawings. Although, strictly speaking, this is an unattainable objective, some suppliers of systems have made substantial progress towards the development and marketing of 'turnkey' systems.

Benefits Available from the Use of CAD

Kaplinsky[9] points out that it is impossible for users to calculate the exact benefits likely to arise from the use of CAD, because the available benefits depend on the organizational context in which it is placed; and gaining benefits in full depends to a large extent on reorganization: it is impossible to forecast the future efficiency of a yet-to-be reorganized design process.

Benefits available from CAD include:

(1) reduction in design lead times;
(2) economies in design and draughting;

(3) ability to design products which are too complex to be designed manually (the outstanding example being in electronics – LSI);
(4) economies in material usage;
(5) combination of CAD with production automation to obtain benefits of integration (CAM).

Why British Engineering Firms Have Adopted CAD

In the late 1960s, there was some interest in CAD in the motor, aircraft and electronics industries. A few firms developed their own, expensive mainframe systems. But, progress in the application of CAD was slow for nearly 10 years, until the late 1970s. From about 1977, interest in CAD increased rapidly in Britain. In that year, the first CAD system was installed in the mechanical engineering industry.

The principal stimulus to this renewed interest was the availability of minicomputer-based turnkey CAD systems from the US. Not only were these new systems cheaper than 'do-it-yourself' mainframe systems, but they were enthusiastically promoted by their suppliers – by means of exhibitions, sales visits etc.

For sales to be made, potential customers – typically middle management – needed to be able to convince top management that CAD was a profitable investment. CAD suppliers provided potential design office customers with a convincing financial case for investment in CAD to present to their top managers in terms of saving in draughtsmen's labour.

Most suppliers, therefore, attempted to sell their equipment on the grounds that it could save their customers substantial costs in terms of draughtsmen's labour. This is reflected in the basis of CAD capital investment appraisals submitted to top managements (table 1).

But few managers had any experience on which to base any realistic cost–benefit analysis. Generally, managers relied on the information provided by CAD salesmen of the benefits offered by their equipment.

Table 1   Basis of cost justification of CAD sample: 34 engineering industry CAD users

Cost justification basis	Number of establishments
Savings in drawing labour	21
Savings in drawing and design labour	3
Savings in estimators' labour	1
No cost justification	5
No reason given	4
Total	34

Source: Arnold and Senker 1982

A number of managers interviewed readily admitted that their justifications were spurious – quite probably because they realized that genuine, quantifiable cost justification data were unobtainable: but that, if they were to obtain the equipment which they felt that their firm needed, they would have to produce some quantitative 'data' to get their proposals past the firm's accountants.

Broadly speaking, our findings were similar to those reported from a series of studies in the US[5]: the data which firms feed into their investment appraisals are often wildly inaccurate, especially when investments involve new technology. Four frequently found sources of error were reported from these US studies:

(1) underestimating the time needed to achieve effective functioning, often by a considerable margin;
(2) overestimating the utilization rate likely to be achieved;
(3) underestimating the need to make adaptive adjustments;
(4) underestimating the tasks involved in securing labour acceptance of changes.

We found that few CAD system purchasers appear to have included any allowance in their appraisals for delays likely to be experienced before gaining full benefit from CAD. In the event, most firms found that there was a good deal they needed to learn before they could achieve significant productivity benefits.

Managing a computer system was often a new experience for drawing office management. In addition to the need to deal with training and industrial relations, the work load, access to the machine and housekeeping all needed to be organized. In order to get good use out of the system, a database of standard components and standard procedures needs to be set up and the instruction 'menu' needs to be adapted to the firm's particular needs. Many users found it helpful to appoint one individual to be responsible for this.

CAD often provided an impetus to reorganize drawing and related activities. At one firm, a new drawing numbering system was adopted for use on both the CAD system and the firm's mainframe computer. This permitted parts list processing to be done on the mainframe computer.

Three main factors tended to slow down the rate at which firms learned to use CAD: industrial relations problems, particularly in the vehicles and aerospace industries; supplier failure to deliver adequate software; and managerial inefficiency, particularly in the mechanical engineering industries. Sometimes there were delays in concluding negotiations between managements and trade union representatives about various aspects such as pay and who was to operate the equipment. One firm had installed a system a year prior to the interview but had not yet put standard parts into their database; it was also severely hampered in the first six months by its failure to appoint a CAD manager.

Most companies treated CAD as a design office investment: few recognized its significance as a strategic step towards CAD/CAM.

CAD CAM

At the time of our interviews in 1981, few firms outside the electronics industry had made much progress linking CAD with 'downstream' production processes. In many parts of the electronics industry CAD CAM has been a technological imperative for years: many highly complex microelectronic components, computers and printed circuit boards simply could not be designed and manufactured in its absence.

Manufacturing printed circuit board artwork and generating microelectronics' masks essentially involve manipulation of a light or laser beam under computer control to produce and superimpose two-dimensional drawings. The manufacture of these highly complex and very precise products can only be achieved using techniques of combining CAD with CAM.

In contrast, linking CAD with production processes in mechanical engineering involves adding other data – feeds, speeds, tool offsets etc. – to the output from CAD graphics terminals.

Obviously, if the output from CAD systems is to be used to drive computer-numerically controlled (CNC) machine tools, the firm has to own or acquire these machine tools, for applications such as diemaking. Generally, however, design and diemaking departments submit proposals for capital investment separately to top management and there are not always adequate procedures to ensure that these requirements are coordinated.

For example, some design offices in automobile companies invested in CAD systems well before there was any consideration of investment in CNC diemaking facilities by their diemaking departments. Although the potential of CAD CAM for rationalizing the process of design and production of car-body dies has long been appreciated, implementation of CAD CAM had proceeded slowly, partly for this reason.[14] Efficient implementation of CAD CAM depends on coordination of software acquisition as well as hardware.

When once drawings have been produced on CAD equipment, those drawings, or digitally defined geometry, then go to the part programming office for programs to be drawn up for use on the CNC machine tools. This may or may not be done on a CAD facility, but is a separate process from creating the drawing in the first place. The part program is then put through a post-processor to generate a program the particular machine tool can read. A wider and more useful variety of post-processors are available for use in conjunction with some CAD systems. If a firm wishes to use its CAD installation to generate programs for driving CNC machine tools (i.e. for CAD CAM) it is obviously helpful for it to invest in a CAD system for which a wide variety of post-processors are available.

Generally, design offices were principally responsible for purchasing CAD systems and, naturally, paid most attention to securing systems which had the most potential for enhancing the productivity and efficiency of the design office itself. As a result, in some cases, subsequent CAD

CAM developments were hindered by the firm having purchased a CAD system for which insufficient post-processors were available.

DCF And Profit Centres: Barriers to Strategic Investment

The traditional method of investment appraisal in Britain is the payback method. Individual departments calculate the costs of a particular investment and attempt to assess how many years it will take for the firm to recover those costs by savings resulting from the investment. These investment appraisals are then submitted for top management approval. In general, top management then approves those investments which will repay their costs within a specified period, say two years. There may be other investments whose costs will take somewhat longer to recoup, but which are approved for special reasons. For example, the use of lifting tackle may reduce risks of operator backpain considerably and may, therefore, be approved by top management even though the investment fails to meet strict financial criteria. Such payback methods are simple to apply and are still widely used: it is claimed that one of this technique's advantages is that it contains a built-in safeguard against risk. If all the money invested can be returned within two or three years, risks are minimized. But the method suffers from the disadvantage that any returns which accrue after the end of the payback period are ignored entirely: various projects may differ greatly in this respect.[11]

In theory, discounted cash flow methods (DCF) can cope adequately with differing time patterns of profits and differing project lives.[7] But they suffer from the disadvantage that they may lead top management to reject proposals to invest in radically new technologies. It may be vital for a firm to set up facilities incorporating new technology in order to secure its competitiveness in new and growing markets.

Yet, because the direct financial return to the new facility may take a few years to materialize, the project may be rejected by top management no matter whether they use payback, DCF or a combination of the two. By their very nature, these methods favour projects which yield a relatively quick return.

In Britain, as in the US, investment proposals generally emanate from relatively junior people – from departmental managers, engineers, supervisors.[6] These people, however able, lack the data needed to take into account the wider benefits of a project – still less to put forward meaningful quantitative estimates of financial benefits outside their own specialist areas.

The use of discounting techniques often supports the case for expanding existing facilities rather than building a new plant on a greenfield site. The initial investment to modernize an existing plant is normally far less and rates of return given by the use of the usual evaluation techniques are often far higher. But, a series of such decisions can lead to 'ponderous outmoded dinosaurs that are an easy prey for the smaller, more modern, and better focused plants of competitors'.[8]

By its very nature, DCF encourages projects which yield quick returns as profits anticipated to arise after several years are heavily discounted. If a firm has to choose between two broadly similar projects, one of which yields quicker returns than the other, then the project rated most highly using DCF techniques is the obvious one to implement. If, however, a project can help to get a company involved in a new technology on which its whole future could depend in a few years' time, it may be vitally important for the company to invest in that project, even if predicted returns are relatively modest.

The profit-centre control system prevalent in Western companies gives managers incentives to strive to maximize the short-term profit performance of a specific profit centre. Managers are also moved frequently from one profit centre to another. These practices tend to discourage managers from promoting projects which depress short-term profitability even if they are expected to yield long-term competitive advantage.[10] Profit-centre control encourages competition between managers. But benefits from such competition may be outweighed by the discouragement it creates to cooperation between divisions which may often be in the long-term interests of the firm.[13]

Automation and Skills

Bela Gold emphasizes that many companies lack sufficient production and engineering capability at high levels.[6] Top managers need more knowledge about CAM to provide adequate support, direction and coordination.

In Britain, deficiencies in engineering training and skills often extend from top management to lower levels of management, to planning, engineering and shop-floor skills.[15] No conceivable investment appraisal technique can compensate for such deficiencies. Firms will not be able to use advanced automation systems effectively unless they take urgent action to remedy their lack of key skills. Investment in appropriate education and training at all levels is extraordinarily expensive: the Japanese experience indicates that it can be very worthwhile.

The FMS (flexible manufacturing systems) Report emphasizes that it 'is no longer sufficient for the specification and justification for new equipment to be made at lower levels, leaving senior management to approve or reject it ... senior management must participate in the analysis process and provide guidance'.[4] This demands that senior management – and others – have the skills to provide appropriate analysis and guidance. They comment that the price paid for the benefits of FMS is high and heavily 'people dependent' and includes a high level of engineering and management capability and commitment, new skills in programming project management etc. Success is determined by the quality and quantity of planning and everyone totally underestimates the amount of planning work necessary for success. Yamazaki invested 100,000 hours of planning work on their successful 18-machine system, far more than anyone else: satisfaction with the performance of FMS systems is far greater at Yamazaki than in US firms which, in comparison, neglected planning.

Effective planning can only be carried out by people with adequate engineering and planning skills. Ball describes the investment in training facilities by large Japanese firms as 'massive': since they regard technically qualified people as their most valuable asset, 'no effort is spared in ensuring that they are properly equipped to carry out their tasks'.[2] Japanese companies second their own key people for short periods to the training department as instructors for the various groups of employees, from production line workers to senior managers.

Automated systems such as direct numerical control (DNC) and FMS make it possible to manufacture sets of components which may be assembled into finished subassemblies immediately they are produced. Potential work-in-progress savings within the automated machine shop may be very substantial, but other savings may also be facilitated. It may be possible to reduce work-in-progress in the assembly shop. More responsive production can permit a plant to offer customers a given standard of product delivery performance with much lower investment in finished goods stocks.

FMS allow for easy modifications of a design at any stage of a product's life and facilitate changes in product mix as demand changes.[6]

Blumberg and Gerwin found that firms in the UK, West Germany and the US all had considerable problems in evaluating such systems, and DCF was of very limited use.[3] Insufficient evidence was available to estimate future net returns, especially if it was proposed to use the new equipment to produce a new product. A critical issue was how to quantify the benefits likely to accrue from enhanced flexibility.

Japanese companies evaluate proposed major investments for their impact on competitive cost position and market share for the whole business over a period of years. They find it easier to initiate investments in projects with a long-term pay off for several reasons: they do not employ the profit-centre control system, and managers' pay is less directly related to short-term profitability: fixed investments are financed primarily by retained earnings and debt and interest rates are lower. These factors make it easier for them to invest in automated plant which may require market share to be increased to ensure that its capacity is fully utilized and that its potential for low cost production is realized.[10]

Japanese decision-making processes, based on consensus, are well-adapted to making the major decisions involved in investing in complex manufacturing systems. Many people take part in the decision-making process, and by their participation, a large number of people feel committed to the decision. This tends to eliminate objections during implementation. Wide participation ensures that fewer aspects are overlooked and tends to reduce the trauma of major changes. This can generate an atmosphere of confidence which facilitates the implementation of bold decisions.[12]

Blumberg and Gerwin concluded that the capabilities of the firms they investigated in West Germany, US and Britain were insufficient for them to cope adequately with the problems of appraising and implementing sophisticated manufacturing systems.[3] They suggested, therefore, that firms should cease to strive towards 'the automatic factory' and concentrate rather on simpler technologies more compatible with their skills and organization. A problem with such an approach is that if some countries – in particular Japan – succeed in using these advanced technologies, then other countries may risk loss of markets if they fail to emulate their more efficient competitors.

Implications

In Britain, DCF capital investment appraisal techniques have been widely used to justify investment in CAD on the grounds of its potential for savings in the costs of employing draughtsmen. But the real justification for CAD is usually strategic – for example, that its use can be a key factor in ensuring that a company is early enough to market with new products to secure its survival. For most manufacturing companies CAD is also an essential first step into CAD CAM, which can help them to secure their long-term futures.

An organization which relies on 'bottom-up' procedures for proposing major new investments in automation is unlikely to secure the most suitable systems for its needs. Assessments can only be made properly in the light of a company's strategic plans. This implies that top management must be heavily involved in the initial selection of major projects for evaluation.

CAD CAM can confer increased flexibility and quicker response on the whole firm. A high proportion of benefits may be conferred on departments outside the one in which the automated facilities are installed. In addition, some such benefits can be difficult to evaluate and quantify by their very nature.

The effective use of CAD CAM can often depend on close coordination between departments and divisions. It also depends on a high level of technical competence at many levels and in many functions throughout the organization – from design office and shop floor to top management.

Economical production of reliable products designed to meet customers' needs forms the foundation of successful manufacturing operations. Success demands the application of considerable engineering and technical skills in a wide variety of functions in planning overall strategy, design and production management and on the shop floor.

But the submission of individual project proposals by departments to senior management for evaluation is no longer a satisfactory basis on which to build a strategy for investment in automation. Top management teams need to acquire the engineering skills necessary to interact with design and manufacturing management and engineers in order to develop coordinated automation strategies.

References

1. Arnold, E., Senker, P., 1982 Designing the future: The implications of CAD interactive graphics for employment and skills in the British engineering industry. EITB Occasional Paper No. 9
2. Ball, G. F. 1980 Report on vocational education and training for employment in engineering in Japan. British Council EITB
3. Blumberg, M., Gerwin, D., 1981 Coping with advanced manufacturing technology. International Institute of Management Labour Market Policy Discussion Paper, Berlin (Reading 4.7)
4. FMS Report, 1982 *Ingersoll Engineers*, p. 47, IFS
5. Gold, B., 1980 On the adoption of technological innovations in industry: Superficial models and complex decision processes. *Omega* **8**(5): 505–16
6. Gold, B., 1982 CAM sets new rules for production. *Harv. Bus. Rev.* Nov./Dec., 88–94
7. Hawkins, C.J., Pearce, D.W., 1971 *Capital Investment Appraisal*. Macmillan, New York
8. Hayes, R.H., Garvin, D.A., 1982 Managing as if tomorrow mattered. *Harv. Bus. Rev.* May/June (Reading 5.4)
9. Kaplinsky, R., 1982 *Computer Aided Design*. Frances Pinter, London
10. Magaziner, I.C., Hout, T.M., 1980 *Japanese Industrial Policy*, pp. 10–11. Policy Studies Institute
11. Ogden, H., 1980 Justifying assembly automation economic methods. *Proceedings 2nd International Conference on Assembly Automation*, pp. 1–8. IFS, Bedford
12. Sasaki, N., 1981 *Management and Industrial Structure in Japan*, p. 59. Pergamon Press, Oxford
13. Sciberras, E., 1982 Technical innovation and international competitiveness in the television industry. *Omega* **10**(6): 585–96 (Reading 4.4)
14. Senker, P., Huggett, C., Bell, M., Sciberras, E., 1976 Technological change, structural change and manpower in the UK toolmaking industry. EITB Research Paper No. 2, p. 17
15. Swords-Isherwood, N., Senker, P., 1980 *Microelectronics and the Engineering Industry: The Need for Skills*. Frances Pinter, London

5.2

Startup Research Presents Purchasing Problems and Opportunities

Michiel R. Leenders and Ross Henderson

Recent research has uncovered a number of major new findings regarding startup. In technologically advanced systems startups:

(1) Take much longer than expected
(2) Tend to be much more expensive than expected
(3) Follow a learning curve pattern
(4) Are predictable as to total length and rate of improvement
(5) Take even longer when they involve major technological advances

These findings, coupled with our already existing knowledge of startup, present a number of major implications for managers involved in the startup process.

Jim Adams[16] posed proudly for news photographers in January 1963 beside a model of the $3 million continuous steel casting machine which he announced would be installed by June 1964, producing at a 200,000 ton per year capacity rate by December 1964, and would add $1.5 million to 1965 profits. He noted that $200,000 had been provided in the capital budget for contingencies. Contrasted to this proud announcement, the record showed, 4½ years later, that the first steel was cast in October 1964, capacity monthly production of 16,000 tons was first achieved in June 1967, startup modification costs totalled $1.7 million, while 80,000 tons of lost production caused a reduction in contribution to profit of $3.6 million during the startup period. This startup of a plant using new process technology had taken 2½ years longer and had cost $4 million more than Jim Adams expected. He felt defensive about the result and would have been much relieved to know that his startup, rather than being an isolated misfortune, was better than the average in such circumstances. Most

Source: from *International Journal of Operations and Production Management*, 1(2–3) 1980. MCB University Press Ltd.

startups using new technology take longer, and cost much more, than expected.

Mike Broda, after encountering more severe startup difficulties than Adams during the first six months of startup at a similar plant in a developing country, was convinced that full capacity production was only a month or two away. He had raised production to 30 per cent of full capacity and felt that the remaining constraints to a 100 per cent production rate could be removed quickly. However, as month after month, and then year after year went by, his goal was ever elusive. Cash was short and a local supply had to be developed to provide a substitute for imported rapeseed oil lubricant. Developing a local repair parts machining operation was tedious. Persuading experts to come around half the world to give advice was time-consuming and cost money. One by one the problems were solved until, finally, the production rate reached 100 per cent of capacity in the 53rd month after the first steel was produced. Broda relaxed with relief. He was not aware that the rate of productivity increase had followed a well-known pattern. Nor did he realize that the production data from the first six months might have permitted him to predict productivity growth during the remaining 47 months of startup with useful accuracy. The pattern known as the learning curve or manufacturing progress function, a method used for prediction, was unknown to Mike Broda as he struggled through 4½ years of startup. But even that startup was shorter than some.

Tom Kruger struggled through even more severe difficulties to start up a continuous steel casting machine which incorporated significantly advanced technology. He was placed in charge after the decision had been made to purchase a new curved mould type machine equipped with automatic controls for tundish level, mould level and casting speed, plus extensive new instrumentation. These equipment innovations were coupled with a product requirement for many analyses of stainless and alloy steels. Some of the most important alloy steels were specified with exceedingly tight tolerances. As a combined result of the technologically advanced machine plus the sophisticated product, virtually no good steel was produced during the first year of startup. Equipment modifications and learning achieved some productivity growth during the second year but Tom Kruger sweated through a full six years before his $8 million new plant was producing, at capacity, 100,000 tons per year. During that period, 270,000 tons of production below capacity levels had been lost, which equate to a lost contribution of $16 million. Modifications and unexpected maintenance added another $6 million to the losses which had to be explained away in the annual report. Tom Kruger, and the company, were disenchanted with the whole idea of this new plant with advanced technology by the time it was operating properly. Not only do plants using new technology take longer and cost more to start up than expected, but where advanced technology is tried for the first time, the startup period may stretch far beyond the worst expectations of any of the managers involved.

Long Startups are Ubiquitous

Companies are finding it increasingly important to their survival and profitable operation, to be able to incorporate and digest sizeable technological leaps forward into a new process such as this, without being stalled for several years by that digestion. Usually, after the marvellous plants incorporating such exciting processes have been built, a ribbon is cut, a bottle of champagne is thrown and the machinery whirs to produce a small amount of product. This event signals the beginning of the long arduous period called startup, which is complete only when production levels out somewhere near design capacity. This may take as long as four or five years, as we have seen with Messrs Adams, Broda and Kruger. And many startups of such new processes are required to deliver the broad array of products in this complex industrial world. Thus, management faces a problem with significant dimensions to increase productivity during startup.

Purchasing managers possess an often unrecognized opportunity for favourably influencing the productivity during plant startup, that painful period of time from production of the first acceptable product until the plant is operating regularly at full capacity. As we have seen, managers frequently discover that the amount of time and money required to bring a new plant to the desired level of production, especially a plant involving new technology, is greater than expected. How long it will take to reach the desired output level is difficult to predict. What action to take to speed the process has often not been clear. However, the findings from a recent research study suggest to purchasing managers in such plant startup situations how they may help to reduce startup duration and its cost, how they may help to secure a useful prediction of startup duration and action which they can take to speed the process.[2] This study builds upon a number of earlier explorations of the subject, as well as upon observations of management practice and startup productivity data.

Many sources show that long, expensive startups are ubiquitous, and generally unexpected. A sampling of corporate reports readily discloses footnotes explaining large write-offs due to long startups, such as the $75 million write-off by Celanese in 1968 for its Sicilian paperboard plant, and Weston's $7.6 million startup write-off in its $9 million sugar plant in 1974. An earlier journal article detailed the unexpectedness of the duration and cost of such startups.[3] Such sources are supported by the research reported here, in which the 30 plants encountered an average startup duration of 2½ years, with some requiring longer than six years.

Theory has evolved which helps to explain, and which may shorten, such long startups. The startups are more susceptible to management control than sales volume and price, which also hinder attainment of planned production levels. Bright recognized the existence of an extensive startup or 'debugging' period for technologically new, machine-intensive systems, 20 years ago.[4] Unfortunately, most research on productivity during startup has been in the labour-intensive air frame and electronic industries. The learning curve has received considerable use and refinement in these applications.[5,6] Authors have applied the learning curve to other labour-intensive operations.[7] With one exception, little has been done in machine-intensive, process manufacture, although some authors have broached the subject.[8,9]

Nicholas Baloff has carried out the only study focused solely on startup in machine-intensive manufacture. He has written at least 11 journal articles on this subject,[10,11] mostly based upon his 1963 Stanford PhD thesis.[12] His work provides solid evidence that increases in productivity tend to follow the manufacturing progress function. Yet such evidence and theory has not been widely used to manage startup.

Actual management practice seems to be a crude dichotomy of theory and conventional wisdom. For public consumption, all managers gather together on the day of first production and announce that the new plant is operating smoothly except for a few bugs which will be ironed out by the end of the next month. Behind the scenes, they plot production, weekly or monthly, on arithmetic charts, always hoping that approximately another three months will bring production to full capacity levels.

Usually, it is many times three months before success arrives. Meantime, the managers exert strenuous efforts to complete a sequence of difficult modifications, each of which increases productivity a little. It is believed that this actual management practice can be improved, based upon both the theory and research findings from the following startup study of a sequence of continuous steel casting machines.

The Research

Productivity data during startup were gathered from 30 Concast machines which were installed during the past 20 years. All machines were designed by one supplier, Concast Inc. of New York, or its parent, Concast AG of Zurich. The 30 machines represent an important portion of over 200 continuous steel casting machines installed in the world at the time the data were gathered. They represent an even larger portion of machines installed by Concast, the major supplier. These machines were located not only in North America, but also in ten countries on four other continents.

A continuous steel casting machine bypasses the ingot mould and primary rolling mill processes to produce endless lengths of solid billets, slabs or shapes from liquid steel. The hot liquid steel is poured into the top of a bottomless mould from a nozzle in a large open vessel called a tundish. Cooling water in the walls of the reciprocating mould cools the liquid steel so that a solid skin forms before it leaves the bottom of the two to three foot long mould. A lubricant, usually rapeseed oil, is forced under pressure between the mould and solid skin, to permit the steel to move downward in relation to the mould. The mould reciprocates constantly and the red hot steel with a liquid core moves downward to a cooling bed. The difficulties of making this process work, without either freezing the steel completely in the mould or allowing the liquid core to break out through the thin steel skin, are evident. The growth in productivity at a new plant tends to be gradual as a result of these difficulties.

The monthly production of steel in tons was gathered from each of the 30 Concast machines for the first 36 months or more of operation. The number of operating hours for each of the months was also collected. These figures were used to calculate a standardized productivity per month which made allowance for periods where production was low due to lack of sales. A steady-state productivity level was chosen for each machine based upon 12 months of production at that level [after] the startup period. The standardized monthly productivity figures were divided by this steady-state productivity level to express productivity as percent productivity efficiency (PPE). The cumulative actual tons of steel produced in each consecutive month were also calculated. The cumulative tons were also divided by steady-state in tons to obtain cumulative tons as a percentage of PPE. These were the data used for the measurement and prediction of productivity progress and have been reported in greater detail elsewhere.[13]

Less quantitative observations were also gathered during the research, often in the form of vignettes of revealing little startup episodes. Both quantitative and less tangible data tended to confirm the three startup experiences at the beginning of this reading: that startups take longer than expected, cost more than expected and are even more difficult to manage where a major technological advance has been incorporated into the process.

Time

Startups usually take much longer than expected by the managers concerned. The expectations of Adams, Broda and Kruger were for substantially shorter startups than they experienced. Those expectations are not precisely predicted by managers in most cases, the research shows. However, statements in press releases and annual reports frequently suggest that full capacity production is expected throughout the fiscal year immediately following startup. This would presume a startup duration of six months to one year. Equipment manufacturers' representatives are often scheduled to remain at the plant for two or three months, which implies an expectation of even shorter startup duration. Financing of many new plants, which budget for operating losses only in the first six months, reinforces the idea that managers anticipate a startup duration of less than one year. Although such expectations are neither precise nor explicit, all the evidence points to expectations of startup taking months, not years.

Actual startup duration observed in the study of 30 Concast plants averaged 2½ years and ranged from eight months for a few well-managed plants up to more than six years. These startup duration measures are much more precise than the rough projections used to form the managers' expectations. Such a large, consistent set of startup data has not previously been available from any one industry using a new technological process. Casual observations, however, indicate that many other processes such as float glass or computer-controlled paper-making machines suffer startups which are just as long. The actual startup duration for Concast machines provides an interesting comparison with managers' expectations.

The actual average of 30 months seems to have been three to five times as long as expected. Plant managers would point out in response that production often reached 50 per cent to 80 per cent by the end of the first year. But tonnages unavailable to customers are still large at these levels. Also, due to high fixed costs and breakeven levels, operating losses were being incurred, in most cases, well-beyond the first 12 months. Regardless of these arguments, the research shows clearly that startup took longer in almost every case than the somewhat imprecise expectations of the managers concerned.

Cost

Not only did they take longer, they cost more. Startups tend to be much more expensive than expected. Jim Adams expected to spend $200,000 for contingent modifications, and to lose about 50,000 tons of production, worth about $1 million in contribution, between June and December 1964. Various reports disclose that contingency funds for equipment and system changes rarely exceed 10 per cent of the budgeted plant cost. Annual reports, as mentioned earlier, often state that profits will be earned from the new facility at full capacity production throughout the fiscal year following the first good production. Breakeven is expected in less than one year. These expectations of costs and profits are usually badly disappointed and often have to be explained away in annual reports, as examples given earlier illustrate.

The actual cost of startup including modifications, extra maintenance, lost and late orders, hiring and firing people and lost contribution, is often comparable to the $5 million which Adams experienced on a $3 million budgeted installation. The lost contribution to profit does not show up in company accounts, of course, but only as missing profits which the company had expected to make. Wearne says: 'Total costs due to delays in achieving full use of plant can be large enough to destroy the economic justification for a project or even cause its abandonment'.[4] In the case of the 30 plants studied, lost contribution amounted to $137 million if each ton was estimated to produce a $20 contribution to fixed costs and profit. Estimates of total actual costs of all the above components for startup are difficult to make, but are, as a minimum, equal to 25 per cent of plant investment and may amount to several times the total investment in difficult cases. An amount of 40 per cent of the plant investment is believed to be a conservatively low estimate of the average total cost for startup with a technologically new process.

Comparison of this 40 per cent with a maximum contingency allowance of 10 per cent shows that startups cost more than expected. Where the total cost of startup exceeds plant investment, the extra cost may be crippling, as discovered in one or two of the plants reported. While these statements cannot be made with great precision since data on costs were not collected formally in the study, enough individual pieces of information were secured to corroborate the truth of the statement: 'Startup costs more than expected.' Fortunately, the progress of productivity follows a regular pattern called the manufacturing progress function, during startup. This pattern is similar to the inverse of the learning curve and it provides a basis for measurements which can be used subsequently for the control and acceleration of startup. [For details, see Leenders and Henderson's full article.]

Difficult High Technology

The research also disclosed that where major technological advances are implemented in the process, startups take even longer. Tom Kruger's six-year startup was undoubtedly extended to such a long term because of the new curved mould and the automatic controls. These features promised lower investment in structure, quicker training of crews and more consistent quality. But their modification to an effective design and learning to operate them regularly took this extensive time. The first eight-strand machine, which promised more production than a smaller machine, encountered the same type of extended startup due to hotter, faster flowing steel at the centre strands, as compared to the cooler steel at the outer strands, which tended to freeze up much more easily. Earlier in the technological sequence of advances in continuous steel casting, an extended effort to cast small, two inch square sections, and so avoid expensive subsequent rolling, was finally abandoned as commercially unfeasible at that time, due both to breakouts and freezeups. The enticing promise of these major technological advances led each of these companies to adopt a major advance to the process first. They expected significant benefits of greater productivity, lower cost or better quality, far in advance of their competitors.

The actual time when those benefits were secured, after a lengthy startup, was so much later than expected that most of the exciting promise was not fulfilled. Kruger's company showed mediocre financial performance for five years and lost the favour of financial analysts and investors. The company starting the eight-strand machine was able to absorb the lack of tonnage from continuous casting by accelerating other operations, but they became much more cautious about being first with a new technology. The company which so enthusiastically commenced production of two-inch-square billets, phlegmatically wrote the project off as a learning experience. The net result in these cases was actual startup very much longer than expected due to major technological advances. These long startups dissipated most of the promised benefits.

An expectation that a major technological advance is likely to cause a longer startup can be deduced from the research. Tom Kruger would have enjoyed much more equanimity during his six years of startup if he had known this. The continuous steel casting technology is nearing maturity now, and long startups such as reported here, due not only to major technological advance but also due to the proliferation of the basic technology to many new locations, are very much less likely to occur. But as major advances are made in other technologies, the expectation of long startups in these instances is likely to be a valid one.

Timing the Purchasing of Technology

A major strategic decision for any organization is whether the time is right to commit the organization to entry into a new technology. This idea has been extensively explored by a number of researchers in the field.[4,15] Henderson's research shows that the very long startup for early adopters may prevent them from enjoying the advantages of early adoption, because not sufficient output is achieved. Early adopters run a high risk that startup will be long and that subsequent modifications in the technology may make the installation obsolete.

From a purchasing standpoint the role of the supplier in this regard is an extremely interesting one. At this stage the supplier is highly dependent on customers to find out what is wrong with the equipment and what needs to be done to modify it. Using the experience of one installation, the supplier modifies the next and the subsequent customer benefits from such modifications. The supplier also uses the startup knowledge from the first customer to assist subsequent customers in faster startup. This key supplier role as the bee that gathers the essential startup knowledge must be recognized. It is not only the technical proficiency of the equipment that is important, but also the supplier's knowledge of how it can be started up and used. How to get the most out of a supplier under these circumstances is one of the challenges for the purchasing manager. Early adoption of a new technology clearly means that relatively little startup knowledge is available.

Early Involvement of Purchasing

Modern literature on purchasing and materials management fully supports the notion of early purchasing involvement in all phases of acquisition planning. The purchase of major equipment should be no exception. Since most of these types of installations tend to be custom produced, requiring close engineering–supplier cooperation, it is important to avoid being locked in with a supplier before all alternatives have been suitably explored. Early involvement in procurement can help prevent premature commitments. The knowledgeable purchasing manager can also make sure that specifications and expectations are realistic in view of our current knowledge of the startup process. Three major areas of procurement assistance involve source selection, contract negotiation and planning for the startup period.

Source selection

In source selection it becomes not only important to have a technologically competent supplier, but also one who is capable of giving startup support. Offshore suppliers may represent special difficulties in terms of language capability and availability of personnel who are willing to spend much time living in a foreign country. Since the startup cost may well be as large again as the original investment, considering lost capacity, wasted product and lost man-hours as well as additional investment required, the initial price of the equipment may become a less significant consideration than at first envisaged.

The idea of purchasing a turnkey operation must be highly appealing. If a supplier can be found who is willing to run the risks and problems of startup, this may be an attractive alternative. Provided that the labour is adequately trained at the time of take-over, an organization can save itself a tremendous amount of worries by placing startup responsibility where it makes the most sense: in the hands of the people best equipped to achieve fast startup. Even with a turnkey operation, however, fast startup is not necessarily guaranteed.

All the evidence suggests that normal supplier promises of quick startup need to be viewed with extreme caution. In the case of continuous casting, the fastest startup took about eight months and the slowest about six years. Supplier and purchaser prior estimates ranged from three to six months.

Startups tend to follow a learning curve, the manufacturing progress function, in two ways. In the first place, a startup will be influenced by earlier startups of similar systems provided by the same supplier. It is possible for the purchaser to check startup experiences with previous installations. This information is usually more easily obtained before a firm commitment is made. Secondly, the startup itself will follow the manufacturing programme function, therefore information gained during the first few periods of operation can be usefully employed to predict total startup time. Unfortunately, the latter information is not available at the time a supplier decision needs to be made. Nevertheless, realistic expectations should be reflected in the contract drawn up with suppliers.

The contract

Purchasers can protect themselves somewhat against long startup delays by insisting that certain understandings be covered by contract. A simple way would be to make sure that major payments are delayed until rated capacity is actually reached. A ten per cent holdback is really not very meaningful under these circumstances; a $20 million investment sitting idle even for one day will incur a carrying cost of about $6,000 assuming a ten per cent capital cost. Penalty clauses may be written in for startups which are slower than predicted by suppliers. A more positive approach can be taken by giving bonuses for quick startup. Consequential damages from slow startups represent a host of difficulties of their own. It is probably advisable to have a separate contract for startup services by the supplier. Normally, a supplier's technical representative will stick around during the first month or so and make sure that some satisfactory output is produced. Then the representative disappears and the organization is left to its own devices. In a startup of as much as two or three years' duration, this is not a particularly happy situation. Since certain individuals within the supplier organization may be better equipped to assist during startup, such contracts for startup services may specify the use of specific persons, rather than leaving this to the supplier's discretion. All the evidence we have tends to suggest that such personal assistance of a knowledgeable individual can be of tremendous value in achieving a quicker startup.

Procurement planning for the startup period

Procurement planning for the startup period must recognize the realities of recent research findings. The materials organization must be ready to support all efforts for quick startup. At the same time the possibility of a long and arduous startup will require that contracts with suppliers for raw materials and supplies may have to be highly flexible. It is unlikely that high initial production estimates will be met and a good understanding of the range of possibilities will be necessary to achieve a procurement plan which covers all options.

Organizing for startup in the purchasing department requires that people and procedures are ready to go. People have to be psychologically prepared for the frustrating period ahead and must also be available to provide the support services as needed. It may be necessary to put a buyer on the startup team. Procedures for emergency purchasing and transport services must be ready to go. Back-up suppliers who have special skills in the technology need to be readied in advance for quick consultation as required. It may be necessary to arrange for special purchasing delegations to members of the startup team, the opening of stores on a 24-hour around the clock basis and the telephoning of orders to specially prepared suppliers. It may also be necessary to prepare special identification tags for receiving and materials handling to prevent startup items from getting lost or delayed within the stream of regular goods and services being received within the organization.

During Startup

During startup the materials organization can do much to assist the startup team in achieving effectiveness. Willingness to roll with the punches, high flexibility and the ability to live through a variety of frustrations are important factors here. A typical example of what may happen during a startup follows.

In the early stages of startup of a technologically advanced machine a serious breakout hazard was disclosed, due to an apparent mould supply problem which occurred in over ten per cent of the hot steel strands. (A breakout occurs when hot liquid steel bursts through the thin shell, just as the slowly moving billet emerges from the bottom of the mould.) The cleanup is annoying, time consuming and delays the next cast. The managers and technical men began to look for the cause of the breakouts and discovered that the four inside walls of the copper mould had bulged inwards. The bulge was located at the metal meniscus; thus, a small billet was formed which was not supported by the 32-inch-long mould, further down below the meniscus. The next problem was to determine the cause of mould distortion.

Hardness tests disclosed that the copper in the original moulds was softer than the 68 Brinell Hardness Number which had been specified. The purchasing agent hastened to request from the supplier replacement moulds which possessed the specified hardness. It was discovered that the supplier had to change his cold drawing practice and the purchasing department ensured that this change would be instituted. Unfortunately, the replacement moulds of specified hardness did not halt the breakouts. The moulds still bulged inwards. The purchasing agent was then required to procure moulds fabricated of welded tube from the same supplier. After a hurried series of specifying and expediting efforts, the welded tube moulds arrived and gave the desired dimensional stability. They did not solve the distortion or breakout problem, however. After extensive additional experimentation and time, the purchasing people involved were advised that bacteria in the water caused slime which was at the root of the mould failure. This in turn required the acquisition of water treatment equipment and chemicals.

During startup the quality of the procurement planning will be severely tested. Are the systems working well, have the eventualities and possibilities been adequately forecast? Has the selection of the people recognized the heavy psychological pressures of continuing and draining negative feedback as option after option fails to produce improved results? It takes a special breed of person to be able to live through such a high pressure period and continue to be at peak effectiveness throughout.

That not all problems are great technological obstacles is illustrated in the following incident.

Several years ago each of the initial attempts at casting steel in the new continuous steel casting machine ended with the steel sticking and freezing in the mould. Low levels in the rapeseed oil tank at each attempt could not provide the necessary lubrication between mould surface and the hot steel billet. Refills at the end of the day could not prevent the tank from emptying overnight and causing a freeze-up in the mould again the next day. At each occurrence, the plant manager's demand for additional oil within hours caused frantic activity in the purchasing department. When the plant manager finally discovered theft of the oil during the nigh., for resale as cooking oil, his demand for ten gallons of laxative by noon hour,

Table 1   Summary of startup research findings and purchasing implications

Research findings	Purchasing implications
1. Startups take longer than expected	Information about startup expectations is vital Supplier assistance is a separate contract Supplies and raw material contracts must be geared to realistic expectations During startup there is a high premium on avoiding delays It may be a frustrating period for procurement personnel
2. Startups cost more than expected	There may be a trade-off between purchase price and final started up cost Early purchasing involvement is highly advisable Turnkey contract at fixed price for guaranteed volume may help minimize purchaser's risk A large part of increased cost may be lost capacity and lost potential revenues as well as carrying costs, compared to the out-of-pocket costs purchasing is normally involved in. Purchasing personnel may resent the 'high spending' at a time of cost over-run
3. Startups follow a learning curve pattern	Information about prior startups may be used for a first approximation

in addition to more rapeseed oil, pressed the purchasing department's ingenuity to the limit. This laxative addition to the rapeseed oil decisively ended the recurring shortages. Although meeting such exigencies may be an unavoidable part of purchasing activities during startup, other, more valuable functions can also be performed, based on recent research knowledge, and these should both speed startup and reduce panic purchasing.

Conclusion

A major startup is a difficult period for everyone in the organization. This research bears little good news in that regard. Table 1 attempts to summarize the major findings and possible implications for purchasing and materials management. It is clear that appropriate preparation and planning by the purchasing and materials group can be of valuable assistance.

Research findings	Purchasing implications
	Procurement group itself can affect the learning curve
	Advance procurement planning regarding supplier back-up, systems and procedures can be valuable
	Different suppliers may show different learning curves
	Contracts can be drawn recognizing the learning curve phenomenon
4. Startups are predictable as to total length and rate of improvement	Actual operating data from first few periods can be used to predict total startup
	Contracts for materials, supplies and services can be readjusted to reflect new data
	Payment for installation could be geared to rate of improvement
	Startup services can be requested for total length of startup
5. Startups take even longer when they involve a major technological advance	Re-examination whether the technological advance is worth the extra startup time required
	All contracts and preparation should reflect the possibility of extra long startup
	Very special supplier assistance may be required
	Anything that can be done to minimize the nature of the technological advance is likely to result in shorter startup

References

1. Henderson, R., 1974 *Plant Startup Productivity*. (Unpublished PhD Dissertation, School of Business Administration, University of Western Ontario, London, Ontario)
2. Henderson, R., 1975 *Plant Startup Productivity: Measuring and Predicting Progress in Continuous Steel Casting Machines*, (limited ed.), University of Western Ontario, London, Ontario
3. Henderson, R., 1971 Improving the performance of capital project planning. *Cost and Management* 45 (5) 33–41
4. Bright, J.R., 1958 *Automation and Management*, Harvard University, Boston
5. Wright, T.P., 1936 Factors affecting the cost of airplanes. *Journal of Aeronautical Sciences*, 3: 122–8
6. Alchian, A.A., 1950 *Reliability of Progress Curves in Airframe Production*, (RM 260–1) The Rand Corporation, Santa Monica, California
7. Andress, F.J., 1954 The learning curve as a production tool. *Harvard Business Review* January–February 87–97

8. Conway, R.W., Schultz, A., Jr, 1959 The manufacturing progress function. *The Journal of Industrial Engineering* X: 5–10

9. Hirschmann, W.B., 1964 Profit from the learning curve. *Harvard Business Review*, XLII: 125–39

10. Baloff, N., 1966 Startups in machine intensive production systems. *The Journal of Industrial Engineering* XVII: 25–32

11. Baloff, N., 1966 The learning curve – some controversial issues. *Journal of Industrial Economics* XIV (3): 275–83

12. Baloff, N., 1963 Manufacturing startup: A model. (Unpublished Doctoral dissertation), Stanford University, Stanford, California

13. Henderson, R., 1978 Measurement of productivity growth during plant startup. *IEEE Transactions on Engineering Management* EM-25 No. 1, pp. 2–8

14. Wearne, S.H., Raine, H.P., June 1976 *Management of the Commissioning of Industrial Projects*, Draft Report TMR 8, Bradford, School of Technological Management, University of Bradford

15. Niland, P., (1961) *Management Problems in the Acquisition of Special Automatic Equipment*. Division of Research, Graduate School of Business Administration, Harvard University

16. Names, dates and some process descriptions used in examples in this reading have been changed to protect confidentiality

5.3

Capital Budgeting as a General Management Problem

Joseph L. Bower

In the introduction to their popular text, *The Capital Budgeting Decision*, Bierman and Smidt offer the following summary of 'a theoretically correct and easily applied approach to decisions involving benefits and outlays through time, that is, capital budgeting decisions'.

> Essentially, the procedure consists of a choice of a rate of discount representing the time value of money, and the application of this rate of discount to future cash flows to compute their new present values. The sum of all the present values associated with an investment (including immediate outlays) is the net present value of the investment. Those investments with the highest present value should be chosen.[1]

The main argument of this reading is that the procedure summarized is indeed a theoretically correct approach to a class of decisions, but that the problem today's large corporations call 'capital budgeting' has very little to do with that class of decisions. In fact, the set of problems corporations refer to as capital budgeting are general management problems. They involve those strategic moves which direct an organization's critical resources toward perceived opportunities in a changing environment. The processes by which resources are committed in turn involve (1) intellectual activities of perception, analysis and choice which are often subsumed under the rubric 'decision making'; (2) the social process of implementing formulated policies by means of organizational structure, systems of measurement and allocation and systems for reward and punishment; and finally (3) the dynamic process of revising policy as changes in organizational resources and the environment change the context of the original policy problem. Management of these processes is a task for general management rather than financial specialists. The importance of this

Source: Reprinted by permission from J.L. Bower, *Managing the Resource Allocation Process*, Harvard Business School, Division of Research, Boston, MA, 1970.

distinction is that recognition of the conceptual form of the problem facing management provides a basis for some tentative new approaches to the problem and a guide to research.

This reading examines the problem of capital budgeting as it arises in its corporate setting. The problem is first considered as phrased and 'solved' by financial theorists, and their solution is evaluated. Then, the results of recent field research are used to rephrase the problem in a new conceptual scheme. Finally this new conceptualization is used to develop the framework and design for the research reported in the complete book.

The Corporate Setting

Today's large corporation is a multidivisional, multiproduct and often multinational giant. Each of these companies annually generates something like 10 per cent of its sales as depreciation and retained profit for use internally and most of this sum is invested in new plant and equipment. Occasionally, the company has recourse to financial markets for additional funds. The implication in terms of a capital budgeting problem is that the average billion dollar corporation spends large sums on its capital programme each year. The attempt to spend such vast sums intelligently occupies a good deal of top management time.

The scope of the problem is relevant because clearly no one manager can be assumed to have the knowledge or the time to generate the detailed programmes to use these funds. And, of course, all large corporations have complex organizations to administer the spending as well as the acquiring of capital funds. The men 'down in the organization', three, four or even seven or eight levels below the corporate president or his appropriations committee are the ones who are involved in creating capital spending programmes.

Consider this description of the capital budgeting process provided by a 'large company' division vice president with general management responsibility for a business. He sits six levels below the chief executive of his company.

> My product department managers carry the principal burden of planning capital requests: we start out with an assumption that there is no limit on capital. This means I am assuming that my managers can develop good projects that meet corporate and division earnings and profit criteria.
>
> Marketing really starts it all off. They plan where we're going to be in the next five and ten years. It's done once a year, but they supposedly keep their numbers up to date. The plans include unit sales, prices, R & D requirements, overhead dollars and people.
>
> We compare the capacity requirements in these plans with existing facilities and the outcome is a requirement for product capacity. The product department managers have to worry about total product requirements – they have to adjust for the peculiar biases of the

individual marketing people. They have to take into account an unwillingness of marketing to fully appreciate the impact on future sales of major cost reductions reflected in price.

The department manager then asks the strategic question: 'Is the top 25 per cent of the business worth more than the bottom 40 per cent?' When he's decided, he pulls together the pieces of information to tell his story to justify the funds he needs.

My job is to correlate each of these pieces into a whole and make sure that product operations and marketing are taking each other into account....

If a project doesn't meet criteria, justification is going to be more difficult. There will always be a segment of business you'll still want to be in, regardless of financial criteria. In other cases, you get together the people involved to make the decision to get out of a business. I think that's my responsibility.

In general, however, I told my group that they should not worry about the approval of their projects. I can sell anything that our groups believe makes good business sense.

The corporation as far as we are concerned is the banker upstairs. They want to know our plans and profits; they provide services and we leave them alone.

Add to this picture the comments of a functional department manager seven levels down with responsibility for putting together five or six multimillion dollar requests each year.

My job is about 20 per cent engineering and the rest operations. Basically, I view my job as spending capital money wisely.... First I try to determine what are the company's needs. What is required in the way of a new facility in terms of cost reduction and in terms of expansion. Ideas are always coming in. Second, I try to train my people. I have 14 engineers and I get them right out of school.... I have to educate my men as to the labyrinth of the plant – really, how to get things done. Every man I have is working on a project and he is spending more than $100,000....

The other side of the capital spending story is getting the justification story together.... We have about 25 jobs going at the moment, ranging from $10,000 to $10,000,000.

Finally, complete the picture with the words of a corporate officer, one of the principals of his company's appropriations committee.

When something gets to the appropriations committee you have one final review prior to the submission to the board of directors. That's all it is really.

Now, I don't think we will get away from the desire of people to sell a project, but we are getting more objective. We are getting the group management to make the review and they have the ability to examine a

project in depth. You know, you are kidding yourself to think that
people on the appropriations committee have the time to read one of
these project proposals thoroughly.

The picture that emerges should be familiar both to corporate managers
and to the readers of a growing body of field research studies. Planning for
capital spending is a process which begins with the operating managers of a
business. They are the ones who define the needs of their part of the
corporation, who make the sales forecasts which justify new capacity, who
review technology to determine what the appropriate design should be,
who evaluate the economics of a strategy and draft requests for capital
funds and, finally, who supervise the design and construction or purchase
of a new plant facility and its equipment. Moreover, the same managers or
their successors are the ones who implement the business plan on which the
capital proposal is based.

The capital proposal which is 'submitted for approval' to an appropriations committee or a board of directors has already been submitted for
approval to numerous levels of division and corporate management.
Although the project definition may not have changed at all from the
conception of its originator 'down in the organization', the request
justifying the project has been screened, revised and politically disinfected
so that it now tells an attractive story in professional tones. Few reserva-
tions and little unguarded optimism remain, although both may have
characterized the underlying analysis performed at lower levels.

It is unnecessary to belabour this point here. The question on which this
reading focuses is 'How does the picture sketched above differ from the
assumptions underlying traditional capital budgeting theory?' We now turn
to the answer to that question. It has two parts corresponding to two
critical assumptions of the theory. First, the theory assumes that for
corporate purposes a capital project can be usefully summarized by a
limited number of quantitative measures. Second, the theory assumes that
choice among project alternatives is the key step in the capital budgeting
process.

The Value of Quantitative Measures

Before beginning an analysis of quantitative measures, it is important to
recognize what is *not* being criticized here, at least directly, and that is the
portfolio model that is usually used to characterize the capital budgeting
problem in economic or financial theory. The portfolio model describes the
firm, or its top management, as having to select each year a portfolio of
capital investments from among many proposals developed by the departments or divisions of the firm and submitted for funding. The model is not
criticized here because in some abstract sense that is what the firm has to
do.

In the reality of the firm, however, selection among projects is an
integral and inseparable part of the strategic process as it is carried out by

the entire general management hierarchy of the corporation. What this means is that the theoretical characterization of a project as a financial security, and the theoretical focus on a discrete and identifiable set of choices made by top management, make for a descriptively inaccurate conceptual framework. The remainder of the argument considers why this structural inaccuracy is a serious weakness.

Traditional capital budgeting theory prescribes that a project be described as the net present value of all cash flows associated with a proposed investment. In fact, the theory is false on theoretical grounds because it ignores the essential effects of uncertainty. More precisely, it could only be correct in a narrow set of special cases. For present purposes it is more important that the theory be broken down on practical grounds for the same reason: management projections of cash flows from different projects are seldom comparable. The degree of uncertainty characterizing the numerical projection in an investment proposal varies with: (1) the kind of project; (2) the kind of business; and (3) the kind of manager. These differences are considered in turn.

The kind of project

As practising managers well know, estimates of the benefits from cost reductions are far more accurate and less variable than those from sales expansions, and these estimates in turn are more reliable than the estimates of the return from new products. A study in one company which compared the results of 50 projects with the forecasts in their respective 'requests for capital' showed that in the last class – new products – the mean of the ratio of actual to forecast net present value of results was 0.1 (as opposed to one mean of 1.0 if the average forecast was accurate). It was not that new products did not make money for the company. Rather, the average project did poorly and the forecasts were no indication of which ones would succeed. As shown in table 1 the mean of this ratio was closer to 1.0 for sales expansion than for new products, and closer still for cost reductions.

Table 1   Discounted actual results compared with the discounted forecast

Type of project	Mean of $\dfrac{\text{PV actual results}}{\text{PV forecasted results}}$
Cost reduction	1.1
Sales expansion	0.6
New products	0.1

The same phenomenon was described succinctly, if wryly, by a manager in another company. 'We're making five per cent on all those 35 per cent projects.'

Other studies indicate that the source of error in estimates of project outcomes is usually the forecast of unit volume or of price. Relative to the uncertainties of the market, investment and operating costs are well-understood. It is also true that managements can and do take steps to bring operating cost into line, and that still another source of error – deficiencies in product quality – shows up in sales volume and/or price.

The kind of business

The objections to a quantitative summary extend beyond the kind of project to projects within a class, for it is certainly true that different businesses involve different levels of risk. The point is obvious in extreme but also in subtler instances. For example, providing fuel on long-term contract for the most modern facility of major metal producer involves risks different from those related to the provision of fuel on contract to a government agency whose funding is subject to cancellation. In the former case, the risk is slight as long as it is reasonable to assume that the newest facility will be the last to shut down. In contrast, many government agencies turn out to be very fickle customers. Thus, while classifying projects helps, it only crudely approximates true variations in risk.

The kind of manager

Finally, the comparability of quantitative measures breaks down most often because the integrity of a set of projections depends on the individuals who provide them and the manager who puts them together. Managers are a highly variable commodity. Two aspects of this phenomenon are evident in the comments of two managers, one the board chairman of a billion dollar concern; the other a division vice president in the same company.

(1) You can't be systematic about people. They're just too variable. For example, some groups in the corporation are such that you just know their numbers are optimistic. In others, you know that they are pessimistic and that if a man says that something is going to be so good, then you know it's probably going to be a good deal better than that. So you have to go behind the numbers and look at the men, at their businesses and at the history of both.

(2) You don't plan on the basis of sales forecasts. Christmas, they're not worth a damn. And any manager worth having can produce numbers that will make a project look good.

What we do is get the people who know about a business together. We try to work out what the nature of the business is, and whether we belong in it given our resources and the other things we can do, and what stance we are going to take. Where we have decided to go into a market, then we just build as much as we can. It's my job to get the dollars from management.

The point is not that most managers are devious and narrowly self-interested. Quite to the contrary. What is true is that for reasons of temperament or design the numerical estimates prepared by a manager reflect his attitude as he tries to fulfil what he sees to be a key requirement of his job: convincing his superiors to fund whatever series of programmes he thinks is best for the company in the particular area of responsibility delegated to him. And, in fact, that is precisely the job his 'bankers' have given him, which leads to the second major point – the theory's emphasis on the act of choice by top management.

The Role of Top Management Choice

The very size and complexity of today's large corporation have necessitated breaking up the job of planning *and of choosing*. While the board of directors of a corporation or a corporate appropriations committee typically reserves to itself the right to make large capital expenditure decisions, a host of critical decisions are made long before a request for funds reaches the board. Among the most important of these decisions are the choice of sales forecasts, the definition of a facility scope and design and the decision to submit a request. The first, as we have seen, critically influences the attractiveness of a proposal; the second determines a project's actual chances of success and the third reflects the decision of a subunit manager to sponsor – to associate his reputation as a manager – with increased investment in a line of activity. That boards and appropriations committees in most companies recognize the extent to which they have delegated their capital appropriating powers is evidenced by the very low rate of project rejections which characterizes their actions. The word 'delegated' is emphasized because it should be noted that delegation is *not* abdication. When boards make a practice of accepting subordinate or subunit judgements, they do it because they are convinced of the qualifications of their subordinates.

An extreme example of what can happen when a top management tries to keep these powers to itself is provided in the case reported by an assistant controller of a divisionalized company with highly centralized control.

Our top management likes to make all the major capital decisions. They think they do, but I've just seen one case where a division beat them.

I received for editing a capital request from the division for a large chimney. I couldn't see what anyone could do with just a chimney, so I flew out for a visit. They've built and equipped a whole plant on plant expense orders. The chimney is the only indivisible item that exceeded the $50,000 limit we put on the expense orders.

Apparently they learned informally that a new plant wouldn't be favourably received, and since they thought the business needed it, and the return would justify it, they built the damn thing. I don't know exactly what I'm going to say.

Notice that the case above is not the self-interested response of a sub-unit manager to an inflexible measurement system. Examples of this latter kind of obstreperousness are increasingly plentiful in the literature of decentralized control and transfer pricing. Rather, subunit management was in this instance responding to what they perceived as corporate interference with their attempt to pursue corporate objectives delegated to them. The response was parochial but not self-serving in any base sense.

The notion that the decisions of subordinates are crucial to the choices presented to superiors, that indeed these subordinate decisions often may constitute the true shapers and initiators of corporate commitment, once stated is obvious. Yet the widespread evidence of this phenomenon is just making its appearance in the literature of capital budgeting and the import of this evidence has not yet been incorporated in new analytical concepts.

Steps in the Investment Process

A discrepancy

It is now simple to describe the origin of an investment project. The first step in a long process which ends with the expenditure of capital funds, begins when the *routine* demands of a facility-oriented job indicate the need for a new facility – when there is a discrepancy between 'what the company wants of me' and the status of existing investment in facilities.

There are three basic versions of this event. In each case, feedback from the measurement system indicates that performance in the job is, or will soon be, inconsistent with the demands of the job.

(1) In the first case feedback indicates that 'costs are too high'. There may be several reasons, although the primary one is often the secular rise in labour and materials cost. The source of feedback is either accounting data that indicate a rise in cost above standard, or the sales force that reports a decline in margin as a result of falling prices. In either case, a need is indicated for research and development which will define a route to cost reduction, and investment is almost inevitably a second step.

An important alternate formulation of this first case is, 'costs may be lowered'. Here the feedback is from engineering or production process specialists whose research and analysis indicate that there is a potential route to lower costs available through major investment.

(2) In the second case feedback indicates that 'quality is inadequate', in the sense that it is not adequate for the market. Normal quality control will

usually assure that quality is meeting specifications, but competition frequently will move ahead of specifications, or specifications will not meet the needs of the customer. The sales force will typically make unambiguous reports if this is the case. On the other hand, an important variant of feedback concerning quality is price. It will often be true that in order to obtain desired unit volume the sales force will have to lower price. While price competition among homogeneous products is the standard reason offered, it is occasionally true that marketing has had to defend the market for a product of inferior quality with price cuts. In this second instance, it may be difficult for management – particularly facility-oriented management – to learn that quality is the true source of the price problem. Facility-oriented managers are not predisposed to admit that they have not done their job well. But when intelligence concerning technology is poor, a report from the division research activity that evaluates company technology favourably may reinforce the facility-oriented managers' view by indicating that 'you can't get better quality for those costs'. In this last instance it is possible for general management to conclude erroneously that their competition has lower profit criteria, a serious strategic mistake.

Again there is an important variation of this case that has its source in research and development work. Reports from technical management may indicate that changes in product or process would improve product quality and hence a facility investment is suggested.

(3) In the third and final case feedback indicates that 'sales exceed capacity' or 'the sales forecast for year X exceeds capacity'. In both instances, investment for expansion of facility capacity is indicated. The message here ought to be unequivocal, particularly in the first phrasing, but in fact both the definition and timing of a 'need for capacity expansion' are critical. This issue will be re-examined later.

It is more important at this stage of the exposition to understand that there is a variant in the third case of special importance: the new product and its near relative the new business. While from a corporate or an external point of view there may be something very special about new products and new businesses, where *investment* as we have defined it above is the result, there is nothing radically different about them from the vantage of a facility-oriented manager's job, except perhaps that they involve more risk. New products or new businesses require new facilities.

Our subject is not the origin of new businesses, but it is worth noting here that when new businesses are entered by internal expansion, they often begin with minor new product extensions of existing businesses, or expanded exploitation of the technology of existing businesses. Reorganization that formally recognizes the new business, typically follows after the first products are introduced to the market. This observation and the literature of new product development both suggest that from the point of view of the investment process, there is nothing fundamentally different about new products.

It is certainly true that in the new product case, there may be no facility-oriented manager to respond to the discrepancy between sales forecast and existing facilities. No-one may happen to have that job and it 'falls between the chairs'. However, the work of Jay Lorsch on new product introductions indicates that organizations that successfully manage the transition from product development to production do so in part by formally creating a facility-oriented job concerned with production of products which fall outside the scope of existing job definitions.[2]

Definition

The preceding discussion is summarized in table 2. Need for a facility is defined by a discrepancy between the information available to a facility-oriented manager and the capabilities of the facilities for which he is responsible measured along the critical dimensions of costs, quality and capacity. The sources for this information are the reports of other specialized activities of the company.

A facility is defined in terms of new fixed investment, a time of availability, a capacity, a product scope of specified breadth and quality and a specified level of product cost. One of the issues that must be examined is what happens to this definition over time as the investment process develops. Given the description thus far, the sources of definition are to be found entirely within the bounds of a discrepancy perceived by a facility-oriented manager. If it is believed that other issues should enter

Table 2   Sources of facility definition

	Discrepancy	Evidence	Source
Costs	too high	input costs have risen	accounting data
		prices have fallen	marketing
	may be lowered		production or engineering
Quality	inadequate	competitive improvement	marketing
		customer need	marketing
		prices have fallen	marketing
	may be improved		engineering
Capacity	insufficient	sales exceed capacity	marketing
		forecasts exceed capacity	marketing engineering development
		new product	marketing general management

into the facility *definition*, such as the product or facility plans of other parts of the company, labour policy, location policy or company financial policy, explicit steps have to be taken to introduce them. Either other managers concerned with these issues must intervene in the process of definition or the framework of the original definer's perception must be changed. Unless one of these steps is taken the original perception of a discrepancy will constitute the sole conceptual framework shaping the final investment project. The corporation has only the choice of accepting or rejecting the project. In fact, once a project emerges from the initial stages of definition it is not only hard to change it, but in some cases hard to reject it. Too much time has been invested, too many organizational stakes get committed and at very high levels of management too little substantive expertise exists to justify second guessing the proposers. A classic and tragic example of the process is available in Schlesinger's description of the Bay of Pigs invasion.[3]

The forces influencing definition

The argument above has one very important implication which needs development. It has been stated that change in the framework of the definer's perception is necessary in order to change his definition of an investment project. While this may seem an empty sort of hope, it only requires rearranging the phrasing of the propositions above to make clear that there are levers which management has at its disposal to influence the definition process. It has been stated that the organization, measures and rewards – what we will call the corporate structure – indicate to a manager what 'the corporation wants of me', and hence play a critical role in shaping the decision rules a manager uses to organize the demands of his job. If management wants to change those decision rules, it only has to change the structure within which a manager works.

Stated most crudely, the definition of a manager's job, the way the performance of his area of the business is measured, the information available to him for purposes of managing his area and the way his own performance is measured and rewarded, define the 'rules of the game'.

The structure of the corporation shapes the way a subunit manager perceives the rules of the game. The rules in turn affect the way he defines deficiencies in the performance of his job. And these deficiencies in turn are the origin and source of definition of investment projects.

Impetus

After a project's initial definition, its progress varies substantially depending upon the organization and procedures of the division in question. In every case the project receives study that in effect constitutes a continuation of the definition process. Engineers make a variety of estimates, forecasts are extracted from marketing activities and eventually a new facility is designed. There is no set period of time for a study. It can take as little as six months or go on indefinitely. Most important of all, it need not involve any managers at levels higher than that at which the project is defined. A project can be defined by a plant manager or an operations manager and never move toward funding.

In fact the rate of progress of a project up the hierarchy of management through various stages of approval to final authorization depends on the impetus given the project. At National Products every capital project of over $250,000 requires the approval of the corporation's executive committee and board of directors. Each project is submitted to the executive committee by the general manager of a division after it has the approval of the group management committee of which he is a member. Projects that have the approval of a division general manager are seldom turned down by his group – although minor modifications are frequently requested – and projects reaching the executive committee are almost never rejected.

The problem facing the definer is, therefore, quite clear. It is also well-understood as a matter of practice. The definer must get his project approved by his division general manager. Sometimes this requires the general manager's personal approval, in other divisions that of his executive vice president and in still others that of a vice president with general management responsibilities. In all cases the source of a project's impetus is the power of a general manager at the officer level of the division. The general manager in question must sponsor the project and shepherd it successfully through the rigours of whatever screening the division in question imposes.

The forces influencing the provision of impetus

Whether impetus is forthcoming depends on two kinds of evaluations made by the general manager in question. The first is his evaluation of the quality of the project. In particular, the general manager must believe the sales forecasts justifying the income projected, otherwise a project is dead. The general manager must also accept the technical aspects of a project, but typically he expects his subordinates to be technically qualified, either personally or in the support obtained during the definition process. It is also at this point that the general manager, or several division officers, may enter and influence the definition process. The extent of influence will depend importantly on the timing of the intervention.

It should be recognized that in emphasizing the market judgement, the general manager is exercising the responsibility delegated to him by making a judgement on the basis of his local expert competence.

The second kind of evaluation the general manager makes is an estimate of the benefits of being 'right' and the costs of being 'wrong'. This estimate depends upon his estimate of what the corporation wants of him in terms of his job history, the history of the product, concurrent projects which he must sponsor and the nature and extent of his responsibility – precisely the same factors taken into account by the definer. Again the manager's perception of these factors is shaped by the structure of the organization, the way in which his job is defined, how the performance of his business is measured, the kinds of information available to him and the way in which he is evaluated and rewarded.

The manner in which these forces operate was made clear in the comments of a division product area manager interviewed during phase 1 of the study. While other managers reached different conclusions depending upon their personal attributes and the structure of their division, they took into account the same relationships among the same factors.

> What it really comes down to is your batting average. Obviously anything cooked up, I have to sell and approve. My contribution is more in the area of deciding how much confidence we have in things. The whole thing – the size, the sales estimate, the return, is based on judgement. I can kill or expand a product based on my judgement. I decide the degree of optimism incorporated into the estimates. You know your numbers change depending on how you feel. The key question is 'How much confidence has the management built up over the years in my judgement?' A guy in my position must think this way. He loses his usefulness when he loses the confidence of higher executives in the company. Otherwise his ideas won't be accepted when he goes up.
>
> I can lose that confidence by being too optimistic or too pessimistic. I can lose the confidence of the people above, and I can lose the confidence of the people below that they and the degree of confidence elicited from them are not going to be ill used.

The point here is the same as that made above. The importance of the role that structure plays lies principally in the fact that structural variables may be manipulated by top management. By the way in which the careers of general managers are advanced or retarded, top management can make very clear its attitude toward the quality of judgement exercised by division general management. When managers who have taken the corporation into exciting new but unprofitable fields are promoted, it is clear that 'creativity and imagination' are important bases for reward. If managers of high return but only moderately growing businesses are passed by, then the importance of return on equity as a goal will be lessened in the eyes of the general managers.

In short, a general manager sponsors a project when he believes it will be in his interest to do so rather than not to do so, given his understanding of 'the rules of the game'. If he does not choose to sponsor a project there are a host of devices at his disposal for saying 'no' including, as common examples: (1) putting higher priority on other activities of the definer and thereby influencing the structure within which the definer works so that he drops the project; (2) asking questions that require further study and thereby are tantamount to indefinite postponement of the project; (3) suggesting directly that 'the timing isn't right', or (4) ignoring the request for review – the pocket veto. Thus, the way in which the general manager in question responds to the structure within which he operates, determines which projects see the light of day and when.

The capital appropriations request

The 'communication' that the corporate management receives is a memorandum from the general manager of a division, in the more or less standardized form of a capital appropriations request (CAR in National Products terminology).

Because a project is defined at two or three levels below the office of the division general manager, it is often a product area manager who provides impetus for the project. There is inevitably a period of negotiation while the ideas of the area manager and his subordinates are translated into a set of 'words' that are acceptable to the division general manager – 'something I can sign and defend', as one general manager put it. The division control manager, or his office, often plays a key role in this process: (1) acting as an intermediary between the general manager and his officers ironing out details of substantive disagreement, (2) introducing a substantial element of standardization and continuity in the format and substance of division capital requests and (3) serving an editorial function screening the prose and analysis, and polishing wherever it appears necessary.

In all divisions of National Products, the CAR is in draft form prior to its submission to division officers. Depending on the quality of the reasoning underlying the project and the style of the CAR prose, there may be as many as six substantial redraftings of a CAR prior to its submission to the group officers, or none at all.

The purpose and effect of the negotiation and rewriting of CARs depend entirely on the pattern and interrelationship of the definition and impetus processes. Where they are closely intertwined, CAR writing is largely an exercise in logical analysis and English composition. Division management has made its thoughts known prior to drafting. The business decisions reflected in the proposal are untouched. Where, however, the definition process is nearly completed prior to the provision of impetus, negotiation can be prolonged and painful. In effect, revisions in the drafting language constitute modifications in business decisions.

Approving projects – The group and the corporation

Once approved by a division general manager, a CAR is routed to the group management committee of which the division general manager is a member. There it is discussed, sometimes thoroughly, sometimes cursorily depending on whether issues raised by the project arouse committee interest. Typically, the most intensive management committee discussion is devoted to major projects demanding a good deal of capital for low-return businesses and to large investments in new business ventures. The extent and quality of discussion also vary with the degree to which the committee has become familiar with the strategic issues of the business in question. Discussion is on a project by project basis. On any given day a management committee may have four to ten major projects on its agenda.

As noted above, most projects submitted for group management committee approval received this approval with only minor changes requested. They were then redirected (the cover letter is changed) to the executive committee. When a project was turned down at the group level, it was generally only postponed for a week until a question or two were answered. In all cases, a division general manager had the right to submit a project to the executive committee without group approval. (No instances of this event were cited, but division officers described the right as 'important'.)

The executive committee (Excom) in its turn reviewed all capital projects before they were sent to the board of directors of the corporation. In fact the committee functioned as an internal board. No instances were found or remembered where a project approved by the Excom was turned down. It is not surprising: the Excom was made up of National Products' president; executive vice presidents; group vice presidents; the vice president, finance; and the secretary. They represented a very substantial aggregation of knowledge and experience. When they chose to discuss a project in depth they were penetrating. Usually they chose to accept most projects with only brief discussion. Comment was reserved for controversial proposals. While the Excom seldom turned down a project, critical discussion of controversial projects sometimes established the criteria by which the business would be judged in the future. Such comments as the Excom chose to transmit to the divisions were regarded by the divisions as important indicators of corporate attitudes toward the conduct of division business.

In short, both the group management committees and the executive committee provided a review of all CARs. The review varied in thoroughness depending in large measure on the extent of the projects' controversialism, but always the result of review was a 'go' or 'no go' response. The definition of a project did not change. In fact, the response was typically 'go' and, as noted earlier, the last level at which projects were turned down with any frequency was the division.

References

1. Bierman, H., Jr., Smidt, S., 1966 *The Capital Budgeting Decision*, 2nd Edition. Macmillan, New York
2. Lorsch, J., 1966 *Product Innovation and Organization*. Macmillan, New York
3. Schlesinger, A.M., Jr., 1965 *A Thousand Days*. Fawcett Publications, Greenwich, Connecticut

5.4

Managing as if Tomorrow Mattered

Robert H. Hayes and David A. Garvin

A Decline in Reinvestment

Raw data on recent capital spending and R&D investment by the private sector [in the USA] tell a tale of modest but steady increase – until, that is, one adjusts for inflation or for changes in GNP and the size of the work force. Then the figures tell a different story. Although gross business investment as a percentage of GNP has remained roughly constant in real terms since the 1950s, the capital invested per labour hour and the share of GNP devoted to net new investment have both declined over the last decade.

Between 1948 and 1973, for example, the ratio of the net book value of capital equipment to the number of labour hours worked grew at about 3 per cent per year. Since then it has increased at only about one-half that rate. Moreover, the growth in the ratio of net capital stock to the number of full-time equivalent employed workers – a figure that adjusts for changes in the hours per week worked by the average employee – reveals an even greater post-1973 decline.

Spending on R&D presents an equally disturbing picture. Viewed as a percentage of GNP, total US investment in R&D fell steadily between 1967 and 1978. In basic research, which involves longer time horizons, the picture is even bleaker. Measured in constant dollars, investment peaked in the late 1960s, then dipped and did not regain its earlier level until 1978. As a percentage of GNP, corporate spending on basic research is today only two-thirds of what it was in the mid-1960s.

American managers are also under-investing in human resource development, especially in critical industrial skills. The average age of experienced tool and die makers, for example, is approaching 50 years. If present trends continue within the next decade this vital reservoir of skills – necessary in a variety of industries and already in short supply – threatens to dry up. Similarly, the Department of Labor estimates an annual demand

Source: from *Harvard Business Review*, May/June, 1982. ©President and Fellows of Harvard College.

for 22,000 skilled machinists during the 1980s, yet only about 2,800 graduate each year from various apprenticeship programmes. Much the same is true for skilled assemblers, forging-machine operators and optical workers.

Taken together, this evidence suggests that business spending on many crucial activities has been lagging badly in recent years. What lies behind this dangerous slowdown in long-term investment?

Searching for answers

There is no shortage of popular explanations. Most fall into one of three categories:

(1)Managerial theories, which blame business itself for the emphasis on near-term profitability that now dominates managerial decision making. Observers attribute this myopia to a variety of causes: the shift to multidivisional organizations, which typically use short-term financial measures as the primary means for evaluating managerial performance; the desire of younger managers for rapid advancement, which tends to limit the time a person spends at any one job; and pressure from the financial community.

(2)Environmental theories, which cite as culprits inflation, high income and capital gains taxes, rising energy prices, constrictive federal regulations, erratic shifts in public policy and other features of the business environment.

(3)Financial theories, which point to the recent increase in mergers and acquisitions as being responsible for the decline in direct investment. According to this view, managers are simply responding rationally to current economic conditions when they purchase inexpensive used assets rather than invest in more expensive – and risky – new assets.

Some of these theories are more persuasive than others. That American managers tend to be more concerned with short-term financial performance than their German and Japanese counterparts is, for example, now widely recognized. So, too, is the effect on investment of dramatic changes in the economic environment, although other developed countries have experienced similar shifts during the past decade without a corresponding slackening of investment. Increases in merger activity and in the funds devoted to corporate acquisitions, however, do not explain this decline in capital spending.

Corporate acquisitions are neither a substitute for direct investment nor a cheap way of acquiring plant and equipment. The bargain-basement character of such activity is an illusion resulting from attention to the wrong set of figures. Typically, analysts cite data that compare the market value of US companies (as measured by the sum of their outstanding debt and equity) with their replacement value to justify the claim that assets can be obtained more cheaply by acquisition than by direct investment.

Government figures do indeed show that market value, measured in this way, has been well below replacement cost in each year since 1972. Most mergers, however, involve acquisition prices substantially above supposed market value because some premium is generally required to entice managers and shareholders to approve the sale. Should a bidding war develop, the acquisition price can escalate dramatically – often to more than double the company's market value.

Nor can the problem be attributed to a lack of capital caused by inflation, reduced profitability, higher taxes or government-mandated nonproductive expenditures. During the past ten years, the inflation-adjusted aftertax return on equity for US corporations has roughly equalled its level in the 1950s. The ratio of shareholder dividends to total corporate operating cash flow, however, was 11 per cent higher in the late 1970s than in the late 1960s – and 30 per cent higher in 1980. The ratio of investment in new capital equipment to corporate cash flow, on the other hand, has generally declined since the 1950s. The problem is not that US business lacks the money to spend; it is simply not spending the money it has in the same ways that it used to.

Is this behaviour evidence of a foolish, but unintentional, mistake on the part of American managers? Not necessarily. They appear to believe completely in the legitimacy of their investment decisions and in the techniques on which they are based. These methods, however, have profound conceptual weaknesses that are not always recognized, and the answers they provide depend on managers' perceptions of the current and future economic environment. This combination of theoretical blind spots and economic misjudgements can often lead a company to short-change its future.

Discounting the Future

The theory is simple: a dollar received today is worth more than a dollar received tomorrow. How much more depends on the current uses to which the dollar can be put. If it can earn 5 per cent interest, a dollar today will be worth $1.05 after a year; if 10 per cent, $1.10. Conversely, at a 5 per cent interest rate, a dollar received a year from now is worth only $1 \div \$1.05$, or 95.2 cents today; at a rate of 10 per cent, 90.9 cents. This determination of a future dollar's present value is, according to accepted theory, the appropriate way to compare future benefits with present costs.

Extending the theory to capital investment is also simple: a company pays a certain amount of money to receive a series of returns stretching off into the future, each of which can be translated into an equivalent amount today. The difference between the amount invested and the sum of the discounted returns determines whether the proposed investment is more attractive than the best alternative use of the funds. Notice that this calculation requires several critical estimates: the size of the anticipated investment, the amount and timing of the resulting cash flows and the rate of return that could be realized if the capital project were not approved and the funds were directed elsewhere. This last figure is generally termed the company's opportunity rate or, more prosaically, its hurdle rate.

Today such calculations have, because of their apparent rationality, gained the upper hand in the evaluation of new investment proposals. Yet these techniques are as subject to misperceptions and biases in application as are other, less formal methods.

Skimping on reinvestment may take other forms than not replacing equipment as it depreciates in value. Managers can, for example, allow the productivity of existing equipment to deteriorate faster than normal by using it more hours per week than before or by replacing it with less productive, and usually less expensive, equipment as it wears out. Or, more subtly, they can replace it with machinery based on dated technology. Similarly, managers can allow spending on R & D, advertising and personnel development to fall below historical levels.

The Theory's Wobbly Legs

The task of most managers is to evaluate specific investment proposals by discounting the estimated cash flows (after taxes and depreciation) from a proposed investment using a hurdle rate that reflects the minimum acceptable return for proposals of that type. Should a given project not promise to generate profits (after depreciation) equal to those available from investments outside the business, it will be rejected.

According to discounting theory, then, a pattern of progressive disinvestment might make perfect sense. Discounting techniques, however, rest on rather arbitrary assumptions about profitability, asset deterioration and external investment opportunities. In fact, we believe that much of the decline in investment in capital stock is the result of misperceptions about the changes that have taken place in these three critical variables over the past decade. American managers have acted as though these variables have been moving in such a way as to make direct reinvestment in their existing businesses less and less desirable. But how have these variables actually behaved?

Cash-generating rate

Many managers are convinced that the ability of their companies to generate earnings is less today than in the past. They blame global competition, industry maturation and intrusive government for the decline; yet according to Burton Malkiel and other leading economists, the overall rate of return on equity for US companies, *after adjustment for inflation*, has remained roughly constant for about 30 years.[1,2] Only during the mid-1960s, when the rate of return rose to double its historical level of about 4.5 per cent, did this pattern change. Even then, the gains were short lived, and earnings soon fell back to their former levels.

Many executives, however, view the rates of return during the mid-1960s as the norm, rather than an aberration. By this standard, things have indeed worsened in recent years. But even though a company's profit margin may have dropped from, say, 10 per cent in 1965 to five per cent today, that 10 per cent figure is not a reasonable reference point for historical comparison.

Nor, in fact, is the five per cent figure always reliable. When managers attempt to net out the impact of inflation on the profitability of their businesses, they usually address only the asset side of the balance sheet. They reduce profits by the amount that inflation has increased the value of inventories and recalculate depreciation on the basis of the replacement cost, rather than the historical cost, of equipment. Rarely, however, do they acknowledge that their long-term debt also declines in value during an inflationary period. By ignoring the debit side of the balance sheet, many managers have overestimated – perhaps by as much as 50 per cent – the decline in profitability attributable to inflation.[3]

Deterioration rate

Even if the perception of a long-term decline in corporate profitability is illusory, there is good reason to believe that the rate of deterioration of capital stock has increased. Inflation is partly to blame, for the cost of many capital goods has risen faster than the prices of the products they make; consequently, it is more expensive to replace the fixed assets that a company employs in its business. Between 1970 and 1979, for example, the price of metal-forming machine tools almost tripled, but the price of all manufactured durables increased by little more than a factor of two. Such high prices often deter reinvestment or limit it to some fixed percentage of annual sales.

Also responsible, of course, are rising energy prices, which have so burdened operating costs that some production processes are no longer competitive. The rapid obsolescence of manufacturing equipment – whether because of high energy consumption, restrictive government regulations or declining efficiency compared with the newer process technologies – can appear prohibitively expensive to remedy. When replacement costs are in the stratosphere, the need to reinvest in capital stock can easily paralyse, not galvanize, a manager's willingness to reinvest in existing businesses.

Hurdle rate

Despite their perceptions of a decline in the profitability, and an increase in the rate of deterioration, of their companies' assets, American managers have not made a corresponding reduction in the hurdle rates they employ in capital budgeting.

These rates are typically quite high, often in the range of 25 per cent to 40 per cent, and there is some evidence that they have been rising over the past decade. A recent survey, for example, shows that about 25 per cent of American manufacturing companies require expenditures for modernization and replacement of equipment to pay off within three years. Ten years ago only 20 per cent had required that rapid a payoff. Shorter payback periods imply higher hurdle rates, just as higher hurdle rates imply a stronger emphasis on near-term benefits.

As with most of the arbitrary numbers that find their way into a company's systems and procedures, these hurdle rates are often used without question, even by executives who profess to be open minded. The chairman of a leading American equipment manufacturer recently described himself as an executive who encouraged his managers to take risks; at the same time, he insisted that all new investments produce a 25 per cent return during the first five years.

Such hurdle rates often bear little resemblance either to a company's real cost of capital (even after appropriate adjustment for differences in risk) or to the actual rates of return (net of deterioration replenishment) that the company can reasonably expect to earn from alternative investments. Again and again we have observed the use of pretax hurdle rates of 30 per cent or more in companies whose actual pretax returns on investment were less than 20 per cent.

How do managers normally defend this practice? First, they claim that an artificially high rate helps protect them from unforeseen reductions in cash throwoffs that are triggered by competitors' actions, unexpected inflationary increases in investment costs, and number fudging by subordinates anxious to have a project approved for personal reasons. Second, they argue that high hurdle rates increase motivation and that difficult-to-achieve targets tend to spur good performance.

As attractive as these explanations appear at first glance, their logic is faulty. Systematic adjustments for risk are quite appropriate when computing present values, but many of the hurdle rates that we have seen contain unreasonably high risk components. Moreover, using such rates as a motivational tool undermines their worth in evaluating investment opportunities. For one thing, they often discourage investment in existing businesses whose risks are known and direct it toward businesses whose risks are less understood.

Such behaviour also reflects a growing preference among managers for acquisitions over internal investments. Despite considerable evidence to the contrary, American managers appear to believe that aggressive acquisition programmes make possible both higher long-term growth rates and greater profitability. Many are so firmly convinced that the grass is greener in almost any industry other than their own that they are far less tough minded in evaluating acquisition candidates than they are in assessing internal investment proposals.

The key assumptions in this approach for analysing investment proposals – assumptions about rates of profitability, deterioration and acceptable return – are highly unreliable and prone to individual bias. Managers may have an accurate sense of their businesses' past profitability, but their belief about future profitability depends heavily on their basic optimism and confidence in the economy. They may know to several decimal points the average depreciation rate for their industry, but they are less likely to know the real deterioration rate of their companies' total capital stock. Even more uncertain is their assessment of the profit opportunities and deterioration rates in businesses other than their own.

Bitten, perhaps, by the merger bug and unwilling to adjust their inflated hurdle rates, many American managers have found reinvestment in existing businesses less and less desirable. Under siege in a changing world, they recall the economic Camelot of the 1960s and believe that it still exists somewhere, waiting to be found outside their corporate bunker. Had they placed less faith in the misleading objectivity of their discounting techniques, they might instead be spending their time and resources reinforcing their own bunker's walls.

The Theory's Blind Spots

Discounting methods are biased against investment in new capital stock in still other ways. Present-value comparisons are especially difficult to make if the projects under review have different lifetimes: when projects are of equivalent length, present-value calculations favour those with shorter payback periods; when projects are of unequal length, those with longer lives often appear more attractive than those with shorter lives. Few investments, however, are intended as 'doomsday projects' for which there is no successor. Managers usually assume that at the end of a current investment's lifetime, another, involving similar activities, will begin. Thus, unless corrected for, discounting's focus on the profitability of initial projects can lead to a series of absurd decisions.

Narrow use of the present-value criterion will, for example, almost inevitably argue for expanding facilities already in place rather than for building a new plant in a different location. The initial investment is normally much less, the returns more immediate. Over the long run, however, a series of such decisions – each backed by its own impeccable logic – can lead to ponderous, outmoded dinosaurs that are easy prey for the smaller, more modern and better-focused plants of competitors.

Consider the experience of one producer of large machinery that opened a simple assembly operation in the 1920s. As sales increased, the plant kept expanding both the size and number of its processes until today the company finds itself with a mammoth and uneconomical complex. Now it is trying to figure out how to break apart a manufacturing operation that appears to have grown like Topsy over the years, although each addition made sense at the time.

Another manufacturer, which recently focused attention on its home plant, discovered a collection of more than 40 multilevel buildings producing an extraordinary variety of low-demand items using equipment that dated back before the Second World War. Rather than undertake the immense task of modernizing this outmoded plant, whose condition was the result of a series of apparently rational investment decisions over a long period, the company reluctantly closed it down.

For similar reasons, the present-value criterion will often suggest delaying the replacement of a piece of equipment by another, more modern machine that performs roughly the same function. The economic benefit of delaying purchase for a year, say, is seldom offset by the efficiencies obtained from using the new machine on comparable activities. Less obvious benefits from increased worker skills and capabilities, new products and a different cost structure are harder to document in advance and so do not fit neatly into a present-value analysis. In fact, to counteract this bias against modernization, some companies are experimenting with a 'sunset law' for capital equipment, under which a piece of equipment is automatically replaced at the end of a predetermined period unless a special review process decides otherwise.

The Logic of Disinvestment

The threat implicit in discounting techniques is not limited to misperceptions, too short time horizons or a bias against major modernization projects. It extends to the very ability – and willingness – of managers to ward off the attacks of aggressive competitors.

Consider, for example, two companies that share the market in a price-sensitive industry. Initially, both use the same production processes and have similar cost structures. A new manufacturing process, however, promises to reduce variable costs significantly. Company A, with a high hurdle rate, rejects the investment out of hand as being insufficiently profitable; Company B, with a lower hurdle rate, decides to buy the new equipment.

Both companies perform similar discounting calculations to weigh the advantages of the proposed investment. They arrive at opposite conclusions because of the differences in the hurdle rates employed and in the importance placed on maintaining competitive vigour. In theory, both should be satisfied with the results.

Company B, once its new equipment is in place, quite naturally proceeds to compete aggressively for market share by lowering prices. Its new manufacturing process, after all, gives it much lower variable costs and requires high production volumes for maximum efficiency.

Can Company A respond? Its outdated equipment places it at a distinct competitive disadvantage. Moreover, its competitor's price reductions have so reduced the profitability of its existing business that the investment required to upgrade its facilities looks even less attractive than before. At the least, Company A will lose market share; at the worst, it could be driven out of the business entirely and, perhaps, be forced to use its remaining capital to acquire another business, one apparently better able to meet its high hurdle rate.

Many American companies today find themselves in a position much like that of Company A. The problem is not that reliance on discounting techniques inevitably leads to inaccurate results but rather that managers can all too easily hide behind the apparent rationality of such financial analyses while sidestepping the hard decisions necessary to keep their companies competitive.

One reason companies so often become trapped in this sort of disinvestment spiral – deferred investment leading to reduced profitability, which further reduces the incentive to invest – is that discounting techniques make the implicit assumption that investment processes are reversible. That is, if one sells an asset, one can always buy it back; or if one delays an investment, one can always make it at some later date with no penalty other than that implied by the company's discount rate.

No company, however, can be sure of recovering lost ground quite so easily. To regain its position, a company may have to spend a good deal more than if it had made the investment when first proposed. As time passes, downward spirals become much more difficult to arrest. Moreover, as the experience of both Ford and Chrysler attests, postponed investments (in downsizing in the mid-1970s) may not be reversible; complete recovery may be impossible.

This irreversibility is partly rooted in the dynamics of human organizations. Companies are collections not simply of tangible assets but of people as well, and the bonds among them reflect understandings and commitments developed over a long time. Such bonds need constant support and reinforcement; once they begin to dissolve, an organization loses its sense of movement and often falls prey to a sense of resignation. Morale sags, performance suffers and employees – generally the best ones – begin to leave. Faced with these circumstances, top management often concludes that a division or product line is unsalvageable and purposely continues the process of disinvestment.

Reversing the disinvestment spiral

It is both difficult and costly for a company to extract itself from such a spiral, for usually no single investment can repair the damage. Instead, simultaneous investments in several projects are often necessary to achieve an acceptable return. If managers evaluated each of these projects individually with no attention to the interactions among them, they might reject some as being insufficiently profitable. Unfortunately, the capital budgeting procedures that most companies follow today do not readily accommodate such interdependencies (what economists call 'indivisibilities') among investment projects. The same logic that got a company into such a predicament can therefore impede its attempt to extract itself.

No company can break out of a disinvestment spiral by relying on the same financial logic that got it there. The only remedy is to understand the shape of that logic as well as the direction in which it leads – and then to take an opposite course. Managers must be willing to reinvest at the very time such action appears least attractive. They must stop pouring funds into refurbishing their images and upgrade their factories instead. They must resist the lure of unfamiliar businesses and mind their own.

Beyond all else, capital investment represents an act of faith, a belief that the future will be as promising as the present, together with a commitment to making that future happen. Modern financial theory argues that under certain 'reasonable' assumptions disinvestment is a logical and appropriate course of action. Today, the future consequences of a disinvestment strategy, as seen through the reversed telescope of discounting, may appear inconsequential; but once tomorrow arrives, those who must deal with it are certain to feel differently.

References

1. Malkiel, B.G., 1979 The capital formation problem in the United States. *Journal of Finance* May; p. 291
2. Malkiel, B.G., 1979 Unraveling the mysteries of corporate profits. *Fortune* August; p. 90
3. Modigliani, F., Cohn, R.A., 1979 Inflation, rational valuation, and the market. *Financial Analysts Journal* March–April, p. 24

5.5

Managing Capital Investment: The 'Total Investment' System

James Morgan and Michael Luck

What is the Investment System?

Our experience in a number of pilot studies suggests that it is wrong to concentrate on the formal decision to accept or reject a proposal. It is more useful to think in terms of the total investment system. This system overlaps many of the firm's activities: sales, production, accounting and others. But, while other systems – such as cost control or production control – are often clearly recognized within the firm, the investment system is rarely regarded in this way.

In reviewing the 'total investment system', we should not define its boundaries too precisely. But, certainly, all the following activities should be considered:

(1) Generation of an idea
(2) Assembly of information
(3) Appraisal – which may result in acceptance, rejection, modification or postponement of a proposal or its referral to a higher level
(4) Assessment of the availability of funds
(5) Implementation
(6) Exploitation of the idea and reporting final results
(7) Follow-up

There can be few managers who do not take some part in one or other of these activities, but it is rare for any single manager to have a deep concern with the full range of them. Each professional function has a different perspective. Accountants, for instance, are concerned with control and problems of liquidity and so on. The relations between some of the components of the investment system are illustrated in figure 1.

Source: from J. Morgan and M. Luck, *Managing Capital Investment: The 'Total Investment' System*. Mantec Publications, Rugby, 1973. ©Tavistock Institute of Human Relations.

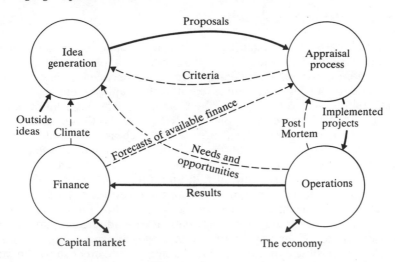

Figure 1   The total investment system of a firm

The investment system has both technical and social components. The technical part – consisting of decision rules and techniques for assessing projects, forecasting the availability of funds, formal control procedures etc. – is clear to see. The social processes, including the development of attitudes and patterns of behaviour, are, perhaps, less obvious but their existence can be detected most easily in the way ideas occur. If experience with a particular project is very satisfactory, proposals to extend it or to introduce similar projects may be expected. For example, firms which have bought numerically controlled machine tools have found that, once their particular advantages have been experienced, other potential applications are appreciated and further proposals result. If, on the other hand, particular types of proposal are frequently rejected at the appraisal stage, people may stop putting forward ideas of those types. There is plenty of published evidence to show that innovation flourishes in conditions which combine technical with social factors.

We have found also that the 'formal' processes of project appraisal, approval etc., are accompanied by 'informal' processes in which many of the real decisions are taken. These informal processes often serve to take into account things which the formal methods ignore – such as, subjective assessments of the many factors which will influence the way a company develops in the future.

A lot of effort has been devoted in recent years to promoting new methods of appraisal. Many companies have adopted these methods or are considering them. Although the arguments for them are attractive, the likely repercussions on and through the remainder of the investment system ought to be considered before opting for a change in its formal procedures.

Formal and informal systems

The 'formal' parts of the system are those which can be specified precisely and which could, for example, be described in a procedural manual. Regular monthly or annual meetings, a meeting between a chief and his subordinate to review the minor capital programme, the system of documentation etc., would be described as formal. The word 'informal' is not used as a perjorative term, but simply to describe those parts of the system which cannot be specified precisely – such as day-to-day contact between managers and their subordinates; the acquisition of information in the normal process of operating the company; *ad hoc* discussions about investment proposals arising out of talk on a totally different subject.

An example of how the informal parts of the investment system provide information and judgements which are not obtained by the formal part is shown by the way proposals for minor capital expenditure are handled in the Universal Stuff Company. Figure 2 charts the company's formal system for dealing with these applications. The numbers on figure 2 show the fate of 100 proposals for special expenditure in their passage through this system. Requests by subordinates are vetted by the divisional manager at a regular meeting. But, in some cases, the formal application is raised without this preliminary. We were unable to find any instances where a higher committee turned down a proposal once it had reached the stage of formal application. The decisions facing the higher committee seemed to be resolved mainly by informal consultation outside the formal system. Thus, the real job of the higher level committees may be to carry out a 'quality control' function; making their influence felt through comment of a general nature, rather than about particular proposals.

With large companies, in which functions are divided, formal procedures are a necessity. If, however, a company perceives its investment system wholly in formal terms, it ignores the capacity of the informal parts of the system to trap information, the existence of which could not have been seen by the formal system and which may, nevertheless, be highly relevant.

The ability to use this kind of information may help a firm to respond positively to its opportunities in the outside world.

A 'Development Activity'

It is just as important to review and develop the investment system as it is to improve production control or standard cost systems.

If the system for managing capital investment is to be most effective, some form of focus is needed to provide a base for review and improvement. It is not so clear who should provide this – the chairman of a project development committee, a development coordinator or committee as a whole or perhaps someone else. Therefore, we will use the words 'development activity' to describe this function. Each company must decide for itself how this activity may be carried out most appropriately

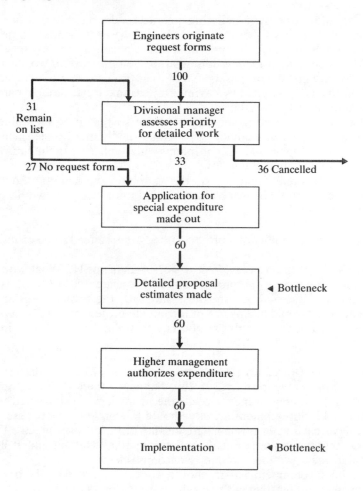

Figure 2   A formal authorization system

and how different individuals should take part in it – for much of the work will have to be done by individuals and not by a committee.

There are a number of broad characteristics which should relate to the development activity:

(1) Its function should include the requirement to stimulate the flow of ideas so that – hopefully – the firm is faced with an increasing number of attractive projects.

(2) It should not take away initiative, responsibility etc. from people who are exercising this already and it should not slow down the rate at which any idea can be put forward or developed. Rather, the development

activity should work by stimulating people to do their own work in a systematic way, guiding them to cover all the necessary points and helping by ensuring that resources are provided where necessary for any investigations. The need is not for a new administrative empire but for working with people who are making proposals so that their projects are not delayed and their initiative is not sapped by having projects taken out of their hands as soon as they start to materialize.

(3) Its way of operating should be set up so that it is adaptable. Its methods and procedures cannot be specified rigidly in advance since they will have to develop to deal with problems which will arise in the future and which we cannot know about now.

(4) The individual members of the development activity should be allowed to develop their own roles in the firm in a way that will best suit their own talents.

To prove the feasibility of a project, three aspects need to be examined.

(a) Feasibility of the process. Will it work satisfactorily? What would be the performance of the process and its associated costs?
(b) Commercial feasibility. Are the returns for the project satisfactory? What effect would having or not having the project have on sales etc.?
(c) Design of the plant. Is this as good as possible? What is the estimated cost?

If any of these three aspects is ignored, it is possible that the firm will take on unprofitable projects. If this happens, then the development activity will be failing to meet one of the needs it was established to fill. Therefore, the development activity should be confident that these three aspects have each been investigated satisfactorily for all projects. This is not to say that it must carry out all the investigations itself but that it should be satisfied they have been properly accomplished.

Thus, the development activity should work in parallel with the proposer of the project, rather than in series with him; taking as much of the load of investigation and analysis as is necessary and useful. If this exceeds the capacity of the individuals concerned, it would be their function to call in other resources to help in this work or else to increase the time scale of the investigation, as they see fit.

This suggests a fifth characteristic.

(5) The development activity should, at least, know of the existence of each project – even if it is not going to do any work on it. Nevertheless, the proposers should continue to be responsible for drawing up the studies of viability for their own proposals. By working with the proposers, the design activity should ensure that the three factors of process, engineering and commercial feasibility have been taken into account.

Detailed analysis of several investment projects suggested that there are a number of stages through which any project is likely to progress before final implementation. These are represented in figure 3. Depending upon size and complexity, not all stages may be followed by every project. Nevertheless, the points in the process at which the development activity should be involved particularly are indicated clearly in the figure.

The departmental manager concerned should be able to deal with small projects which are obvious in the sense of having a very high return and are confined within one department. However, the development activity would expect to know of the existence of such projects. And, where a small project is likely to have a very good pay-off, there is a good argument for informing the board of this – both to ensure the proposer gets a 'pat on the back' and also to ensure that the project does not fall by the wayside.

On the other hand, projects which are expensive, those which are uncertain in that they rely heavily on intangible benefits or on things like sales or market considerations, innovations etc., those which may be difficult to assess and those which are highly interdependent with other projects and overlap many departments should all be referred up to the board too.

A further principle related to delegation is that delegated decisions must be sufficiently substantial to the people concerned for them to take seriously the task of going into the pros and cons of the case. Also, the company must know how much of its resources are likely to be committed in this way. As an example, our analyses showed that the investments made by one company split fairly naturally into three groups:

(a) Must be considered by the board.
(b) Must be considered by the development committee (plus a main board director).
(c) Could be considered by the local manager.

Sums under £3,000 accounted for only five per cent and sums over £10,000 accounted for 80 per cent of the total investment. The development activity should establish figures like this and check whether delegation causes any change in the trend-using, perhaps, statistical controls.

It is possible to take a fairly systematic view of the cost of project estimates. This could lead to a logical procedure, whereby the exceptions are identified and reasons for these are sought. At the same time, a record would be kept of the cumulative estimated cost of proposals compared with the cumulative actual costs – either over a period or as a moving average. Here, it is important that the estimate which is taken into account should be the initial one, on the basis of which the firm originally committed itself, rather than the most recently revised figure, which may have been made for the purposes of engineering control. And there seems to be little reason why similar procedures should not be adopted in respect of returns.

Although capital costs of plant are established quite rigorously, there is little evidence of similar rigour being applied consistently to the estimation of future benefits. In particular, they did not seem to be very well-articulated and the value to the firm was not stated explicitly when future

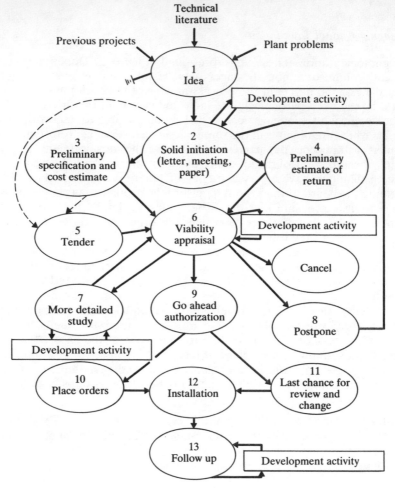

1. The idea for the project may originate from many sources, such as technical literature, problems in the plant, high running costs etc., which can seldom be pinned down exactly
2. Solid initiation occurs when the idea is put forward in a positive way by writing a letter to a director, circulating a paper or holding a meeting
3. A preliminary specification and cost estimate is then called for. This is often done by the chief development engineer
4. At the same time, a preliminary estimate of return is made – usually by the person who carried out stage 2
5. In some cases, tenders may be asked for
6. When estimates of cost and return have been made, a visibility study can be carried out
7,8,9. As a result of stage 6, the decision may be made to postpone, go ahead or make a more detailed study
10. If the decision was to go ahead, then orders are placed
11. There may be the situation where a final approval is made and important modifications introduced
12. The equipment is installed
13. There may be a follow up to see how successful the project has been and make changes to improve its performance if necessary

Figure 3   Stages in the development of a project

benefits were in terms of increased capacity or commercial advantage, or when projects were adopted for reasons of safety or amenity.

Usually, a good idea exists of the things that could affect the outgoings. However, the same principle of building up from smaller elements as is used when estimating project costs seems relevant. This would provide the best way of getting an organized feedback of the actual returns compared with what was estimated. This will help to improve future estimates of return.

For many proposals, of course, the returns from investment will be difficult to translate into straight cash terms. For example, they may be justified on grounds of improvement in product quality. Or they may be needed to meet developments being offered by competitors. However, in all these cases, it may be possible to make some quantitative statements about the improvements and it may be necessary for the development activity to develop units of measure for the effects which are not easily quantified – quality, delivery, flexibility within the shop, effect on customers and on competitors and so on. In time, experience with the use of such measures may suggest how they can be translated into cash terms.

Another characteristic of the development activity is:

(6) The development activity should satisfy itself that, for any project, key points relating to the benefits from the project have been specified before the go ahead and, after the project, that the proposers go back to see what has happened at those key points.

For example, in a proposal for a new boiler, key points might be the total weight of steam raised, the specific fuel consumption and maintenance time spent on the boiler. For a project closer to the market, key points might be the output tonnages achieved of different types of product, the reactions of customers etc.

In working along these lines with the proposers, the development activity can be expected to build up a body of technical experience which will contribute usefully to the project. Then it will be sought actively as a collaborator as well as a provider of extra resources where necessary. And the development activity must become aware of possible projects at an early stage. Then it will be able to urge those which are likely to have most effect upon profits and also spot connecting links between different projects in different sections of the company – for example, where one project may release space or machinery which could usefully be taken up by some other activity within the company.

(7) This early warning will be achieved most effectively if the development coordinators make a definite point of getting out and about and keeping in touch with all parts of the company, so that any formal channels of communication are supplemented by quick and informal routes to the development activity for the exchange of information.

The progress that has been made on projects which are outstanding so far, the rate of implementation of projects and the state of the queue of projects waiting to be worked up in any way or waiting for appraisal should all be reviewed regularly. And priorities between these should be established for any work involved. Furthermore, regular meetings should provide opportunities for the development activity to introduce any new topics which come up – for example, by discussion of ideas arising from visits to other works or from suggestions made by colleagues. Regular meetings will also give opportunities to set up *ad hoc* groups to work on particular problems and to receive reports from those groups which have been set up already. With major projects, the members of the development activity should be so involved in the groups that they, themselves, will be engaged in preparing reports, as well as taking part in the experimental engineering and other work.

However new developments are managed, there is a need to keep an eye on the cost of the system:

(a) Direct costs of development engineers, draughtsmen and other people working up ideas.
(b) Indirect costs of senior management time and effort.
(c) Benefits and costs of projects accepted and rejected.

(8)   Although we cannot yet identify these components accurately, we consider that the development activity should be collecting data on them and so reporting to the board whether the company has adequate performance. Nobody else will have the combination of a broad view with the time and ability to go into detail in order to relate them correctly.

As part of the job of working with other project proposers, the development activity should take responsibility for monitoring and reviewing the system for managing capital investment. This will involve the following practices:

(9) Developing more systematic methods for assessing and for measuring future benefits.
(10) Keeping an eye on the implications for cash flow of the expected costs and returns from the present portfolio of projects.
(11) Pressing for the consideration of alternatives as widely as possible and trying to stimulate ideas from others within the firm, perhaps by conducting tests, by providing data or by making analyses.
(12) Encouraging the consideration of interdependence between projects and also the consideration of possible consequential effect.
(13) Developing a logical structure of delegation.
(14) Keeping an eye on the balance of projects and reporting to the board on the overall level of profitability of new projects coming forward to ensure this is not declining. Also ensuring that the balance between projects which will increase profitability and those which will directly increase amenity only is preserved at a level with which the board agrees.

Seven Questions for Management

Management can choose many of the features to make up its own investment system. But what features should it choose to get the best total system? How can it decide what changes to make now in the present system?

A comparison with other companies may be difficult. Some use complicated techniques and procedures. Others have approaches which are simple to the point of naivety yet stay in business. Of course, there is no single best. The best system for each company is the one which is most appropriate for its own characteristics. For example, Universal Stuff is engaged in the same process industry as Crawshays, but it is much bigger. Therefore, more formal ways of handling information about possible interactions between projects and of dealing with 'soft' information about other aspects of the business may be more appropriate in Universal Stuff than in Crawshays.

The diversity of the components which constitute the investment system makes it difficult to review the whole thing systematically. Our way of carrying out such a review is to try to answer seven major questions:

(1) What are the pressures which actively stimulate the flow of new ideas?
(2) What is the method of appraising proposals and is it appropriate to the firm's circumstances?
(3) What are the pressures within the firm to achieve profit improvement rather than mere survival?
(4) Where in the management structure does the maximum of relevant information exist, and are decisions taken at this point?
(5) In what way are capital investment proposals related to the long-range plans of the company?
(6) What control is there over implementation of projects?
(7) What sorts of post-mortem does the firm carry out and do these really enable it to learn from experience?

A review of these questions should encourage the company to ask whether each point receives enough emphasis.

It is interesting to note that our three companies would give very different answers to these questions. This suggests that each might start looking for improvement differently from the others. Attention may be given to some aspects at the expense of others – for example, so much engineering effort may be going into detailed estimates that none is left for nourishing new ideas. Or the outgoings on projects may be checked very closely without comparable effort being made to investigate returns.

We are going to consider some of the alternative answers to each of these seven questions and the problem of choosing between them. For example, activities to stimulate the flow of ideas for investment can include: setting up a special department, special responsibilities for particular individuals, working parties and so on. In some companies, appraisal using simple calculations may be most appropriate. But, in others, the best appraisal process may be to use mathematical programming techniques for more formal analysis of profitability, risk and liquidity.

The point is that there are many aspects of the total investment system in any company and these interact with one another. Management is under pressure from time to time to make changes in respect of one or other of these aspects. In responding to these pressures, management should consider whether the same effort might give greater rewards if applied to other, perhaps less fashionable aspects, and how the proposed changes could have repercussions in other parts of the system.

What are the pressures which actively stimulate the flow of new ideas?

Ideas for change exist in any firm but it is difficult to stimulate the flow of new ideas and accelerate the good ones without at the same time so formalizing the system that their generation becomes inhibited. One possibility is to insist that every proposal should quote other ways of achieving the results at which the proposed investment is aimed. Universal Stuff has recently adopted a system of approval in which each person making a proposal has to establish an extensive set of alternative ways to achieve the same effect. By concentrating attention on the need to improve particular activities rather than seeking a particular piece of equipment, the firm hopes to make people regard themselves rather than others as the source of ideas for the future.

At the more basic level of generating completely new themes, the research department is seen by many firms as the main source of ideas. However, attention should be paid, perhaps, to strengthening other sources of ideas. This may include: setting-up some form of 'development activity', an 'information centre', 'management by objectives programmes', 'brainstorming' and/or other methods of management discussion. Basically, all of these serve to make connection between different parts of the firm easier and more fruitful. The importance of doing so is emphasized by the studies which Burns and Stalker made in the electronics industry: the firms which were most successful as innovators were the ones in which members of the research staff could go most easily into all other departments. The problems for top management are whether a sufficient variety of ideas comes up from different parts of the firm and whether the ideas can be trapped in the form of timely and workable proposals.

What is the method of appraising proposals and is it appropriate to the firm's circumstances?

The extensive published literature on investment problems is concerned mainly with appraisal, i.e. with the problems of deciding whether a project is acceptable, or of choosing between alternative projects or sets of projects. In real life, however, many investments are undertaken without the benefit of formal appraisal in economic terms. And yet the firms concerned survive with apparent profit. It is evident from this that a case for any form of appraisal (and for the time and expense involved) only exists to the extent that it succeeds in improving overall profitability. The basic case then, is that by applying a formal method of appraisal, a 'better' set of investments will be chosen than would otherwise be the case.

There is intuitive appeal for a simple rule or formula which will attach to an investment proposal a number indicating its 'profitability'. By comparing this number with a preset criterion, it is hoped to 'weed out' the less profitable proposals when they arise and, thus ensure that the performance of the firm is better than it otherwise would be. But the matter is not so straightforward. The case for investment cannot usually be reduced to a monetary one independently of:

(1) The existing activities of the firm
(2) Other investment opportunities
(3) The sources of capital
(4) The uses to which income may be put

These considerations are too rich to be compressed into a single number and, in theory at any rate, the methods of budgeting and appraisal should take formal account of:

(1) Preferences for income at different times
(2) Preferences between return on capital and aversion from risk
(3) Dependence relations between projects
(4) Uncertainty of outcomes
(5) Covariation between outcomes of projects and with existing activities
(6) Physical restraints on capacity to accept combinations of projects
(7) Uncertainty of financial provision for opportunities which have not yet arisen
(8) Optimal scale and timing of projects
(9) Lending and borrowing

Methods incorporating several of these features have been described but no method in practical use has them all. Most treat important desiderata in a perfunctory way. Our studies suggest the available theory is not implemented, partly because of a realization that it does not deal with the problem fully and partly because managers do not find it natural to provide information on such matters as risk and interdependence in the manner that would be required. Most of the firms with which we have dealt rely on the informal system to recognize and deal with interdependence between projects. An alternative might be to formalize and improve the flow through the firm of 'soft' information about the plans currently being canvassed, so that people are alerted to possible interaction.

Assessing risk is another difficulty. This can be dealt with by requiring originators of proposals to make explicit statements about possible contingencies or by 'buying' information – for example, through research. But, in none of the companies we have studied, were the possibilities of errors in estimates or the risk of nonfulfilment taken into account in the formal assessment procedures. The main reason for top management being involved in the appraisal process may be, therefore, to relate it to their knowledge of the future plans for the business and their subjective assessments of risk.

Financial considerations, such as the need to preserve certain 'normal' ratios on the balance sheet, may constrain the total investment system as much as the need to preserve liquidity. In General Gadgets, a conservative policy of financing investment wholly out of retained earnings meant that valuable opportunities were being missed.

But, the constraints acting on the investment system may not all be financial. In Universal Stuff, the availability of engineers to work up schemes which had been agreed in principle was the bottleneck which really limited capital investment. In General Gadgets, lack of staff to commission and 'debug' new plant also led to delays and difficulty in implementation.

There are a number of problems to be resolved in designing an appraisal system even if the economic factors are dealt with in a satisfactory way. One such problem is including in the assessment the cases which are not judged against direct profitability criteria, such as necessary spending on safety or welfare items to improve unpleasant working conditions etc.

Another problem is who should make the assessments. Should it be a special central department or should it be the originator of a proposal? In favour of a central department are the arguments that more complex mathematical and other techniques may be used if they are relevant and a more critical approach can be brought to bear on some of the assumptions. However, it may engender a 'them and us' situation in which, though part of the formal system, it is excluded from the informal and is consequently ineffective.

In summary, the problem in designing the appraisal aspects of the investment system is one of making compromise between simplicity and completeness, in choosing how much of the more complex theories to implement and how to rely on the 'informal system'.

What are the pressures within the firm to achieve profit improvement rather than mere survival?

It is not obvious that conventional management information – related, as it usually is, to past performance and existing standards – gives an incentive to profitable innovation, or whether its warnings of impending difficulties are sufficiently timely. Standard costing systems, for example, are usually directed towards questions of pricing and of control rather than to improvement beyond the standard costs which obtain with existing equipment. Indeed, there is some evidence that administrative controls which concentrate on checking the details of individual activities tend to suppress innovation. The study in American hospitals, by Rosner, showed how innovation was slowest when there was most control over medical staff activities and fastest where there was looser control over activities but a quick feedback of results. These relationships could be expected to be even more pronounced in organizations which are more tightly controlled than hospitals, where medical staff activities are usually fairly free of administrative interference.

Profit improvement, therefore, rather than bare survival requires a positive incentive to innovation in managers at all levels. One currently fashionable way to do this is 'management by objectives' planning. This may be easier if each production engineer or manager can have a clearly defined part of the plant within which he is required to set for himself definite targets for profit improvement. However, this clarity of definition is not possible if production processes are highly integrated and a less compartmentalized view of life is required.

Are decisions taken at a point in the management structure where the maximum of relevant information exists?

We can distinguish between two sorts of information: information about a project and its details, and information about a firm as a whole, its plans and finances.

Some form of budget is needed to allow for adequate consideration of each proposal in the setting of the total plan and expected future cash position of the company. There were considerable differences between the three firms. The virtue of General Gadgets' highly specific annual budget is that wide interdependence between projects can be taken into account and the most profitable investment programme chosen. On the other hand, the rather lengthy budgeting process seems to have a stultifying effect on junior and middle managers who have to spend a lot of time revising and altering their estimates. A continuous review, as in Universal Stuff, may be more flexible and allow the benefits of 'good' projects to be taken quickly.

Top management do not have enough time to go into the details of each proposal and, for this reason alone, some measure of delegation is needed. Apart from this, junior and middle management – if they are able – need to be allowed to make some decisions themselves. Projects which are small, which show a very high return and which are confined within one department, can clearly be dealt with by the departmental manager concerned. One such project in Crawshays showed an annual saving of £2,000 on water costs for an investment of £300 on a pump.

Some companies delegate authority by setting aside a lump sum to be spent at the discretion of each manager. Others set a figure up to which a manager can approve individual projects. Another way is to specify the type, the degree of risk or the degree of interdependence of projects which a manager can approve. Any decision rule such as this should be examined so that a company can know how much of its resources it is likely to commit in this way. It should also be examined from time to time to see what side effects it may have.

In what way are capital investment proposals related to the long-range plans of the company?

Top management needs to know the effects of any investment proposals on the future cash flows of the business. Will there, for example, be years in the future when the aggregate cash flows may cause concern with regard to financing or capital structure? More generally, what will be the effect of the future commitments which are now being accepted on the physical, financial and human resources of the company? Will plans to open up new works or to develop new markets require so many men with particular experience that it will be difficult to find them? How can the right man be sought out and given the appropriate experiences and training? Do current plans take into account possible developments by competitors and possible changes in the market and in technology?

The investment system must generate, select and implement projects which capitalize on the distinctive competence of the firm and cause it to develop lines of business with profitable and expanding futures. The realization by the companies operating passenger liners that they are more engaged in the luxury holiday industry than their traditional business of scheduled transport provides an illustration to this point. The way in which the boundaries to its activities are perceived by the organization as a whole will certainly affect the course of its future development. As Miller and Rice have pointed out, it is a matter over which senior management has at least some control.

The development of a firm depends on the actual outcome of factors that are necessarily uncertain at the time of decision. Whatever the economy and whatever competitors may do in the future, it is desirable that the firm's current projects should be capable of leading to other profitable developments, rather than to dead-ends. The extent to which the investment system gives rise to projects which are appropriate in these terms is an important measure of its effectiveness.

What control is there over implementation of projects?

In many situations, after a proposal has been approved, new information becomes available which would indicate desirable changes in the project. In the extreme case, the project itself may be shown to be irrelevant if a lengthy budgeting process causes delay between the original conception of a project and its full implementation, competitive opportunities may be lost and the full benefits of the project missed. There is, therefore, a need for some control over implementation so that advantage can be taken of any new information or changes in conditions and so that the benefits from the project are realized promptly.

This kind of control over implementation could be achieved in several ways: by simple control procedures using the normal management chain, by setting up specific commissioning teams or by making it the specific responsibility of some individual or department to check progress. This last type of control often exists over cash outlays: it could without difficulty be extended to cover the attainment of programme targets. But this may mean a change in the attitude with which these tasks are approached.

What sorts of post-mortem does the firm carry out and do these really enable it to learn from experience?

Only if it knows how individual projects have prospered and why, in some cases, success is not achieved, can a firm learn from its own experience to improve the processes of conceiving, estimating, appraising and implementing projects.

An effective system of post-mortem must depend on adequate documentation and written justification of spending proposals. There is a danger that overelaborate documentation could deter people from producing realistic estimates but some effort must be made to record information in such a way that organized comparisons are possible. Since price inflation has a confusing effect on money figures, it may be better to break estimates down so that the basic units of output: quality, man-hours etc., can be specified for each component of the proposal. Calling for such key points can help in controlling implementation, as well as making the results of post-mortems more useful in improving future estimates of cost, project timing, market reaction etc.

Whether the organization of post-mortems should be based on the normal management hierarchy, or should be made a function of some particular staff department, will vary from firm to firm. In some cases, it may be better to review a small proportion of projects in depth rather than all of them superficially. The projects to be reviewed should be chosen by setting up 'control limits'.

Existing data and reviews of completed projects, supplemented by subjective assessments, can give information about the degree of error which may be expected. Where results fall outside the expected margins of error, more detailed investigations can be made and the cause of variation may be taken into account in future proposals.

The arguments here are identical with those for statistical quality control. Indeed, a post-mortem system should do for the investment system what quality control should do for the production system.

The Performance of the Total Investment System

Since the outcome must to some extent be uncertain when an investment decision is made, there will be risks that some accepted proposals will turn out to be unprofitable. Also there will be risks that some rejected proposals would have turned out to be profitable. An efficient system is one which balances the risk of accepting too many projects which turn out to be unprofitable and the risk of rejecting or failing to generate a large number of projects which would have been profitable if accepted.

In the long run, the performance of a firm's investment system may be evident from the growth or decline of the firm. However, there may be other confusing factors and we may want to assess the performance of the investment system without waiting till the end of the long run. One approach is by comparison with the requirements of capital budgeting theory or with the practice of other firms who seem to handle capital investment effectively. The components of the investment system can be assessed quantitatively against the seven points for review described previously. Most published material about quantitative assessments concentrates on one component, appraisal, and the effect of different techniques for choice within a given set of possible investments. This limitation is arbitrary. Quantitative methods can be used in assessing other components of the investment system – for example, the optimum degree of effort to put into making more precise estimates.

However, to evaluate the investment system in financial terms, broad assumptions are made about nonfinancial factors which must be taken into account. One assumption, for example, is that decision makers within the investment system operate in a predictable way and their variability will not greatly affect the comparison of systems. Some comparative assessment of the investment system is needed, therefore, and we must define:

A good investment system as one which generates and implements profitable proposals for a particular company. This requires that the investment system must be able to tolerate changes in the conditions in which it operates and be sufficiently adaptable that changes in market conditions, in government policy, competitors' plans or in technology, do not leave the firm without the capability of proposing and accepting appropriate projects.

Note

This reading reports work carried out in the Institute for Operational Research, now Centre for Organizational and Operational Research, which is a unit of the Tavistock Institute of Human Relations in London.

5.6

Identifying and Quantifying the 'Company-wide' Benefits of CAD within the Structure of a Comprehensive Investment Programme

P.L. Primrose, G.D. Creamer and R. Leonard

Introduction

Senker wrote that firms normally justified the costs associated with CAD by relying on labour savings from employees such as draughtsmen and estimators.[1] Conversely, while other authors have suggested that the advantages of CAD extend beyond the confines of the drawing office, these advantages have not been identified in sufficient detail to enable them to be used for the financial justification of a potential system. It should be noted, however, that companies who invest in major capital projects without a detailed financial appraisal run the following risks:

(1) They might invest in a project which is incapable of generating an adequate return on capital.
(2) They might invest in a project which does not represent the best potential application, with the project offering the greatest return remaining unidentified.
(3) They may refrain from investing in a project, even though such an investment would be more advantageous than simply continuing with current practice.

The costs associated with the design function often represent less than one per cent of the total cost of sales, therefore, by erroneously concentrating the appraisal of CAD on savings in this single area, industrial relations problems are needlessly created within the drawing office. Conversely, by considering the effects of CAD on the nondrawing office areas of the

Source: Unpublished paper, 1984.

company, the potential advantages of CAD will be shown greatly to exceed the salaries of a 'handful of draughtsmen'. It is only by correctly identifying these 'company-wide' benefits of CAD, that a rigorous investment appraisal can be carried out, thereby enabling the company to concentrate its resources on the areas which will maximize the advantages of the system.

Identification of Costs and Benefits

Whilst Massey and Millward have listed the advantages of CAD/CAM in general terms,[2] they have refrained from suggesting how these advantages might be quantified. In order to derive a 'scientific approach' to CAD evaluation within the present work, all the potential items which affect the appraisal of CAD have been identified and defined in such a way that they can be both quantified and the danger of duplication avoided elsewhere in the analysis. The objective when compiling the lists was to identify every factor which could have relevance to a company. However, it is recognized that certain parameters will not be applicable to a specific application. Yet by having a comprehensive list, a company considering CAD may discover areas of saving which otherwise would have gone unnoticed. Although the overall list displays some similarity to the collective suggestions of previous authors, the resulting analysis of company benefits is very different. Thus, the contentious term, 'reduce drawing office labour' has to be specified as both 'reduction in number of existing draughtsmen' and 'avoid recruiting extra draughtsmen'. Although the effect of both terms is a saving in labour costs, the first includes redundancy costs while the latter avoids recruitment costs.

Costs of CAD

Initial costs

(1) Hardware
(2) Software
(3) Installation (including building alterations etc.)
(4) Consultancy costs (may include customising software)
(5) In-house project team (may include customizing software)
(6) Database development
(7) Operator training
(8) Lost time during transition (may include subcontracted work)

Running costs

(1) Maintenance contract
(2) Insurance
(3) Running costs (i.e. electricity)
(4) Consumables
(5) Software updates
(6) Training updates (i.e. for new staff)
(7) System management
(8) Labour shift premium (if two-shift operation is introduced)

Items such as the cost of capital, government grants and tax rates have not been included in this list. They will, however, be dealt with later. Similarly, although 'downtime' is suggested as an important cost, it is not really a cost but rather a loss of productivity. Therefore, if downtime is allowed for in the labour balance, it cannot be included as a separate cost. Likewise, as will be discussed later, in certain circumstances some factors may become costs rather than savings.

Benefits of CAD

Drawing office savings

(1) Reduce the number of existing draughtsmen
(2) Avoid recruiting additional draughtsmen
(3) Reduce clerical labour in drawing office
(4) Reduce or avoid subcontract design work
(5) Take on subcontract work
(6) Eliminate model making by use of three-dimensional design
(7) Reduce outside graphic design work (for marketing, service department, publicity etc.)

Increased sales from reduced delivery times

(8) Reduce design/documentation time for customers' orders
(9) Improved drawing quality reduces production delays (e.g. easier assembly)
(10) Eliminate incorrect ordering of components, thus reducing production delays

Increased sales from other causes

(11) Company can quote more reliable delivery dates
(12) Faster and better presented quotations
(13) Company image improved by having CAD
(14) Orders would be lost if company did not have CAD design facility
(15) New products can be introduced more quickly

Reduced stock levels

(16) Improved drawings reduce production lead times, hence reduce work-in-progress
(17) Component standardization allows a reduction in finished stocks
(18) CAD may avoid ordering unwanted components

Reduced production costs

(19) Improved drawings/documentation reduce production costs (e.g. easier assembly)
(20) Reduced scrap and rework
(21) Production efficiency improved by eliminating 'stock outs'
(22) Component standardization enables larger batches to be produced
(23) Design optimization reduces production and material costs

Cost control

(24) Unprofitable orders eliminated by improved estimating
(25) In-house cost control improved by better estimating and quotations

CAD–CAM link

(26) The purchase of separate systems for NC programming avoided by CAD
(27) Linking NC programming to CAD reduces programming costs
(28) CAD aids company-wide information system
(29) CAD avoids the need for other expenditure, e.g. expanding the drawing office building.

Description of Benefits

Although it might be imagined that certain factors are insignificant, within the context of the wide range of systems available and the varying needs of companies considering CAD, items should not be excluded from the list simply because an individual has found them to be unimportant in the past. Similarly, while many of the factors are self-explanatory, some require clarification. For example, because the preparation of items such as service manuals and publicity literature is ordered directly by marketing from specialist firms without the drawing office being involved, graphic design is listed under a separate category to drawing office subcontract. A number of factors lead to improvements in delivery dates, with this consideration being of major importance in the current economic climate. It has been suggested that in some industries, companies without CAD may realistically be excluded from the market.[2] Similarly, a two-fold benefit results from CAD enabling a company to launch new products earlier. Firstly, the contribution of the new product to overhead recovery will start at an earlier date, while the advantages arising from being 'first in the market' will generate greater total sales over the product life. Although it might be argued that CAD enables a company to design products which were not feasible using traditional techniques, the choice in the evaluation is not between 'CAD + new product' and 'No CAD + no new product', but between CAD and a suitable subcontract bureau where the work could be done.

One of the benefits claimed for CAD is that value engineering becomes easier, enabling components to be 'standardized' and designs 'optimized'. This results in savings in the categories listed as 17, 22 and 23. Similarly, if manual parts lists are incorrectly drawn, problems will be generated in production, resulting in high levels of stock. Yet another major potential benefit for CAD is that of improving the quality of estimating for quotations. By using CAD, a company can avoid orders which would have proved unprofitable; equally important, the CAD procedures will prevent 'excessive' quotations being sent, with the subsequent loss of 'profitable work'. An accurate method of estimating, based on CAD procedures, enables a company's order intake to be biased towards those quotations which give the greatest profit. Similarly, improved estimating enables a company's internal cost control system to be improved. For example, variances tend to be ignored with manual estimates because it is assumed that the fault lies in 'estimating'. However, if estimates are 'known' to be accurate, variations in production costs can be investigated, identified and corrected. Figure 1 depicts a CAD quotation system operating within a company associated with this work, as described by Creamer and Leonard.[3]

CAD/CAM Links

Considerable literature exists relating the benefits of linking CAD with CAM, with detailed examples of CAD/CAM applications.[4] However, caution must be exercised when considering CAM as a direct financial benefit of introducing CAD. Whilst certain CAD systems now have the capability to enable NC machine programs to be produced more efficiently than by manual methods, it is equally possible to buy a relatively inexpensive dedicated computer, specifically designed to produce tapes for NC or CNC machines. This specialist computer is likely to cost only a fraction of the price of a large CAD system and, more importantly, the NC tapes produced will operate the machine tools in more efficient conditions than a corresponding tape from the CAD system.

At a recent conference on integrated manufacture,[5] the participants were unanimous in the view that many years will elapse before the concept of a company-wide integrated computer system becomes a practical reality. Therefore, while both CAD and production MRP systems rely on the creation of large databases, the make of computer, its operating system, programming language and database management system are all likely to be different, thereby making the systems incompatible. Thus when integrated CAD/CAM finally becomes a reality, companies may be faced with rewriting the data currently stored on a range of databases, at the expenditure of considerable extra cost.

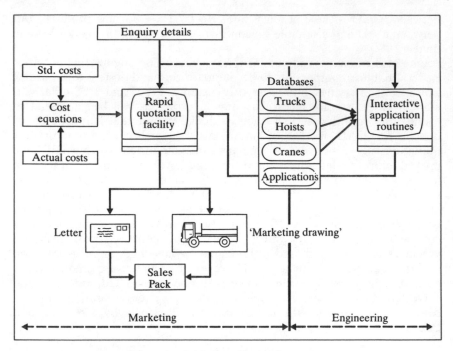

Figure 1   CAD-based quotation system

Investment Appraisal

Nondiscounting appraisal techniques, such as payback or accounting rate of return, are condemned by accountants because they do not take account of the timing of cashflows or cashflows beyond the payback period. The problems of using payback are particularly acute when considering CAD because up to three years can elapse between the implementation date and the full benefits being achieved.[4] Since the objective of the analysis is to provide a financial methodology which is acceptable to accountants, DCF techniques are used to evaluate the return on investment, measured as net present value (NPV) and/or internal rate of return (IRR).

A company considering CAD may wish to appraise a range of different systems but to attempt this by traditional methods, using discount tables, would be both tedious and prone to error. In fact, it is as a result of the practical problems normally associated with DCF that many firms still use payback. Fortunately, the problems previously stated can be overcome by the use of the programs now to be described.

Computer-based Evaluation Technique

The computer program for financially evaluating CAD systems takes account of all the costs and benefits listed earlier. Items such as working life, government grants and the cost of capital add significantly to the 45 factors previously listed. In practice, however, many of the factors are 'nonapplicable' to a particular evaluation. For those factors which are relevant, the cash flows might be on an annual basis (e.g. labour saving) or a single sum (e.g. start-up costs) or a combination of both. Figure 2 shows how, by a simple yes/no decision, factors can either be passed over or, where relevant, multiple entries made. Savings will not usually result from a single item but, as shown in figure 3, will comprise multiple questions to quantify each factor. Once the check list sheet corresponding to the questions asked by the VDU has been used to identify relevant factors, a complete set of data can be inputted in less than five minutes, with the corresponding results being instantaneously displayed.

Quantification of Data

When companies invest in machine tools such as CNC, they are able to identify savings and accurately estimate the value of new production. This leads to the belief that investment appraisal is an exact science, with this view being reinforced by accountancy literature, which concentrates on complex mathematical techniques. In practice, investment appraisal is only an aid to decision making, with the results reflecting the quality of data used. For example, previous CAD evaluations have concentrated on labour saving in the drawing office, with a 4:1 productivity improvement being incorporated within the calculation. However, Dawson claims that a 2.5:1 improvement is more realistic, while Gott even suggests that any productivity gain will be marginal.[6,7]

Within the analysis described in this reading, the lists of costs and benefits enable each factor to be correctly quantified. The problem of quantification is twofold, namely (1) estimating the magnitude of the values and, (2) ensuring that only incremental cash flow data are used. For example, when trying to identify, and subsequently quantify, savings in areas such as production or sales, the help of management should be elicited to define the effect of factors such as: drawing and documentation improvements. For this case, the question to be asked should be of the form: 'If we reduce the throughput time of paperwork in the drawing office by X weeks, will this reduce product delivery time by X weeks and, if so, will this lead to an increase in sales?' If the answer is yes then the subsequent question is, 'What per cent increase in sales will result from a reduction of X weeks in the average delivery time?' Because of the nature of sales and production it is difficult to measure cause and effect, thus the answer to the above question is unlikely to be 'Yes 1.25 per cent'. More probably, it will be of the form: 'We think so, it is probably about 1 or 2 per cent'. Although the answers will not be specific, it is possible to establish upper and lower values. Therefore, by appropriately defining costs and benefits, data can be entered for optimistic, pessimistic and mean values, resulting in three DCF returns. It then becomes management's task to establish the probability of the results being achieved and thus whether to proceed with the investment.

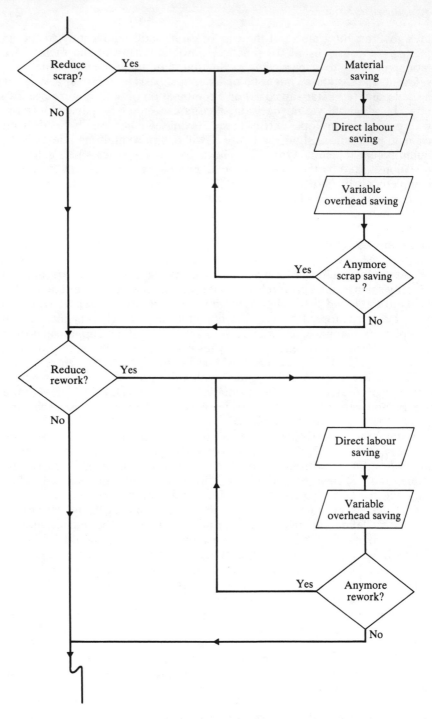

Figure 2   Selection of relevant factors and multiple entry of savings

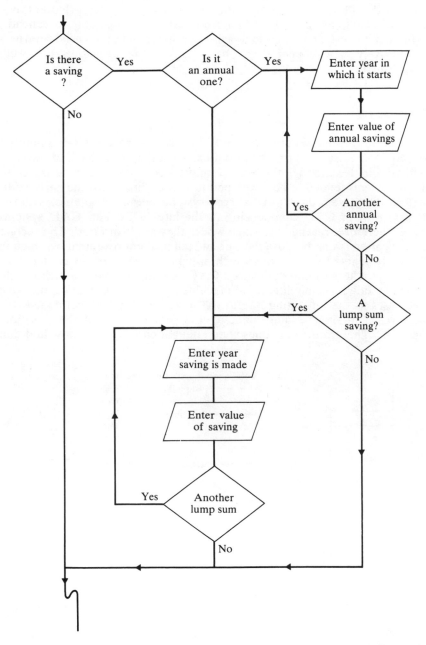

Figure 3   Multiple questions to quantify a relevant factor

If the check list has been comprehensively designed, and the questions correctly phrased in the interactive input, the data used, as shown in figure 2, will be in the correct format. Thus the increase in sales, previously discussed, will result in a contribution to overhead recovery. (Note: this is not the same as savings in material, labour or variable overhead.) Although the use of a program to achieve a statement of financial returns is a relatively new concept. the basic accounting principles are well-established.

Working Life

When DCF techniques are used for CAD investment appraisal, a number of problems exist when trying to establish what the working life of a system will be. Firstly, although the term computer-aided design can be traced to 1960, systems tended to be developed in-house, thus it was the early 1970s before the present day 'turnkey'-type systems became available, with the real growth of CAD commencing in the late 1970s. Few CAD systems have, therefore, reached the stage where they are 'worn out'. The second complication is that because the rate of technical improvement was high in the early years of CAD, firms which bought systems are now contemplating the purchase of more advanced CAD systems. Franks and Scholefield describe how, as the life of a machine is extended, the capital cost decreases as the 'operating inferiority' increases, thus the combined cost achieves a minimum value, namely, the optimum life of the machine, figure 4.[8] Therefore, companies which bought early systems now find that

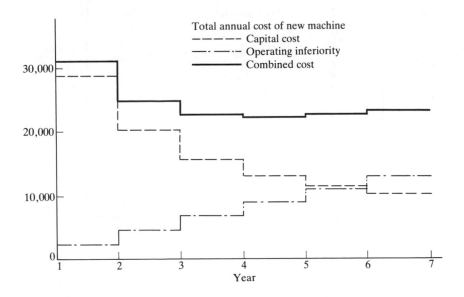

Figure 4   Change in total cost showing optimum life

the difference between what they have and what is now available, e.g. the 'operating inferiority', is so marked that a replacement can be justified on the grounds of increased savings. Although improvements will continue to be made, these are likely to be at a reduced rate, therefore, companies who are buying CAD today may retain their systems for their full working life, which may amount to 10 years. Figure 5 illustrates the effect of working life on the internal rate of return. For this figure, a resale value of 10 per cent has been used. Little difference occurs, however, if this is reduced to 1 per cent, representing only scrap value.

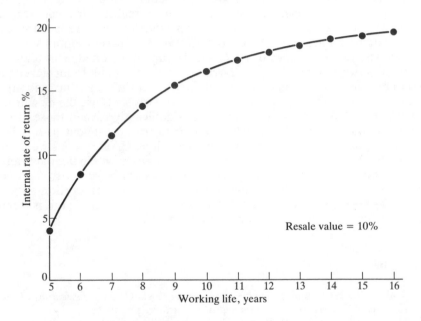

Figure 5   Effect of working life on IRR

Use of Computer Program

The evaluation program has been used to appraise CAD investments in firms which range from a 'one-off'-type company to a 'large batch' manufacturer. In all cases, the potential savings from nondrawing office areas proved to be much greater than from the drawing design departments. As a consequence, the corresponding return greatly exceeded the figures that companies had previously regarded as optimistic. In a company which already had a major CAD system, the original justification was based on saving draughtsmen's salaries, thus efforts had purely concentrated on optimizing the design process. However, by showing that the potential benefits of increased sales and reduced production costs far exceeded any savings in the drawing office, the company is now seeing CAD as a company-wide system.

Conclusions

The lists of costs and benefits, given in this reading, enable a company considering CAD to identify those aspects of its operation which will exhibit the greatest benefits. In addition, because the lists have been structured to prevent individual costs and benefits from overlapping, the corresponding sums can be quantified within a realistic financial evaluation. Although savings in factors such as production costs, reduced stocks and increased sales are subject to uncertainty, by using optimistic and pessimistic values, a range of DCF returns can be obtained. This allows a more realistic appraisal to be carried out than would be achieved by concentrating on the single factor of savings in drawing office labour – where errors of 50 per cent are common. Similarly, because the magnitude of the potential savings, company wide, significantly exceed those in the drawing office alone, it is much easier to justify investment in sophisticated, expensive systems. Thus companies selling CAD systems would be well-advised to incorporate a company-wide program within their appraisal of a potential CAD customer so that all benefits can be correctly assessed. In addition, CAD suppliers should devote a significant proportion of their R & D efforts to functions outside the drawing office because this is where the greatest gains will ultimately become manifest.

References

1. Senker, P., 1983 Some problems in justifying CAD/CAM. *Proceedings of 2nd European Conference on Automated Manufacture*, pp. 59–66
2. Massey, I.C., Millward, W.J., 1983 The financial justification of CAD/CAM systems. *Proceedings of DES '83 Conference*, pp. 101–12
3. Creamer, G.D., Leonard, R., 1983 The design and application of an advanced CAD method for customer quotations. *Proceedings of 6th Annual Design Engineering Conference, Birmingham*, October
4. *A Guide to CADCAM*. The Institution of Production Engineers
5. Integrated Manufacture – A Total Concept I. Prod.E. Conference, October 1983
6. Dawson, F.T., 1980 CAD – A user's perspective. *CAD Journal* 12(3): 127
7. Gott, B., 1983 CAE as a business. *Electronics and Power* 29(1): 75–77
8. Franks, J.R., Scholefield, H.H., 1977 *Corporate Financial Management*, 2nd edn. Gower Press

5.7

The Decision to Adopt New Technology – Effects on Organizational Size

Martin K. Starr and Alan J. Biloski

Introduction

Production and operations systems can be categorized by work configuration. Job shop (or batch production systems), which utilize relatively inexpensive general-purpose equipment, constitute about 75 per cent of manufacturing configurations in the USA. This percentage undoubtedly would be increased substantially if service and office systems were included along with maintenance functions in manufacturing. Flow shops (or serialized production systems), epitomized by a dedicated, paced conveyor belt, require more pre-engineering and higher fixed costs in trade-off for lower variable costs. That the trade-off is hard to justify is exemplified by the relatively low percentage (25 per cent) of manufacturing configurations that have been serialized into flow shops. Even automobile assembly lines in Detroit use many batch processes to feed components to the serialized assembly conveyor. A totally serialized manufacturing system (termed automation by Del Harder of Ford Motor Company in the 1940s) requires very costly special-purpose equipment and fixed transfer lines, resulting in no flexibility. Such levels of fixed automation never proved practical.

Major developments in computer control now have produced the potential for a new type of work configuration which reduces variable costs while improving the flexibility of the process so that output variety can be increased.[1,2] This is often called flexible automation or more recently, flexible manufacturing. The cost structure of fixed automation, which utilizes unchangeable transfer paths between special-purpose equipment, can only be justified when market demand is sufficiently great, over a long enough period of time, to allow the fixed costs to be amortized. Flexible

Source: from *Omega* 12(4), 1984.

manufacturing systems (FMS), on the other hand, provide computer control of expensive, programmable work centres as well as of the direction and rates of transfer lines. Thus, in addition to numerical control (NC) of workstations, the fixed conveyor belt constraint of mass production lines is relaxed.

One of the major obstacles, however, in using FMS is that little is known about how to coordinate activities to obtain a relatively constant flow of marketable output. Furthermore, the production cost structures are more complex than have previously been experienced. The same observations apply to new technological effects on service, office and maintenance operations. It is hardly surprising that research efforts are growing rapidly in an effort to understand the benefits and costs of FMS, as well as flexible office systems (which we call FOS). According to our analyses thus far, the conceptual basis of FMS applies equally well to FOS.

The past year has seen an acceleration in the publication of FMS literature. By and large, these articles can be categorized into one of two general areas: (1) descriptive articles that detail a particular FMS and describe its production performance and (2) technical papers which focus exclusively on the development of (a) analytical models[15] and (b) simulation studies.[13] The underlying emphasis of all treatments, however, is to highlight the reductions in variable production costs promised by FMS.

Methodology and Description of Technology

This work represents a completely different approach in that it aims to uncover the effects of FMS technology on organizational size. The investigation has proceeded along two diverse, but complementary, paths. Nonlinear breakeven analysis has been used to establish a theoretical framework for the problem. The effects of FMS on important variables are identified, and related to the organizational size at which profit is maximized. The findings have been supplemented by empirical data provided by the General Dynamics, Boeing and Northrup Corporations.[8,11]

Objectives: Consideration of Quality and Cost Factors

To understand the organizational size issue, we found it necessary to include other factors in our theoretical model in addition to the variable production cost savings which have been demonstrated by earlier writers to play an important role in the decision to adopt flexible manufacturing. By understanding these additional influences, FMS advocates will be better able to overcome institutional roadblocks and achieve successful implementation of this important new technology, when it is feasible. Avoiding FMS failures is implicitly the other side of the coin.

Experience has shown that FMS are adopted for reasons other than cost-effectiveness (i.e. output variety increases and improvements are likely to occur in product quality). While accounting cost considerations may remain the ultimate arbiter of a new proposal, it has to be recognized that the intangible cost aspects of FMS also deserve serious consideration. These may, in fact, be overriding under various circumstances.

Evaluation difficulties arise because existing accounting systems have been designed for traditional operating configurations and cannot accurately measure the total system's performance of a radically different manufacturing method. For example, the random processing nature of FMS is almost equivalent to continuous production while traditional job shops are labour intensive. Since most justification schemes are direct labour oriented and assume that what exists today will last forever, there is a strong bias towards the status quo. It could be argued that this is true of all innovations affecting quality and variety, but it is especially relevant in the case of FMS since conventional accounting systems are unable to quantify the potential benefits offered by flexible automation.[7,9] Given this situation, and the need to express all proposals in accounting terminology, the general approach has been to translate high fixed cost flexibility into quantifiable, but indirect, measures such as shorter lead times and reduced in-process inventory. This gives an incomplete picture, however. Adequate comparison of these vastly different alternatives requires a comprehensive strategic evaluation of a company's operating systems, product portfolio and marketing environment. The ability to rapidly and inexpensively change the output mix is a powerful advantage that heightens organizational responsiveness to competitive and/or market changes. A balanced presentation also requires consideration of the negative intangible aspects of new technology. The problems of new equipment, faulty control logic and insufficiently trained personnel can lead to serious cost overruns. The contrasting effects do not necessarily cancel each other however, as the latter can easily be foreseen and countered by fostering good vendor relations and early development of in-house expertise. The net result is that accounting-based FMS evaluations are not likely to be sufficiently objective.

FMS product quality improvements are a second major intangible factor. The substitution of computer-controlled equipment for human workers must necessarily increase the precision and standardization of the output. This increased process control is manifest in two ways. Customers perceive the resulting improved output performance due to higher consistency between parts as a desirable quality improvement. In addition, the manufacturer achieves greater reliability for his products so warranties can be extended and made more generous. This total package of improved manufacturing and customer service increases the product's value, where value is defined as quality per unit price. Price then diminishes in importance for the purchasing decision and the elasticity of demand decreases. The improved product attributes will also stimulate marginal customers and increase the demands of existing consumers. These twin effects are graphically depicted by the increased slope and shift to the right of the FMS production demand curve in figure 1.

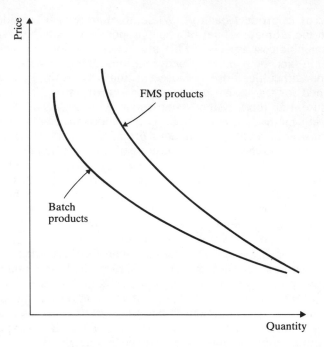

Figure 1   Demand curve structure for batch vs. FMS production

Theoretical Approach

Linear breakeven analysis has been used for capacity planning since the turn of the century. It appealed to managers who wanted to include the many factors that affect the 'bottom line'. For larger organizations, the linear constraint does not apply. Market saturation is one cause of nonlinear revenue performance; overtime cost is another. Thus nonlinear breakeven analysis (BEA) seems an ideal theoretical model for evaluating FMS.

Applying nonlinear breakeven analysis, one may examine the effects of process innovations on optimal organizational size by ascertaining the output volume which generates maximum profit, before and after the work configuration change. The model illustrated by figure 2 defines the revenue and cost functions of a firm.

Our profit model is straightforward, but only one of many possible models to describe non-monotonic profit. For example, we have treated the total cost curve as being linear, which is accepted as a reasonable approximation in many situations. But, it is especially suitable for FMS because of their high-level mechanization and absence of worker-induced variability.

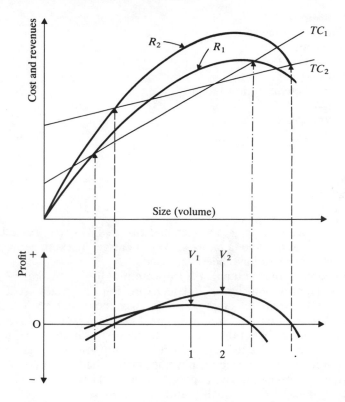

Figure 2   Nonlinear breakeven model. Subscript 1 = non-FMS production, 2 =
FMS production

$$\Pi = \alpha S - bS^2 - c - dS \tag{1}$$

where

$$R = \alpha S - bS^2$$
$$TC = c + dS$$

and

$$\Pi = R - TC$$

The derivative yields optimal output volume.

$$d\Pi/dS = (a - d) - 2bS_{opt} = 0 \tag{2}$$

This leads to equation (3) for optimal size.

$$S_{opt} = (a - d)/2b \text{ for max } \Pi \tag{3}$$

Π total profit
S output volume (size)
R = revenue
TC = total cost
a = unit price
b = market saturation factor
c = fixed cost
d = variable costs

Considering FMS, we find that –

a increases
b decreases
c increases
d decreases

Whereby, we state that, for maximal profit, FMS requires a larger S, greater investment, and a more comprehensive marketing programme than might otherwise have been surmised.

It can be seen that optimal size (represented by S_{opt}) is independent of fixed cost (represented by c), but is a function of variable costs (represented by d). From our empirical data, we know that FMS technology requires a substantial investment in fixed costs. When this investment is made, it is with the belief that a major reduction in variable costs (d) can be obtained. There is increasing evidence to support this belief. Quality effects (which to our knowledge have heretofore been ignored) show up in the revenue component. The decrease in price elasticity (due to improved consumer perceptions of quality and increased variety) is reflected by a lowering of the saturation factor (b). This change in b can be pictured as a reduction in the discount factor which allows the achievement of further increases in sales volume. The net result is that a larger spread between the breakeven points occurs and optimum profitability requires an expansion of the organization. Indeed, user experience has corroborated this important result. The implication is that if growth is a necessary consequence, the window between FMS and dedicated production narrows.

The appropriateness of breakeven analysis for a multiproduct company merits some discussion. Job shop production is unsuitable for such an analysis as the multiple products, each with their own set-up cost, produce a discontinuous step function for the total cost curve. FMS have set-up costs that ideally approach negligible values. This restores the smoothness to the total cost curve and makes the situation amenable to breakeven analysis. Thus, we can compare the high-volume, single-product production process with the high-variety FMS process using this model.

Marketing Considerations

The lack of an explicit marketing cost factor in the breakeven analysis is based on the assumption that such costs are part of overhead. It is believed that production overhead reductions should approximately cancel marketing increases. Examples are shop supervision (reduced since direct labour is less), plant facility costs (higher machine utilization means less factory floor space is needed), lowered scrap and rework costs and lowered prototype part development costs.

User experiences appear to corroborate the various conclusions drawn from the model above. Companies able to provide a varied offering of products increase their market share both by offsetting boredom of individual consumers and by appealing to a larger number of specialized segments of consumers. But, while variety has marketing advantages, it is accompanied by higher costs and increased complexity. Quality and variety increases were assumed to operate in tandem externally to decrease the price elasticity of consumers. The internal effects are much more subtle. Expansion of the product line primarily affects the variable marketing costs. Quality improvements are reflected in higher fixed costs of production, but increased variable testing costs also may result.

As an organization moves from a 'make to order' basis to one that incorporates the additional ability to manufacture a wide variety of new parts, a restructuring of marketing strategy and sales force behaviour may be necessary. For example, job shops normally are demand driven – marketing instructs production which parts are needed to fill customer orders. As output expands under FMS, the buildup of inventories will give rise to a reverse information flow as production instructs marketing which products to push. The size of the marketing department may have to grow along with its capability to achieve fast turnover for a variety of items which management produces to fully utilize its newly acquired, highly capital intensive equipment. Finding customers for new products may be more difficult than anticipated which could result in larger inventories. So a balanced production–marketing planning strategy must be put in place. In a well-designed organization, increased inventories should be a temporary phenomenon, and not a fundamental characteristic of FMS.

Group technology is a method of classification and coding that identifies and sorts related parts so that design and manufacturing activities can take advantage of their similarities. If group technology is used as the marketing emphasis, salesmen can be organized to sell families of specialized components. Personnel would focus on aggregating parts which share similarities in design and production. This perspective insures minimum transition costs. External marketing economies result from reduced customer search and sales targeting activities. Internal savings accrue from a reduction in the number of production line shifts.

Alternatively, modular production concepts may be utilized to focus on products. The principle of modularity is to design, develop and produce the minimum number of parts (or operations) that can be combined in the maximum number of ways to offer the greatest number of products (or services).

The ramification of using FMS in conjunction with a modular production philosophy is that marketing's responsibilities will increase many-fold. For example, while FMS may allow production planners to manufacture a certain number of new parts, the number of new products assembled from them may be orders of magnitude larger as a result of the combinatorial magnification.

Other considerations revolve around the distribution and demand patterns which characterize the new output. Large changes are possible and may be desirable if they serve to insulate the company from cyclical or seasonal fluctuations of demand for its predominant product line. Benefits from product diversification will, however, be accompanied by increased costs. The expenses of establishing and operating new, and perhaps unfamiliar, distribution channels could be substantial.

In any case, marketing personnel in an FMS environment cannot maintain a passive, order-filling job shop mentality. Instead, they must assume an active, aggressive search to locate buyers for the expanded output. A consequence of selling this increased variety often will be higher marketing costs. This will cause the slope of the total cost curve to increase (i.e. d increases somewhat but remains below the pre-FMS level) and acts as a brake on the expansion predicted by our nonlinear BEA.

Managerial Implications

Up to now, our discussion has assumed that organizations will adapt their size to operate at the point of optimal profitability. The assumption that firms strive to attain this optimum is open to debate however.[6] The advantage of limited liability that incorporation confers carries with it the creation of an independent business bureaucracy that is divorced from the firm's owners. That corporate bureaucracies pursue their own objectives is illustrated by the recent spate of 'poor' acquisitions which aim at lowering takeover risk by reducing profitability. Large corporations are especially susceptible to such behaviour since the various managers have relatively minor personal stakes in the profits of the firm. Empirical research has shown that these very firms are the ones most likely to be the carriers of innovational changes such as FMS.[3,4]

These behavioural factors relate to the predicted growth in organizational size. By and large, this growth will come about via the opening of new markets. Market penetration requires different managerial skills than those used in business expansion. Launching a new product calls for ambition, imagination and good timing if entry is to be successful. The limited amount of available managerial resources may not be sufficient to undertake such a tremendous effort. Production responsibilities may also be increased as the computer controls of FMS elevate a significant fraction of manufacturing supervision from the shop floor to the executive suite. Consequently, management may decide to produce a satisfactory rather than optimal level of output. Advancing a step further, the foreshadowing of these increased marketing and production efforts may lead top management to reject FMS, as it needlessly increases their burdens.

The earlier discussion on quality and cost factors revealed how traditional accounting-based evaluations are biased against FMS. Since the effects of flexible manufacturing extend beyond production into marketing, new product planning, labour relations etc, FMS changes the way a company does business and can only be evaluated from a top management point of view. Adequate consideration therefore requires a comprehensive review of a company's products, markets and objectives. The fear or inability of corporate officers to undertake such analyses, as evidenced by the tremendous growth of strategic consulting services, must also be taken into account. (Revenues from strategic planning consultancy in 1980 alone are estimated at up to $1 billion.)

The desire to be profitable against international competitors (who are moving rapidly ahead with FMS investments) may counter a large corporation's tendency to satisfice instead of profit maximize. This leads us to conclude that large corporations will innovate as shown by a partial list of current FMS installations: Allis Chalmers, Bendix, Caterpillar, Chrysler, John Deere, General Electric and Hughes Aircraft.

We raised these issues in order to illustrate that FMS proposals are likely to encounter more than the usual resistance to change because of their wide-ranging organizational effects. These will (and should) only be overcome if a very good fit between production requirements and flexible manufacturing exists.

Work Force Adjustments

Great savings in labour expense result in the switch from job shop production to flexible automation. A General Electric FMS for locomotives in Erie, Pennsylvania, demonstrates this potential. The system incorporates 13 machine tools and has reduced staffing requirements from 64 to 8 workers.[12] The West Germans have been particularly aggressive in moving to exploit FMS technology. The $50 million Messerschmidt–Bolkow–Blohm (MBB) complex in Augsburg was completed in 1981 after 10 years of planning and construction. This system ranks as one of the most advanced machining centres in the world and represents the only significantly automated machining plant within the entire airframe industry. When compared with previous manufacture on stand-alone NC equipment, the automated system has allowed for a 52 per cent reduction in personnel. Half a world away, an equally impressive FMS is busy churning out industrial robots and small machine tools. The $84 million Fanuc plant in Fuji, Japan, commenced operations in December, 1980. The 29 automated machining cells operate with only 1/5 the personnel requirements of a conventional facility.[11]

These dramatic cuts in staffing are largely responsible for the variable cost improvements and are the prime motivators behind flexible automation of batch production facilities. When viewed from the workers standpoint however, these reductions are devastating. The social and moral effects of job displacement – indeed, job elimination – must be addressed. What is the ultimate fate of this tremendous pool of skilled personnel? The challenge of finding alternate employment that will use the full potential of these workers is one that has not even begun to be considered. Managers charged with implementation of FMS should be aware that failure to reach equitable solutions will only exacerbate union resistance to further automation.

FMS Design Concepts

The trade-off of set-up cost levels and output variety provides further interesting insights. In conventional job shop operations, machine tools are only cutting for a fraction of the available time. The balance is spent in manual operations such as fixturing (attachment of the part to be machined onto a rigid support), alignment of the assembly on the machine tool, selection of cutting tools and speeds, etc. The tremendous labour intensity of this process, together with lost production time are collectively referred to as set-up costs and serve to restrict the output variety. Figure 3 reveals the inverse relationship between small, but significant, set-up costs and output variety.

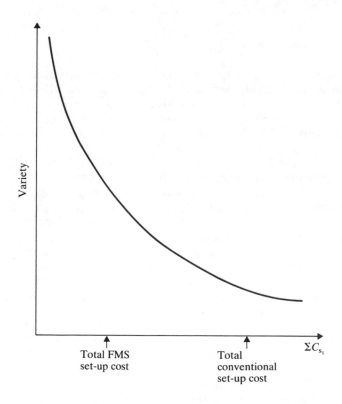

Figure 3   Set up cost – output variety trade-off

As the set-up cost levels approach zero the plant can continuously alter its output mix to match demand. This situation produces maximum output variety. If the set-up cost level is finite, then optimal combinations of products must be found. Solving this combinatorial problem represents a great opportunity to achieve a novel method of FMS design. This can be viewed as the feedback loop depicted below.

A	*Objective A:*	Determine profit margins for each part
B	*Objective B:*	Establish constraints of part demands and machine resources
C	*Objective C:*	Maximize objective function [Max$\Sigma N_i\Pi_i$–programming function]
D	*Objective D:*	Evaluate scheduling characteristics (i.e. job waiting and machine idle times)
E	*Objective E:*	Evaluate shadow prices on constraint functions above. Adjust resource levels to maximize profit

The optimal mix can be determined from an LP-type formulation.[14] An objective function could be constructed using the profit margins for each part type (i.e. $\Sigma N_i\Pi_i$, where Π_i = profit margin for ith part and N_i = number of ith parts produced) less a programming cost function. The programming cost function would reflect the increased control expenses that accompany large variety. Note that profit margins are set outside the feedback loop. This allows managers to determine the effects of margin adjustments on the production mix. The objective function would be maximized subject to the following constraints:

(1) Demand levels for each part.
(2) Machine resource limitations (these would be derived from an assumed system – i.e. 40 hours of five-axis lathe cutting available).

The shadow prices from the machine availability constraint can then be used to determine resource utilization rates. Additional machining centres or facilities may then be added and the LP repeated. After a number of iterations, an excellent fit between parts requirements and system capabilities should result. It is worth noting that such an approach cannot be used to determine the best job shop output. Earlier workers have demonstrated that this problem is NP complete – hence, not amenable to solution by optimization techniques.[10] This can be attributed to the fact that costs depend not only on the part mix, but also on the sequence of production due to the discontinuous set-up costs.

We believe that our procedure would be superior to current FMS design practices which generally follow a two step sequence:

(1) A family of parts having similar production requirements is grouped together.

(2) A system capable of manufacturing this output mix over a particular time span is designed and built.

With the parts mix fixed, engineers and production managers subsequently spend an inordinate amount of time 'fine tuning' the control logic and system hardware to maximize utilization and productivity. Recent work of Buzacott and Yao reveals a flaw in this method however.

One of the intriguing results from the use of analytical models has been the close agreement between stochastic models which summarize part processing requirements by probability distributions and the simulation models which capture all the detailed part information. This suggests that there can be effective control without collecting all available information from the system.[5]

Their findings indicate that as long as parts are approximately similar, fine tuning will lead to only marginal productivity improvements. Therefore, matching the system as closely as possible to the parts mix is not nearly as important as selecting the optimal mix. Our hypothesis is that larger gains in profitability could be realized by exercising more care in the selection of the production mix.

References

1. Barash, M.M., 1980 Computer integrated manufacturing systems. In *Towards the Factory of the Future* (Edited by Kops, L.). American Society of Mechanical Engineers, New York
2. Barash, M.M., 1982 Computerized manufacturing systems for discrete products. In *Handbook of Industrial Engineering* (Edited by Gavriel, S.). Wiley, New York
3. Benvignati, A.M., 1982 Interfirm adoption of capital goods innovations. *Rev. Econ. Statist.* p. 330
4. Braun, E. 1981 Constellations for manufacturing innovation. *Omega* **9**(3): 247 (Reading 3.2)
5. Buzacott, J.A., Yao, D.D.W., 1982 Flexible manufacturing systems: A review of models. Department of Industrial Engineering, University of Toronto, Working Paper 82–007
6. Child, J., 1980 *Organizations: A Guide to Problems and Practice*. Harper and Row, New York
7. Church, J., 1982 Flexible integrated simulation tool (FIST). IIT Research Institute, 10 West 35th Street, Chicago, Illinois 60616
8. Closuit, E., Wamba, P., Rykels, S., 1982 Manufacturing technology for advanced machining systems. First Interim Technical Report, March, 1982; Second Interim Technical Report, June, 1982; Third Interim Technical Report, September, 1982; Fourth Interim Technical Report, December, 1982; Boeing Military Airplane Company, Advanced Airplane Branch, PO Box 3707, Seattle, Washington 98120
9. Fox, K., 1982 Cincinnati Milacron simulation scheduler model. Cincinnati Milacron Inc., 4701 Matburg Avenue, Cincinnati, Ohio 45209

10. Garey, M.R., Johnson, D.S., Sethi, R., 1975 The complexity of flowshop and jobshop scheduling. Technical Report 168, Computer Science Department, Pennsylvania State University
11. General Dynamics Corporation 1982 Manufacturing technology for advanced machining systems. First Interim Technical Report, March 1982; Second Interim Technical Report, June 1982; Third Interim Technical Report, September, 1982. Air Force Integrated Computer Aided Manufacturing Program (ICAM). These documents were obtained from Mr. Chester D. Beerid, General Dynamics Corporation, Box 748, Fort Worth, Texas 76101
12. Iverson, W.R., 1982 Flexible automation woos manufacturers. *Iron Age*, February 19, 48
13. Pratt, C.A., 1982 Simulation tools for manufacture. *Simulation* October
14. Stecke, K.E., Solberg, J.J., 1983 Formulation and solution of nonlinear integer production planning problems for flexible manufacturing systems. *Management Science* **29**(3): 273–88
15. Stecke, K.E., Solberg, J.J., 1977 Scheduling of operations in a computerized manufacturing system. NSF Grant APR74 15256, School of Industrial Engineering, Purdue University, West Lafayette, Indiana 47909

PART 6
New Technology and the Organization of Work

6.0

Introduction

The final part concentrates on the work-force-related issues of implementing new technologies. The term work force is used in a general sense to refer to all levels and categories of employee within a production unit. The case study associated with this section is reading 6.1 (Snackco) which is a research-based case study set in the food industry but fictionalized to protect the identity of the firm and work force in conditions which are still sensitive. The case study centres around the introduction of more integrated and higher speed process equipment which cuts across existing patterns of work organization and of job demarcation. One of the consequences was conflict within management and between shop floor groups but the case study indicates the potential for overcoming these and channelling them in a constructive direction through early – and genuine – consultation and by negotiation. It also indicates the possibilities for choice in technical matters that are taken up in some of the part 6 readings and the constructive potential that is available in the experience of the shop floor work force. This relates to the accommodation of changes to plans that may become necessary as experience grows, and to the development of more satisfactory solutions than may emerge from a solely technical approach.

Rosenbrock in reading 6.2 reflects on the probable direction of technically oriented approaches. He is concerned at the persistence of Taylorite attitudes and approaches to the organization of work, particularly within research and development work. This is despite the manifest spuriousness of the 'scientific' basis of Taylorite hypotheses which are in reality ideological and which can often be shown to be inimical to the long-term efficiency of production systems. A Taylorite perspective may find new technologies providing new opportunities 'for eliminating human skill and for subordinating people to machines'.

Rosenbrock, (here concerned with microelectronic technologies) sees the acceptance of Taylorism as being related to a view that accepts as knowledge only that which is derived from formalized scientific activity. In contrast he sees the great body of knowledge as being derived from experience and suggests that there are two interdependent types of knowledge – explicit scientifically derived knowledge, and tacit knowledge and skills, of which the latter is by far the larger category. While rejecting the proposition that skills should be artificially preserved, Rosenbrock says that approaches to technological change which consciously seek to utilize tacit knowledge and skills are more likely to yield acceptable results for all those concerned.

The issue is taken further in reading 6.3 by Walton who considers the issue of social choice in the specific case of information technology. Rejecting the notion of technological determinism which is implicit in many programmes of technological change and development, he points to the adverse consequences of systems that deskill and demotivate those who are left with the task of operating them. He points out that not only can choices in technological design often be made without sacrifice of economic purpose but that the possibilities are extended by the low cost and flexibility of modern computing systems. That such choices are not made is because the processes of design and planning are generally abstracted from consideration of their human effects. This is left to be somehow sorted out through consultation and other means when installation takes place and policy options are closed, or at best, limited. Walton advances a number of suggestions for guidelines to improve the identification of choices and their development through a dynamic interaction of design and implementation.

A concrete example of the range of different possibilities of using given technologies, and of the consequent benefits is provided in reading 6.4 by Hartmann, Nicholas, Sorge and Warner. They stress that 'it is inadequate to consider a technology as given and to observe the effects which follow from it'. Clearly the extent and nature of choice will vary from one technology to another. In the case of the utilization of computer numerically controlled (CNC) machine tools, they see a considerable degree of choice and show as do a number of other research studies that rather different routes have been taken by British and by German companies. The different routes are essentially a matter of corporate and departmental choice and policy.

Important among the factors influencing the choice of routes are: (1) those elements of national cultures which may shape patterns of work place organization; and (2) differences in the levels and continuities of skills in the two populations. The differences are reflected in the location, on or off the shop floor, of planning and programming tasks. The more skilled German operators are more likely to be involved in these tasks than their British counterparts. However, since considerable differences in CNC utilization are also found within the two countries it would seem that ultimately contrasting choices reflect differing calculations of economic success. One of the tasks facing those (like trade union groups) seeking to shape decisions on technological change is to demonstrate in acceptable terms the benefits that may be yielded by different options.

Trade union groups are sometimes regarded as providing a major, if not the major obstacle to technological change. The reality seems to be that with a few notable exceptions this is a Fleet Street fostered mythology reflecting Fleet Street's own 'problems'. These 'problems', perhaps coincidentally, provide a highly effective barrier to would-be new entrants to the industry. Yet work place groups *can* provide hurdles that have to be overcome if technological change is to be successfully accomplished. Hurdles are, however, likely to be found at any point within a production unit – not just on the shop floor. Keen in reading 6.5 examines 'social inertia' and other factors within organizations which may impede technological innovation or 'damp out' their intended effects. Looking at changes in information systems in particular he identifies a number of causes of social inertia – causes which contribute to experience of change as rarely dramatic but more typically as an incremental or 'muddling through' process. In essence his framework of explanation rests like that of earlier authors (like Bower) on the gap between what actually happens and perceptions or policies of change based on assumptions of rationalistic decision making. In practice, complex organizational 'political' processes tend to shape change. In explaining this, he draws on the pluralist model of organization. Despite assumptions of common purpose, different groups within management and elsewhere perceive the issues in different ways and may adopt tactics of counter-implementation where they believe proposals either to be inimical to the 'real' interests of the organization or to be threatening to the 'legitimate' interests of the group. Keen discusses a number of possible 'counter-counter-implementation' approaches, essentially resting on the acceptance of gradualist or 'tactical' approaches to achieve strategic redirection and the mobilization of coalitions in support – a reflection perhaps of Japanese 'consensus building'.

Rothwell's study of the impact on new technology on supervisors (reading 6.6) brings together the issues of choice in the selection, application and use of new technologies, and the ability of interest groups to deflect change. She found that even though supervisors may play a central part in some of the stages of implementation, they are often one of the last groups whose role is considered in detail. In such cases their impact on the planning of change is likely to be marginal and they are unlikely to receive adequate training to enable them to cope with the new issues they may face. In addition, technological change has considerable potential for an unforeseen erosion of supervisory roles, status and authority. Yet Rothwell's case studies also demonstrate that there are considerable possibilities for the development of the supervisory role and the enhancement of both the authority and credibility of supervisors. This depends upon a clear analysis of supervising roles as part of the early planning process and may lead to a shift in emphasis towards the enhancement of aspects of supervision such as the technical role or that of 'people motivation'. However, such analysis can also bring into question the necessity for retention of the supervisory role.

Reading 6.7 provides a further element of international comparison (Britain, France, the USA and Germany) – contrasting experience of movement towards FMS systems (also considered in reading 5.7). The starting point of the reading is the role of manufacturing innovation as a strategy for responding to environmental uncertainties. Thus while Gerwin and Tarondeau chose to look at what, in its essentials, is much the same technology in all four cases, they found that FMSs were introduced for rather different markets and other reasons. They also found variations in the strategies for FMS adoption and implementation and of the problems arising during implementation. The range of problems they identify provides an instructive list. They identify four strategies for coping with implementation uncertainties: the development of a human and technical infrastructure capable of supporting the innovation; installation in assimilatable stages; the involvement of operating managers in adoption and implementation; and the use of skilled operators in proving the new equipment.

In the final reading, Hatvany, Bjørke, Merchant, Semenov and Yoshikawa consider some of the broader social and technical questions raised by the extending application of 'advanced' – i.e. highly automated and integrated – manufacturing technologies. Apart from the very considerable issues of the training and skill levels demanded by advanced systems they also examine the implications of the greater adaptability that may be necessary. One related factor is the level of job satisfaction that, increasingly, has to be designed into the production system as a matter of necessity. They also point out that cultural differences may shape the extent to which adaptability becomes a problematic issue. Also, the shift to advanced systems may imply that some sectors of the population are regarded as more employable than others. In addition it raises new issues of industrial location because of the much smaller numbers required to operate and manage conventional production systems.

Snackco: Negotiation over the Introduction of High-speed Equipment

Robin Williams

Introduction

This case study is an example of negotiation over technological change conducted not through a formal technology agreement, but by the elaboration of 'conventional consultative procedures'.

The technological change studied trebled labour productivity in the most labour intensive part of the production process. Labour inputs were not substantially altered, but management sought to overcome industrial relations problems and improve efficiency of operations by a profound change in occupational roles (in particular the demarcation between production and maintenance workers and unions).

Characteristics of the Industry

The case study was selected from the food, drink and tobacco industry. This industry was ranked second only to electrical engineering in terms of rate of uptake of microelectronics in production processes (Bessant, 1982). The food, drink and tobacco industry can be broadly characterized as follows:

(1) It sells primarily to the public through retail outlets – with a greater role for distribution, advertising and marketing in the enterprise.

(2) Much of the industry is concerned with producing a limited range of standard products, through very high-volume production.

(3) Although there is often a long tradition of trade union organization (and this was the case in the firm studied), for production workers at least, this is not based on craft skills. The bulk of production jobs are defined as semi- or un-skilled (like materials handling). Though there may be areas of

higher skill, they are highly process/industry specific. Related to this, is the fact that the industry employs a high proportion of female labour in production as well as administrative jobs.

(4) Though the industry was highly labour intensive until the 1940s, there has since been an increasing degree of automation. Many sectors of production could be characterized as flow line production (especially raw materials processing) and current changes were increasing the integration of finishing and packing processes where the bulk of remaining labour is employed.

The effects of the recession on sales in these industries has been less dramatic than in other sectors. However 'consumer loyalty' to traditional brands is decreasing. The advertising and other costs of defending old brands and launching new ones were extremely high. Companies sought to minimize production and distribution costs to provide the funds to defend their sales in a declining market. A number of leading companies have embarked on major restructuring and reinvestment programmes, involving significant job loss.

Industrial Relations at Snackco

In 1981 Snackco in its three divisions employed approximately 15,000 workers broken down into groups and represented by unions as follows:

(1) Management	800	
(2) Staff	3,600	Mainly Snackco Staff Association, followed by ASTMS and some members of staff sections of general unions.
(3) Factory Supervisors	750	ASTMS, and staff sections of two general unions.
(4) Craft Engineering	1,500	AUEW (fitters) 450; EETPU (electricians) 450; and five other craft unions.
(5) Production	8,500	In three industry and general unions.

Negotiations for wages and conditions of employment were conducted on a central company-wide basis. Other issues were handled at plant level with a provision to refer disputes to the national procedure.

Though there are uniform wage scales for manual and clerical staff, negotiations are conducted separately for the above groups of workers.

Wage rates in the company are high, and there are excellent welfare facilities. Manual basic wages in 1983 ranged on a ten-point scale from £100–£150 per week with most workers in the middle range. The company combined a paternalistic image with a highly professional industrial relations team.

There was no formal 'Redundancy Agreement' at Snackco. However, the company offered 'very generous' redundancy payments including the option to retire on company pension at 50 (the retirement age for men at Snackco was 60, five years before the state retirement age).

There were additional financial incentives for workers to accept change. In particular an agreement in 1978 provided for 'change payments' – lump sum payments to both displaced and remaining workers where jobs were eliminated through technological and organizational change, that had given some workers between £10 and £400 in certain circumstances. In addition a bonus scheme yielded a wage supplement for increases in cigarette output per head. This was about 10 per cent of basic pay which was periodically consolidated into basic pay rates. However, payments in recent years had been lower or nonexistent due to the declining sales and surplus labour.

The Introduction of High-speed Cigarette-making and Packing Machinery at Snackco

This case study focuses on the introduction of high-speed making and packing equipment (HISE) into Alpha Division, one of three divisions of Snackco. This was the most important part of a £30 million investment programme initiated in 1979.

Apart from its ability to operate at up to three times the speed of conventional making and packing equipment, the HISE equipment was in a form very similar to its predecessor, with many complex mechanical parts, although it incorporated microprocessor-based monitoring and quality control features, leading to claimed greater reliability and less operator involvement.

Two divisions of Snackco (Alpha and Beta) were introducing the same HISE equipment. However a radically different pattern of implementation emerged. At Beta the HISE machines were installed alongside conventional machines in the separate making and packing department in several of the plants (figure 1).

At Alpha a decision was made:

(1) to introduce all its HISE equipment into its modern 'Sunrise' factory;
(2) to link making and packing equipment in the same department;
(3) to use this opportunity to conduct extensive job redesign in particular involving an integration between the traditionally separate maintenance and production functions.

The Structure of Alpha Division

These technological changes coincided with substantial changes in corporate and management structure.

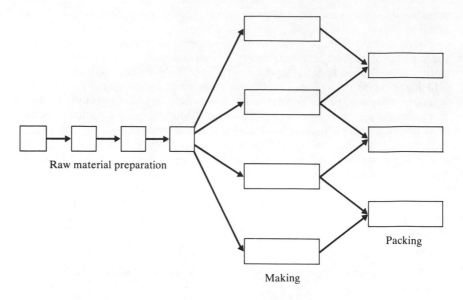

Raw material preparation

Making

Packing

Figure 1   Simplified diagram of production process

In 1970 when the Sunrise plant was opened, Alpha Division had 9,000 workers in three older factories in the same provincial town, as well as one in Scotland. By 1978 the two older provincial factories had been closed and their work transferred to the Sunrise plant, which by then accounted for 65 per cent of Alpha's production. In 1982, faced with further decline in markets, management announced the closure of the Scottish factory (with a loss of 500 jobs) by the latter half of 1983. Subsequently the remaining provincial factory was progressively run down to a third of its capacity as a prelude to consolidating all snack production at the Sunrise plant by spring 1984 with less than 3,000 employees.

Alpha Division had maintained separate managerial structures for engineering and production functions. Thus craft maintenance engineers at factory level reported not to the factory manager but the chief engineer who was in turn responsible to the technical director of Alpha Division. Similarly, line management in production reported through the factory manager to the production operations manager who reported to the operation director.

Subsequently substantial changes in management structure took place following the centralization of administration at Snackco head office. The separate Alpha and Beta boards were dissolved and the management structure was severely pruned (between January 1982 and March 1983 the numbers of Alpha managers were reduced from 220 to 120). When the new structure came into force, in November 1983, the factory manager for the first time took over direct responsibility for the engineers on the site.

Certain functions were transferred to Snackco headquarters – for example the machinery evaluation section, leaving just a service engineering function at Sunrise concerned with maintenance and minor amendments to the technology.

These changes, which were beginning to be implemented during the case-study period, created considerable internal friction and resistance among management circles.

In 1983 the Sunrise plant employed about 2,000 workers. The size of the different departments and their approximate contribution to total employment are as follows:

(1) Primary production	– preparation	5%
(2) Secondary production	– Making and packing (divided approximately equally between making (conventional equipment), packing (conventional equipment), and the high-speed making and packing department	50%
(3) Engineering (maintenance)	– maintenance of plant and equipment (craft engineering workers)	13%
(4) Ancillary manual operations	– despatch, transport, canteen	15%
(5) Management and staff		17%

The plant operates on a two-shift system with a 35 hour week.

The Traditional Technology, Organization and Work Force in Preparation Making and Packing

Primary raw material preparation

This had become highly automated over the previous decade. Automatic quality control and process control had been progressively increased. Raw material preparation was virtually flow production and was computer controlled. This had affected relatively few jobs, since most labour had been displaced from raw material processing 20 years previously. Before this, the area had been the most labour intensive area.

Secondary making and packing

Making and packing had traditionally been conducted in separate departments and employed the vast bulk of manual labour. Both processes had utilized complex mechanical equipment, designed on the same basic principles, for over 20 years. The processed raw materials are piped to the individual making machines where they are moulded and cut. The finished product is then loaded automatically onto trays which are taken to the packing machines. The snack-bar making machinery can handle a limited range of product specifications without major adjustment.

The packing equipment is more complex in principle and operation than in making. The product is bundled and foil wrapped around it. Then a packet is built around it from cardboard blanks. The packets are then wrapped in cellophane. Packing may be integrated with 'crating', i.e. wrapping the packets into larger packs etc. Packing machinery is less flexible than making machinery and is harder to adjust (e.g. for different size packets).

The main technological changes have been in materials handling. For example, the introduction of automatic loading of products into trays for transfer to the packing machines eliminated two 'crew girls' on each making machine a decade ago.

The manual functions involved in making and packing can be differentiated in order of status and 'skill' level as follows:

(1) Unskilled: Transport of materials, cleaning,
    (largely men)   'ancillary' operations.

(2) Semiskilled: 'Feeding and watering' – i.e.
    (largely women)   loading raw materials on to the
                                        machine, Quality control –
                                        inspection of product (particularly
                                        in making).

(3) Skilled – narrow band skills: Low   – Stopping and
    (largely men)   starting machines,
                                                   removing
                                                   blockages.

    Defined as production   Medium   – Operating
    activities.   adjustment to
                                                   machine.
                                    High   – Diagnosis of faults,
                                                   and rectification.

(4) Skilled – broad band skills: Low   – Servicing machine.
    (all men)
                                    Medium   – Routine
                                                   maintenance.
    Defined as craft   High   – Repair and
    engineering/maintenance   overhaul of machine.
    activities.

The packing jobs are considered more difficult and have higher status and pay than making. The traditional method of recruitment was for production workers to start with low-skilled jobs. Their first move into 'secondary areas' would be to 'making jobs'. Those with ability and experience would be promoted to packing and from lower skilled to higher skilled jobs. There was a traditional demarcation between production and craft engineering (maintenance) workers. Maintenance workers had responsibility for machinery breakdowns and overhaul/repair. Other aspects of setting and maintaining the machine during production remained with skilled production workers – the packing mechanics. Under this demarcation, maintenance workers were not allowed to run the machine and production workers could not deal with problems that involved 'gears and bearings'. In the making department, maintenance workers had a much greater involvement with responsibility for some fault rectification and repair whilst the machinery was in production. In 1982–83 the operating arrangements in making were reorganized. Routine production maintenance and fault rectification were given to making mechanics, supported by 'tool carrying' supervisors known as section leaders.

There had been very rigid gender-based segregation of jobs, with women concentrated in the semi-skilled crewing jobs (type 2 above) on low to middle grades (but not the lowest 'labouring grades'). Formal barriers to recruitment and promotion were eliminated over a decade ago with the 'Equal Opportunities' Agreement. Some jobs had become 'unisex' and a few women were beginning to enter the more skilled jobs. However the old sexual division of labour was largely intact.

The Development of the HISE Proposals

High-speed making equipment was available as long ago as 1972 while packing machines became available in 1976. Even at this stage, one manufacturer had already experimented with linking making and packing machinery. The Alpha Division, technical development area began technical evaluation of this system over several years. This combination (maker, mechanical link, packer) was subsequently introduced when the HISE area was established in May 1980, but had to be withdrawn 12 months later because of high 'down time' in production conditions (although it had performed adequately in the technical development area with very high levels of manning and technical support). Alternative manufacturer's high-speed equipment which had been evaluated in the intervening period was acquired and introduced.

After reviewing the 'economic arguments' for recapitalization, in 1978, the Snackco Board decided to buy 24 pairs of HISE making and packing machines for Alpha (and a similar number for Beta). Divisional management were given responsibility to determine the implementation strategy, operating arrangements and industrial relations policy. A management study group was set up to consider these issues for Alpha Division. This included many younger managers on the grounds that they would be the ones who would end up running the new arrangements and were able to operate more creatively than older managers, freer from traditional assumptions and less disposed to defend their 'empires' than senior managers.

The study group started with a three-day brainstorming session away from the factory. They identified the fundamental problem as being the demarcation between production and maintenance and that the introduction of HISE would be a battleground between different affected groups. The group set its objectives as identifying and eliminating the worst practices (viz. demarcation) and reinforcing the best practices like the identification between production workers and machines.

The deliberations and decision of the group were as follows:

(1) Location of machines: it was considered that if HISE was introduced in a fragmented way alongside existing machines, existing work practices would be taken over. It was decided to set HISE in a self-contained section to encourage new attitudes. Making and packing machines would be installed side by side in 'combination units'.

(2) Organization of production: total flexibility of skill and labour allocation was sought. It was considered necessary to retain the distinction between 'high', 'medium' and 'low' skilled roles on the grounds that not all workers would be capable of the more skilled tasks. Work groups would be installed to operate several combination units on a self-relieving basis. Although initially they would only operate one type of machine, the ultimate objective was that workers would be capable of operating both packing and making machines.

(3) Maintenance: it was decided that the HISE area should contain all the skills needed to carry out rectification and adjustment for production as well as to deal with breakdowns. In the longer term these skills would be sought from craft engineers. However it was initially necessary to draw on existing skills in production as well as engineering areas (with appropriate retraining). This was not just a technical, 'resource utilization' decision but reflected also 'political' and industrial relations considerations. It was anticipated that packing mechanics would resist their displacement from front-line maintenance functions. A grade of 'machine specialist' was created which merged craft engineering and production mechanic functions. To give technical and diagnostic help, a 'trouble shooter' grade was created (this also provided a niche for selected section leaders).

Engineering management argued that craft engineering workers would not take orders from production foremen (the senior supervisor of a shop). A role of 'technical foreman' was created between these two levels, with appointments given to existing supervisors from maintenance areas. These 'technical foremen' reported to the 'production foreman'.

By January 1979, by a process of 'logic and accommodation' a set of occupational roles and responsibilities had been outlined, virtually identical to the ones eventually introduced.

Another management group was established to work out the detail of implementation – manning levels, grading, training, hours of work, timing and rate of change, redundancies and the strategy for negotiating change. This detailed advance planning was conducted in order to eliminate the technical problems of implementation. Management wanted to be sure that the proposals would work *before* opening negotiations with unions, to avoid being diverted by union objections.

So far, knowledge of the proposals had been tightly restricted (because of anticipated opposition from the unions and sections of management). A presentation was made to the rest of local management which provoked a 'considerable reaction'. However the study group were gradually able to overcome these objections to sort out 'real' problems and to identify objections that were based on sectional interests or prejudice.

Finally discussions were held with Snackco management to seek approval for these arrangements. By October 1979, the go-ahead had been given and management were ready to approach the trade unions. A negotiating strategy had been developed.

Management Strategy

The problem management faced was of how to negotiate a change which affected workers in four different bargaining groups (production, craft engineering, supervision and staff) and had a particular impact on three influential groups of workers:

(1) packing mechanics (who were the top of the hierarchy),
(2) craft engineers (who had sought to establish a separate identity in 1974), and
(3) section leaders (front-line, tool-carrying supervisors).

Management sought to bring these groups together to cooperate in a single hierarchy. Management believed that the new arrangements could not be introduced without the consent of the work force, and therefore paid particular attention to the question of consultation.

The principles of the management strategy for negotiation can be summarized as follows:
(1) To set up a joint forum to make its presentations about the changes to all the unions (supplemented where necessary by direct negotiations with the individual bargaining group). This was known as 'the composite group'.
(2) To 'share' the misery (and the benefits) as evenly as possible between all groups. Each existing group of workers would be offered some HISE jobs, with opportunities for upgrading where possible.
(3) To encourage trust between management and union by 'even-handed' operations, by only committing themselves to matters they were willing to implement and by being as open as possible.

The Union Strategy

Given the economic situation in the company the shop floor union membership at Sunrise did not feel that a strategy of resisting the technology was viable. If they accepted HISE they would lose jobs. If they resisted, the technology would go elsewhere in the company and their future might have been further threatened (especially in a multiplant company with surplus capacity which could switch production between factories). Given this state of affairs, and that the union nationally was not directing and coordinating the union response, the problem for the unions was not of *whether* the change should take place but *how*.

Between the different bargaining groups at Sunrise, relationships were 'reasonable'. After the initial presentation, the unions realized the dangers of groups being played off against each other and seeking sectional advantages at the expense of others. The unions therefore supported the formation of the 'composite group' and sought to extend its role in consultation. They also sought to extend involvement to the white collar groups (staff and supervisory). Furthermore, an agreement was reached that each group should have the right to veto the responses made by the union side at the composite group.

The Conduct of Negotiations

Management made its first presentation to the unions in September and October 1979. Consultations continued over a period of 12 months.

The initial response of the unions was one of extreme concern. At the first meeting, the craft engineers for example, said they wanted no involvement as machine specialists, but later retracted when the probable reduction in engineering jobs (which was proportional to the number of machines) was shown to them.

At the second meeting, unions presented a barrage of objections to the management's proposals. Management took a gamble and set a challenge to the unions, that if the composite group could work out a set of better proposals for running the new equipment they would listen.

The union side went away and considered this. After two weeks they came back and stated that they could not produce a better alternative. The union side had spent a whole day discussing possible operating arrangements together. Each section had been able to develop a solution that would protect the interests of their particular group. The unions found that management had worked out many of the technical and political problems. The unions did not have the time or the resources to develop fully worked-out proposals, and realized that if they had been able to conduct such an exercise, in a way that balanced the sectional interests, their solution would have been broadly similar to the management proposals. The only alternative scenario was the continuation of existing working arrangements. However the union side could see real benefits to their members from the new arrangements. Whilst the benefit for maintenance engineers from management's proposals was mainly a reduction in job loss, it also led to a direct involvement in production. For production workers, there were opportunities for increased gradings and job enhancement. The unions decided to 'hear out' management's proposals.

Management had been uncertain about the response they would meet from the unions. When the initial responses from the unions were less hostile than feared, management began to elaborate their proposals. Management emphasized their resolve to implement the changes, and that there could be no 'half-way house'. Management's conduct of the meeting was more one of *selling* the ideas than of *negotiating* proposals.

At no stage did the unions formally accept the introduction of HISE. Gradually an air of inevitability emerged. The discussion shifted from whether to introduce change to how to introduce change.

The main issues for consultation focused on selection of workers for the HISE section (which involved ancillary questions of pay, status and negotiating rights). For example, management had wanted to recruit HISE machine specialists equally from amongst craft engineers and packing mechanics. The unions were concerned that the making department had been excluded and insisted that they be given a share of the jobs. Eventually it was agreed that machine specialists should be drawn on an agreed ratio from all three groups, with three from engineering, three from packing and two from making sections.

In return for allowing engineering workers to be involved in production jobs, it was agreed that five mature apprenticeships should be offered to production workers to be trained as craft engineers (again the union insisted that this should involve making as well as packing mechanics).

The changes involved particular difficulties for the craft engineers to be recruited as machine specialists, since they stood to lose their independent status and negotiating rights as well as overtime earnings without any increase in basic pay. On two occasions this led to severe arguments when engineering managers gave their workers a different interpretation of how this would be resolved than was presented at the composite group.

Although formally gradings were to be determined under the Job Evaluation (JE) Scheme, and thus could not be sorted out until the job descriptions were finalized, management made it clear from the outset (in breach of the JE procedure) that machine specialists would be on the top manual grade (grade 10 – the same as the craft engineers).

It was agreed that both making operators and packing machine 'front girls' should be eligible for the leading operative position on HISE. It was expected that this would be a grade 7 job. In the event it was evaluated as grade 6, which involved downgrading for the ex-making operators – some of whom therefore went back to their old jobs. A major change regarded selection for jobs – which had traditionally been based on seniority. Management sought agreement for selection solely on the basis of aptitude. This was accepted (to management's surprise) on the proviso that length of service was one aspect of the selection criteria.

The problems of loss of earnings for engineering and other workers was eased during the implementation period by enhanced 'experimental pay' of £5 per week. The unions were not very happy about the proposed work force levels for HISE equipment but eventually accepted them with minor modifications. The unions felt that efficiency would be impaired with such small crews and felt that management would be obliged to revise them.

The Relationship between Union Negotiators and Work Force

The negotiations between management and the different bargaining groups were very complex. The negotiators saw 'a lot of sense' in management's proposals. However many of these cut across traditional job demarcations and sectional interests. While the negotiators accepted the need for a compromise solution between the interests of the different groups of workers, the rank and file union members continued to view the situation 'parochially'. Negotiators found it necessary to accept certain issues like the erosion of demarcation, that would not necessarily be approved by the membership. In certain situations the negotiators were forced to come to informal understandings that were not made public. In others, it was necessary to ask management to make its ruling to resolve a difference rather than be seen to concede voluntarily.

Despite a high level of union membership, a much smaller proportion of members participate in union activities – as is the case in many companies in this sector. However, the negotiators did hold meetings with the wider union membership. At one stage a mass meeting of all workers was held. The membership agreed to 'go along with' the proposals although most were not happy. The union was heavily involved in discussions about implementation of the change and about job redesign, but the work force as a whole, of necessity was not party to these detailed deliberations. However, regular progress reports were made by union representatives to members – particularly those groups directly affected by changes.

The Role of Company-wide Procedures and the Experimental Agreement

A 'Manpower' Monitoring/Experimental Agreement provided for national monitoring of locally negotiated changes. This had been set up in response to concern over earlier mechanization and automation. Suspicion about the long-term implications for employment and company structure of equipment changes had led to a series of disputes relating to new equipment in the late 1960s. As a result in 1971 the company agreed to set up the 'Experimental Committee' as a subcommittee of the National Negotiating Committee for all production workers at Snackco. This had the task of monitoring the (local) introduction of new or modified equipment and working methods/manning. In practice the agreement enabled management to get carte blanche acceptance of (1) trials and (2) the introduction of equipment changes that did not affect working arrangements/manning levels. At the same time unions were able to monitor changes in equipment and working methods. When the likely impact of technology, restructuring and market reductions were announced to the manual unions in 1980, a new agreement expanded the role of the Experimental Committee. The committee was renamed the 'Manpower' Monitoring Committee (MMC) and its tasks included consideration of proposed changes in employment levels and disposition, means of responding to these (such as retraining and redeployment and consultation over redundancy if this became necessary). However management at no stage considered introducing HISE under these procedures. The changes were so far-reaching that comprehensive agreement from the work force was considered necessary through conventional negotiating procedures. Initial presentations about intended equipment purchases and the implications of these and market changes for employment levels was presented to the National Negotiating Committee (and MMC).

The final phase of the exercise was a formal trial of work force numbers for each combination of making and packing machine, under 'experimental' conditions, during which Work Study carried out extensive observations. Trade union agreement to end the experimental period in December 1982 marked the start of HISE being accepted as standard machinery together with the proposals for work force numbers.

The National Negotiating Committee (and the national union organizations involved) decided that it was up to the local union committees to negotiate the detailed work force arrangement for their plants. Subsequently, some criticism had been voiced by the unions at a national level of what had 'been conceded' at Sunrise. Union representatives at Sunrise and Alpha felt aggrieved by this, since they had kept the national union organization informed of developments and had asked for advice, but had been told to work it out for themselves.

Implementation

By April 1980, management were able to put a proposal to the unions for initial implementation of two sets of HISE machines on site, and these were installed in May on an experimental basis. However there were severe technical problems with these and after a year it was decided to replace them. The alternative equipment that was brought in was installed straight on to the shop floor in August 1981 (two brands of packers and one brand of making machine).

Further changes in the programme of equipment introduction were made. It had been originally intended to introduce some medium-speed packing machines (packing 5,000 units per minute as opposed to 2,700 on conventional equipment) which were slower than the HISE making machines selected (making 6,000 units per minute). These could not be linked because of their different rates of output. However during the implementation process, the manufacturers launched faster machines. Since both of these makers and packers were capable of running 7,200 units per minute, it was again feasible to consider linking making and packing equipment. Alpha managers worked with the making and packing machine manufacturer to develop a mechanical link between the two, which could if necessary be operated in the conventional way (i.e. manual transfer of automatically loaded trays of snack bars from the maker to the packer) thereby avoiding problems of balancing output rate and of losing production from one machine if the other had broken down or was switched off.

It had been intended to install the medium-speed machines first, then gradually introduce the faster ones. In the event the faster machines were introduced more rapidly. During this period the number of HISE machines ordered was reduced from 24 to 18 makers and packers. Equipment was installed gradually. The changes in machine type caused considerable delays. Some people had to be retrained from medium- to high-speed packers and to familiarize themselves with the machines anew. The last of the 18 pairs of machines were due to be installed by the end of 1983.

Elimination of surplus labour did not cause major problems for management – because of the age profile of the work force and the high redundancy payments there were always more volunteers for redundancy/ early retirement than were needed. The numbers being displaced due to restructuring and market changes were much greater than job losses due to HISE. Employment at the Sunrise plant has remained constant at about 2,000. However, this has been due to consolidation and transfer of personnel from other factories. The introduction of HISE reduced direct labour from about 650 to about 250 and indirect from about 400 to 100. These figures relate to two shift operations.

The biggest job losses occurred amongst female quality attendants – only 15 of whom got jobs as HISE operatives. Most had long service and retired. There was some increase in the number of women in higher grades – e.g. leading operatives. However the proportion of women in the work force fell due to the disproportionate reduction of lower graded jobs (which were traditionally womens' jobs).

Selection of workers was conducted on the basis of aptitude and attitude (including previous output and absence records) and was carried out with the assistance of line management. Particular attention was given to selecting maintenance engineers for machine specialist jobs who would fit in with the requirements of production (e.g. being tied to a machine). Despite this, three of the ex-craftsmen decided to return to engineering jobs because they could not adjust to the new requirements.

Training

The HISE crew were simply taught 'what they needed to do' to run the new equipment – which was not substantially different from running the conventional equipment.

The biggest change was faced by the machine specialists. Those recruited from maintenance engineering lacked knowledge about materials and the production ethos (a production worker might 'bodge' a machine to keep it running where a maintenance engineer would strip it right down, which was inefficient). To prepare them, they were given a month's trial as a crew member on conventional machines (feeding and watering duties). Production workers had fairly shallow engineering skills and were given a month's training at a skill centre to learn basic engineering skills and concepts. After this, all groups went on an in-house training course on maintaining and operating the HISE equipment. The biggest jump was for the making machine operators who had done very little maintenance work in the past. However they all coped with the change.

The Results of the Change

Most groups had adapted well to their new roles. There was a strong group identity in the HISE area. Morale was high and cooperation was good. One problem arose from the selection of 'trouble shooters' which had not proved effective. They had been drawn from section leaders and had no engineering background. They had lost their supervisory responsibilities but were receiving more money than the machine specialists. The machine specialists had more direct experience of the new equipment (and one-third had maintenance engineering backgrounds), and rapidly acquired greater expertise than some of the trouble shooters were able to do. Replacement trouble shooters were therefore being recruited from amongst the machine specialists as required.

The major problem was that of production efficiencies. Efficiencies on HISE equipment were very variable. Whilst some machines were operating at their intended efficiencies, most were not achieving this. The most recently installed kit had even lower efficiencies, which was attributed to design and teething troubles (especially on the faster machines) and the fact that they were at the beginning of the 'learning curve' of operation.

At the same time, there were growing problems on conventional production lines where efficiency was falling. There were a range of reasons for this:

(1) More frequent brand changes and changes in plant layout. This meant that 25 per cent of machines were being changed over each week.
(2) Some of the replacement workers needed to keep these lines going were inexperienced in the jobs they now held (e.g. the making mechanics) while many of the more experienced staff (e.g. the section leaders) had retired, if they had not gone to the new lines.
(3) The selection of the most able employees for the HISE area had left less productive workers on the conventional equipment. There had been some loss of morale among those who stayed behind, who saw their jobs as being less secure and having lower status/pay. Some were being downgraded (such as the machine operators who became operatives).

Wastage on packing was very low. Wastage on 'making' in HISE areas was, at 3.5 per cent, below half that on conventional equipment (7.9 per cent).

The low production efficiencies on the new and old machines, combined with transitional labour surpluses, meant that the output per head had increased only marginally during the period of the case study. Production costs for Alpha Division were anyway generally higher than for Beta because the Sunrise plant was on shifts and paid a 17 per cent shift premium. This matter began to attract more detailed scrutiny by management. They considered changes in arrangements for maintenance with more preventative maintenance, and a greater linking between engineering workers and machines in an area. Further 'fine tuning' of the arrangements for work force members was also being discussed – in particular the possibility that an increase in crew and maintenance levels on the new machines might increase production efficiencies enough to yield an increase in output per head.

Assessment of the Negotiation of Change

The unions had not been involved in decisions about choice and purchase of equipment. Once a decision had been taken they were notified – which was as much as two years in advance of introduction. Once equipment had been received, the unions had easy access to see it (e.g. during evaluation) and were able to assess the problems of operating the equipment – this aspect of getting information did not present a problem. However, the unions were not provided with costings. The unions would have liked more involvement and consultation over corporate development, and had asked to see management's long-term plans. Management had told them that there was no 'master plan'. At the same time union activitists were divided in their thinking about how useful such information and involvement would be. Some argued that at the end of the day, they would still face the same market pressures and that it was beyond the union prerogative to attempt to influence management on such issues.

Management attributed its success in negotiations partly to its ability to plan change in advance, to convince the union representatives that the change was inevitable. However they emphasized the need to win the commitment of workers to change – not least to gain their cooperation in meeting production targets – hence their concern to gain the trust of the work force and not favour particular groups. Management's commitment to 'openness' was not the same as providing unlimited information. Care was exercised in the timing and presentation of information. In some cases early release of information was sought to encourage 'acclimatization' to new ideas, in other situations information was restricted. Management were deliberately vague on certain issues – in particular, the provision of precise job-loss figures to the work force and union representatives during discussions of the introduction of HISE.

However, a major factor underlying the successful handling of these dramatic changes, recognized by union and management alike, was the wealth of the industry. Although competition was increasing, the company remained very profitable. Very high inducements could be offered for the acceptance of change (both to those made redundant and those staying behind). In addition, the company was able to support the burden of a certain amount of overstaffing, and to delay its response to technological change. These circumstances might not exist in other companies (or the same company in future years).

As noted earlier, Beta, the 'sister' division of Alpha, had introduced HISE dispersed into different plants, but alongside conventional equipment. Making and packing had not been integrated. The equipment had been operated on the basis of existing job design and work force number arrangements (which were broadly the same as those for conventional equipment at Alpha). Whilst Beta was able to negotiate introduction much more rapidly (in 16 months) they had not been able to tackle the production/engineering relationship. Their managers were 'slightly envious' of what Alpha Division achieved. Snackco were reviewing the benefits of the Alpha approach and were considering the possibility of applying it (or some elements – such as linking making and packing) to Beta.

It is interesting to note that a different company making similar products had opened a 'greenfield' factory with HISE equipment. They had selected labour with no experience of the industry, in an area without a militant tradition of trade union organization. The staffing arrangements they had adopted were very similar to those achieved at Sunrise (in particular the linking of maintenance workers and production workers around groups of machines). This integration was carried further so that work groups of about 50 employees carried out a very wide range of production and administrative functions.

In the case of Snackco, management were able to plan detailed working arrangements well in advance (which appears to be a consequence of relative certainty of knowledge of production in this sector, and management's monopoly of that knowledge). Linked to these factors (in particular that management were not dependant on 'shop floor' expertise) and given the different traditions of union organization in this sector, long-term planning of management's industrial relations strategy was possible. Faced with this, the opportunities for the unions to alter management proposals were very constrained, and union representatives lacked the information and other resources to challenge management. However, management still had a considerable need to win shop floor cooperation to achieve its production goals. Thus anticipated union and worker responses were central to management planning of change.

Reference

1. Bessant, J., 1982 *Microprocessors in Production Processes*. Policy Studies Institute, London.

Note

This reading draws on research conducted at Aston University Technology Policy Unit, in particular *Trade Unions and Policies for New Technology*, a study funded by the UK Economic and Social Research Council (ESRC), and *The Role of the Parties Concerned in the Introduction of New Technology*, the British contribution to a study funded by the European Foundation for the Improvement of Living and Working Conditions. The opinions contained in the reading do not necessarily reflect the position of the ESRC or the European Foundation. The help and support of Fred Steward in the conduct of the research is gratefully acknowledged.

6.2

Can Human Skill Survive Microelectronics?

H.H. Rosenbrock

Introduction

Everyone who is engaged in the development of new technology – that is of computers and communications and the allied developments based upon microelectronics – will be aware that its intention is to supplant human labour. From one perspective this offers us a benefit which can be taken in several ways. We can produce a greater quantity of goods and services with a given human effort. Alternatively, we can have the same quantity of goods and services with increased leisure, or we can have some combination of these benefits.

From another perspective, which is increasingly emphasized, the development brings losses of various kinds. Rapid change brings dislocation and disturbance which have a human cost. Increased consumption brings increased pollution and environmental damage. Increased leisure can reduce the feeling of participation in society which comes from the contribution made by work. Increased labour productivity increases the rewards of ownership as against labour, and creates problems of the distribution of wealth – problems in which the major burden is borne by the unemployed.

These negative aspects have caused some to reject entirely our technologically based society, and the view can be respected. To be consistent, however, it must then reject the gains as well as the losses. And since the history of humankind is one of continually (if not continuously) advancing technology, some line must be drawn, up to which the advance of technology was acceptable, and beyond which it was not. It is difficult to see any defensible basis on which such a decision could be made.

An alternative is clearly to examine whether the development of technology can be carried on in a way that does not bring such damaging

Source: Conference paper published by Forschunginstitut für Mikroprozessortechnik, 1984.

losses. This reading is addressed to one rather narrow but nevertheless important aspect of this question. New technology offers continual opportunities for eliminating human skill and for subordinating people to machines, and these opportunities have been amply exploited in the past. Is this process inevitable, or is it possible to have a technology which is not antagonistic to skill?

Taylorism

The spirit in which technology is developed today remains very nearly that which was codified by F.W. Taylor in his 'scientific management'.[1] It is true that in studies of work, carried out by social scientists, Taylorism finds almost no support. It is also true that many companies have experimented with non-Tayloristic forms of work organization. Nevertheless, when one visits the research and development laboratories in which new technology is conceived and brought forth, one finds a climate of technological endeavour in which the effect of technology upon people is ignored. Often it is assumed that systems will operate without people, and the difficulties which ensue when they do not are patched up when they arise.

The essence of Taylorism is an emphasis upon explicit, scientifically authenticated knowledge, at the expense of skill and the tacit knowledge upon which skill is based. Skill, to Taylor, is something that belongs to the past, and is destined to disappear before the advance of scientific knowledge. Skill is something that belongs to the worker, but he cannot be given the scientific knowledge that replaces it: '... in almost all the mechanical arts the science which underlies each act of each workman is so great and amounts to so much that the workman who is best suited to actually doing the work is incapable of fully understanding this science.'[1]

This quotation implicitly defines the kind of work that is envisaged. It is work that is peformed in ignorance of the purpose for which it is performed. This indeed is made fully explicit: 'Under our system the workman is told minutely just what he is to do and how he is to do it: and any improvement which he makes upon the orders given to him is fatal to success.'[2]

Trust in the skill and ability of workers is therefore to be withdrawn, and replaced by a minutely specified routine in which all contingencies are foreseen. Yet the practical consequences are easy to predict. By rejecting the pre-eminent capability of men and women to react to the unforeseen, and to overcome difficulties through a sense of purpose, systems are made rigid and inflexible. The quality of work suffers, together with the motivation of workers.

All this is well-recognized, and many would say that Taylor's attitudes no longer represent current thinking. In some areas of management that is no doubt true, yet a very different view will be obtained by attending any technical conference concerned with the development of new technology, say of robots or of advanced computer systems. There Taylor's vision reigns unhindered. Similarly, as in a recent conversation, one can find eminent production engineers who will defend and support the quotation from Taylor, (above), as the appropriate spirit in which new production systems should be developed.

Why does Taylorism persist so strongly in the technological community? My belief is that although Taylor had no serious claims to being a scientist, he had hit upon an essential element in our scientific culture – the culture in which all of us, scientists or technologists or not, live and work. In this culture, only one kind of knowledge is fully accepted, and that is the explicit knowledge which arises from scientific activity.

In most practical activities, explicit scientific knowledge is only a small part of what needs to be known. Much of what is needed is tacit knowledge, in Polanyi's sense: things that we can do but cannot readily explain.[3] Anyone who has taught a son or daughter to drive a car will recognize how one has to learn to describe what one can do easily without description.

Much of the activity of a designer or a manager or a shop floor worker is of this kind. Indeed, in any practical activity, the explicit, verbalized knowledge about the activity has first to be made implicit before it can be used. No amount of familiarity with the theory of driving a car will allow one to drive a car until the skill has been acquired by experience. Designers, or managers, similarly acquire their skill by practice. What Taylorism does is not to abolish the need for tacit knowledge and skill, but to divide it into such small parcels that each can be acquired in a short length of time. As Henry Ford said of his foundry workers '... 95 per cent are unskilled, or to put it more accurately, must be skilled in exactly one operation which the most stupid man can learn within two days'.[4]

An Alternative

The Tayloristic view of technological development can be represented by the following analysis. At the present time, there is a certain area which is explicitly understood: we can think of it as a central sun of explicit knowledge. Surrounding it is a corona of implicit knowledge and skill without which the explicit knowledge cannot be applied.

The Taylorist would see the process of technological change as one in which the sun expands continually at the expense of the corona until it has entirely disappeared. All knowledge has been made fully explicit, and skill is no longer needed. It has become possible to tell the worker 'minutely just what he is to do and how he is to do it', and all the worker has to do is to follow these instructions.

In fact, as was pointed out above, the corona has not disappeared. Workers, even when they have been told what to do and how to do it, need to acquire the ability to do the work by practice. What has actually happened is that the corona has been impoverished by fragmentation, so that each fragment can be learned in a short time. This is possible only where there are many workers, and the sum of all their fragmentary skills is still considerable, and essential.

A truer representation of technological development could therefore be described as follows. The sun of explicit knowledge has expanded as before, but now the corona has also expanded proportionately. An aggregate of skill and of tacit knowledge, greater than before, is needed to make the explicit knowledge useable. Computers, satellite communications or integrated circuits, are based upon explicit knowledge, but can only be made and used by a corresponding development of skills. Science does not replace skill, it calls new skills into being.

What is important, however, is that this second analysis should represent not only the aggregate, but also the individual experience. That is to say, if we take some area where a significant personal skill exists in workers, then the development of technology should allow an equivalent skill to be maintained into the future. Note that the skill will not remain the same. It will change as technology changes, but should remain comparable in extent.

Can such a development be competitive with the Tayloristic route? There is every reason to believe that it can, because it makes better use of the specific human abilities: to respond to the unexpected and to pursue a desired objective in the face of difficulties. By contrast, Taylorism builds in an inflexibility and inability to respond to disturbances by rejecting the initiative which is needed to meet them.

Does technology have the flexibility to allow us to follow such a route? Again it seems that the answer is yes. When we consider in detail an area such as manufacturing technology, it becomes apparent that there are many opportunities for skills to change without being fragmented or destroyed.

Possible Objections

There are several difficulties in pursuing the line of development proposed. Long experience in Tayloristic styles of management may have made it difficult for many managers to adopt a different style. Yet the interest in Japanese styles of management suggests that this difficulty may not be insuperable.

Another difficulty is that the proposed route is not without its problems for workers. As technology changes, some existing areas of skill will become outdated. Even if they are replaced by equivalent skills in new areas, the process of change may have its inconveniences. The young will adapt more easily than the old, and if the change is too rapid, many may fail to adapt in time.

Finally, the route that is suggested is believed to be as productive as the Tayloristic alternative. This means that it will do nothing to preserve jobs. Historically, the increase in labour productivity has been accompanied by the creation of new needs, and therefore of new types of job which did not exist before. It has also been accompanied by a reduction of working hours, from around 80 in the early nineteenth century to around 40 now.

Both of these remedies have now become doubtful: the second because pressure from workers for shorter hours has largely ceased, while the unemployed have no organized representation. The creation of new needs, on the other hand, has now become more difficult for many reasons – environmental among others – and seems unlikely to cope with the rapid increase in labour productivity.

Nevertheless, it seems to me a mistake to resist technological change for the sake of preserving existing jobs. Such an attempt can succeed for only a limited time, and when it breaks down, all control over the subsequent rapid change will have been lost. At least, if we accept a more continuous and gradual change, we may hope to guide it in a more desirable direction.

The problems just mentioned illustrate the point that was made at the beginning. The problem of fragmented and deskilled work is only one part of a much larger whole. It is a part about which one can be more optimistic than about some others, but a good solution in this area would still leave many other problems unresolved.

References

1. Taylor, F.W., 1911 The principles of scientific management. Reprinted in *Scientific Management*, 1947 (Harper)
2. Taylor, F.W., 1906 *On the Art of Cutting Metals*, 3rd edn. American Society of Mechanical Engineers, p. 55
3. Polanyi, M., 1966 *The Tacit Dimension*. Routledge
4. Ford, H., (in collaboration with Crowther, S.) 1923 *My Life and Work*. Heinemann

6.3

Social Choice in the Development of Advanced Information Technology

Richard E. Walton

Extensive automation of white-collar work has become possible because of two types of technological advances. The first is an explosive growth in computer power per unit of cost – in the order of tenfold increases every four or five years. The second advance is in telecommunications, making it possible to achieve unprecedented movement and integration of electronic information. Consider also two economic facts: (1) the annual growth rate of capital per employee in offices has lagged behind that in manufacturing, and (2) office overhead costs have risen rapidly in recent years. These factors combined to make an extraordinary variety of new applications economically feasible. Experts regard information technology as the most dynamic sector of technical innovations.

Implications for Work and People at Work

The first proposition is that the new information technology has profound implications for the nature of work performed by clerical, professional and managerial personnel. The potential impact on the work place may be greater than any earlier wave of new mechanization or automation to hit industry. Thus, the human stakes are high.

The new technical systems differ from those of prior generations, particularly because their relationship to human systems has become more pervasive and complex – and more important. Earlier systems utilized large computers, performed a limited number of separate functions, relied upon batch processing and were tended by special full-time operators. The newer technologies utilize a network of large and small computers and embrace many activities within a given system, often crossing departmental boundaries. Managers, professionals and clerical personnel are required to

Source: from *Technology in Society*, 4, 1982.

interact *directly* with computer terminals, often as an integral part of their responsibilities. And, because these systems are on-line, the relationship between the user and the system is more immediate. Thus it is not surprising that the newer systems have the potential for affecting more employees in more ways than ever before, and for influencing work and communications patterns at higher executive levels than previously.

The studies conducted by the author and his colleagues have covered a number of different applications, including the following three:

(1) Electronic mail terminals were placed on the desks of thousands of managers and support personnel in a large firm. This innovation affected the nature of vertical and horizontal communications, access to executives at different levels, and decision-making processes, and it modified somewhat the contents of jobs of those who used the tool.

(2) A procurement system was installed in a large company, embracing buyers and their clerical support, as well as personnel in the receiving and accounts payable departments. The system made it possible to monitor more closely the performance of purchasing agents, changed the interdepartmental patterns of accountability for errors and created more tedious clerical work.

(3) A telephone company automated its local repair bureaus, employing information technology to test phone lines automatically, to monitor the status of all repair orders in the bureau and to provide telecommunication linkage with service representatives who received subscriber complaints at a new centralized office. Before automation, these service representatives were located at the local repair bureaus.

Looking more closely at this application, in the *repair bureau*, the new technology reduced the number of personnel and decreased skill requirements. The 'test man' is a case in point. In the past, the test man's job was a professional one, with a high-status dress code of 'starched white shirts and ties'. Mastery required innate ability and experience. Today the testing function is becoming increasingly automated, and the test man's skills and knowledge have become technologically obsolete. The persons who held those positions, therefore, have suffered psychologically and economically.

The new system also dramatically affected personnel in the new *centralized answering facility*. The service representatives felt that they had become physically and informationally isolated from other steps in the process of satisfying the customers whose complaints they take. The central facility takes complaints for local bureaus in several states, and the answering personnel neither learn what happens to a particular complaint nor know the people in the bureau to which they pass along the complaint. Service representatives cannot determine the status of repair work and, therefore, either cannot respond to customers who call back or have to provide customers with meaningless promises about delivery of service. This has led to tension and mutual fault-finding between the service representatives and the repair personnel. In this, and many other respects, the technical system helped to produce 'unhealthy' jobs – jobs which failed to meet normal human needs for knowledge and control of the work place. The result was employee alienation and defective problem solving.

Not all of the human side effects of these and the other systems studied were negative; this point will be discussed shortly. But the negative human consequences that were found were significant – and largely predictable. The following behavioural generalizations describe some of the common organizational consequences of office applications of the new microprocessor technology.

If the technical system decreases skill requirements the meaning of work may become trivial, and a loss of motivation, status and self-esteem may result. This was a common occurrence. In some circumstances, those who suffered counterattacked the system.

If the system increases specialization and separates the specialty from interdependent activities, then jobs may become repetitive and isolated, and fail to provide workers with performance feedback. Such jobs produce alienation and conflict.

If the system increases routinization and provides elaborate measurements of work activity, job occupants may resent the loss of autonomy and try to manipulate the measurement system. The fact of measurement itself can put excessive pressure on individuals and can strain peer relationships.

Impact of Technology Varies and Can Be Influenced

The second proposition here is that technological determinism is readily avoidable. Technology *can* be guided by social policy, often without sacrifice of its economic purpose. Information technology is less deterministic than other basic technologies that historically have affected the nature of work and the people at work.

True, the side effects described above were generally negative, but sometimes the *unplanned* consequences are positive. In each of the areas listed below the effects were not inherent in the technology. The directional effects resulted – to an important degree – from particular choices made in design or implementation.

(1) Work systems based on the new technology often require less skill and knowledge, but sometimes these new systems result in more jobs being upgraded than downgraded. System design can influence that outcome.
(2) The technical system can increase the flexibility of work schedules to accommodate human preferences, or it can decrease flexibility and require socially disruptive work schedules.
(3) New systems often contribute to social isolation, but sometimes they have the opposite effect. Similarly, they often separate an operator from the end result of his or her effort, but occasionally they bring the operator into closer touch with the end result. Seldom are these planned outcomes, but they *can* be.

(4) These systems sometimes render individuals technologically obsolete because of changed skill and knowledge requirements, but they also open up new careers.
(5) New technology can change the locus of control – toward either centralization or decentralization.
(6) New information systems can change – for better or worse – an employee-typist into a subcontractor operating a terminal out of his or her own home.

The problem is that those who design applications and those who approve them currently make little or no effort to anticipate their human effects. Thus positive organizational effects are likely to be accidental, as are negative ones.

Why is computer-based technology becoming less deterministic, allowing planners more choice?

First, the rapidly declining cost of computing power makes it possible to consider more technical options, including those that are relatively inefficient in the use of that power.

Second, the new technology is less hardware-dependent, more software-intensive. It is, therefore, increasingly flexible, permitting the same basic information-processing task to be accomplished by an ever-greater variety of technical configurations, each of which may have a different set of human implications. For example, one system configuration may decentralize decision making; another may centralize it. Yet both will be able to accomplish the same *task* objectives.

Trends Favour the Exercise of Social Choice

The third proposition is that a number of factors could produce an industrial trend in which human development criteria would be applied to the design of this office technology. The author has not yet observed such a trend. But a social revolution affecting work in the manufacturing plant gathered momentum during the 1970s, and the most natural extension of this social revolution to the office would be a movement to seize upon this new office technology and shape its development.

Managements – and unions, where workers are organized – are increasingly acting to modify the way blue-collar work is ordered and managed. And the changes are explicitly in the interest of promoting human development as well as task effectiveness.

The work improvement movement began in the United States and Canada and in some European countries in the early 1970s, after several years of sharply increasing symptoms of employee disaffection. Symptoms included costly absenteeism and sabotage, and the media labelled the general phenomenon 'the blue-collar blues'.

Over the past decade attention has gradually shifted from symptoms to solutions. Work reform in plants throughout the United States and Canada has grown steadily, and the trend appears to be taking the path of the classical '*S*' growth curve. Today the rate of growth in these experiments continues to increase annually, suggesting the steeper portion of the curve is being approached.

Work reform reverses many practices launched with the industrial revolution in which tasks were increasingly fragmented, deskilled, mechanically paced and subjected to external controls. The current trend combines specialized jobs to create whole tasks, integrates planning and implementation and relies more on self-supervision.

Implementing Social Choice

The idea that technology has a social impact certainly is not new. Social scientists have long argued that technology can dramatically affect individuals, institutions and society as a whole. Managers who introduce new work technologies have long appreciated that there will be organizational side effects. But this knowledge has had little influence on the introduction of new work technology.

In the past, considerations of the human impact of innovation have led merely to efforts to overcome workers' resistance. These efforts have emphasized implementation methods, including communication and training, and employment assurances. But efforts to ameliorate the impact should increasingly extend upstream to the design stage itself and affect the design of hardware, software and management operating systems.

In the past, where human criteria have been considered in the design of work technology they have centred on narrow factors, such as ease of learning, operator fatigue and safety. The criteria should be extended to include a broader array of human needs – for autonomy, for social connectedness, for meaningful work, for effective voice.

But, in order to exercise social choice in the significant sense that has just been described, one must break new methodological ground. Some ways in which this should be done are as follows:

(1) Organizations need explicit normative models, by which designers can judge what human effects are to be considered good, bad or neutral, and which ones are especially salient. An organizationally specific model would be based both on general knowledge about human development, and on an understanding of the particular circumstances of the company.

(2) Designs should not be approved until an 'organizational impact statement' has been prepared and reviewed. The first step would be an examination of the requirements for a proposed technical system. This would clarify the first-order social consequences of the system – how it changes the degree of specialization, locus of control or skill requirements. The next step would be a prediction of second-order consequences, such as motivational effects, social conflict and human development. This would require the perspectives of behavioural disciplines not currently involved in systems projects.

(3) One needs practical methods for involving those who will eventually use and/or be affected by the system. While 'user involvement' in systems development has been a widely endorsed concept for more than a decade, in practice users seldom report that they have been meaningfully involved.

(4) System development should be approached as an evolutionary process. This contrasts with a more typical assumption that the design can and should be completely conceived before implementation. This methodological recommendation is based on the finding that the human impacts of complex information systems are *dynamic*, in the sense that their effects change over time; for example, some initially negative reactions disappear as tasks are mastered, and some initially positive reactions decline as novelty wears off. Complicating the picture is the fact that effects are *reciprocal* in the sense that the employee will react to the technical system; for example, user reactions may affect the quality of inputs to the system and, in turn, the functionality of the system.

(5) The final recommendation here is that significantly greater effort must be devoted to evaluation of the operational system, and this evaluation must comprehend social effects as well as economic and technical achievements.

These methodological proposals have an additional implication: management should assign a fraction of every development budget to be used to explore the human implications of these systems, and then it should act upon the knowledge derived from these explorations.

People are only beginning to learn how to exercise social choice in the course of technological development. There are still relatively few instances in which designers have paid explicit and comprehensive attention to potential impacts on human systems. In Europe there is growing experience with trade unions which have insisted on being involved in evaluating new computer-based technology before it is installed in the work place. Two Cornell professors have developed a model for design and implementation of work-processing systems that attend to social dimensions.[1] In the United Kingdom, Enid Mumford and her associates have developed a participative approach to the design of systems which affect clerical groups.[2] These are pioneering efforts, and though their achievements may be instructive, they are by no means definitive.

Conclusion

To summarize: applications of the new information technology should be guided by human-development criteria; they can be so guided, and now there is a decent probability that they will be. If this new work technology is to be shaped by social criteria, it will be necessary to gain new implementation 'know-how', and a rich field will be opened for basic and applied research.

The design and implementation of advanced information technology poses major organizational problems, and these problems must be dealt with. These innovations also represent the most important opportunity available in the 1980s for the introduction of constructive changes in clerical, professional and managerial work.

First, a few pioneering organizations, and then a larger number of progressive ones will exploit this opportunity. The introduction of this technology offers the chance to rethink the organization and management of professional and clerical work in the office that is analogous to the way greenfield plants created an opportunity to pioneer new approaches to managing factory work.[3]

The 1980s will be a period of trial and error as man learns how to exercise social choice in systems design. Academic institutions can contribute to the analysis and dissemination of these experiences, but only if some managements, systems developers and unions choose to lead the way in this uncharted field.

References

1. Lodahl, T.M., Williams, K., 1978 An opportunity for OD: The office revolution. *OD Practitioner*, December
2. Mumford, E., Weir, M., 1979 *Computer Systems in Work Design*. Associated Business Press, London
3. Walton, R.E., 1979 Work innovations in the United States. *Harvard Business Review*, July–August

6.4

Computerized Machine Tools, Manpower Consequences and Skill Utilization: A Study of British and West German Manufacturing Firms

Gert Hartmann, Ian Nicholas, Arndt Sorge and Malcolm Warner

Introduction

The impact of new information technology on organization and manpower depends on where and how it is used. A statement such as Rada's to the effect that 'electronics ... will substantially condition industrial and service activities and the socio-political structure' is at least potentially misleading, for its consequences may be more due to the specific culture prevailing in such societies.[3] It is thus advisable to exercise great caution when examining any prediction of its effects.

Since the 1950s, metal cutting has undergone major technical advances. It has, in the eyes of the layman, been 'automated'. The development of numerical control (NC) of machine tools has been followed gradually by the introduction of computer numerical control (CNC) of such machines, involving the use of microprocessor technology. The microprocessor increases their capabilities from that of merely registering control information and monitoring step-wise applications to providing programming and other control facilities right on the machine. An alleged consequence of both these technical developments has been greater 'deskilling' of shop floor workers, according to writers such as Braverman and others.[1] We argue that this has not been equally true in advanced economies and we have looked at British and German experiences to question this view and theories built upon it, which emphasize 'deskilling' as characterizing the 'labour process' in the twentieth century.

Source: from *British Journal of Industrial Relations*, Vol. XXI, No. 2, 1983.

The theoretical significance of the 'deskilling' debate recently discussed critically in the volume edited by Wood cannot, however, be divorced from its practical implications regarding manpower utilization and training.[6] Moreover, we feel that national differences are likely to affect work organization, and hence the relative distribution of skills to be found. We therefore hypothesize that the integration of the new technology into enterprises follows quite different routes in Britain and Germany. British companies will in turn train and use noticeably less skilled workers than German ones, and the difference will be particularly visible in production, as opposed to maintenance jobs.

It may be further hypothesized that the reverse of deskilling, may result if 'changing technology (automation) induces a tightening up of selection criteria' because more skilled labour is required on the shop floor as Windolf suggests.[5] We also conjecture that a greater utilization of skilled labour may be anticipated in the future, and stronger in Germany because of a relatively greater supply of trained manpower there at all levels, as indicated by Prais.[2] In Germany, over half the primary work force has gone through a formalized apprenticeship.

Previous machine-tool applications controlled by paper tape programming prepared in the planning department (such as NC) have been to date essentially geared to specialized, homogeneous mass markets, inflexible automation and an erosion of craft skills. In the labour market, there has also been a move towards information processing, administration and different kinds of clerical work, particularly visible in the British experience. Recent directly programmed machine-tool applications, (such as CNC), in our view, may reverse this trend at least in the manufacturing sector; there is now increasingly a focus on craft skills and the levelling-out of the growth of indirectly productive employees very clearly seen in German firms. This consequence basically follows the organizing tradition of the respective national work cultures. It is misleading, we believe, to examine technological change outside this 'societal' context; rather, we see it interacting with it. In so far as the study has implications for organization theory, it argues that technological change needs to be seen as having much more open social implications and leaving organization designers more options than hitherto assumed.

Research Findings[4]

National institutions and cultural practices

If CNC exercises different effects according to technology and machine type, it also adapts to different 'societal' environments, namely to *national institutions and cultural practices*.

The German companies, for example, distinguished less than those in Britain between specialized functions and departments for production management, production engineering, work planning and work execution functions. Similarly, there was a consistently greater use of shop floor and operator programming in Germany; programming is seen as the nucleus around which the various company personnel, the managers, engineers, planners, foremen and operators, are integrated. The differences are reinforced by the CNC dimension. Furthermore, the greater separation of programming and operating in Britain ties in with the increasing differentiation of technician and worker apprenticeships, whereas technician training in Germany invariably comes after craft worker training and experience. In addition, whilst in Britain the planning and programming function confers white-collar status, it is much less common in Germany, where blue-collar workers are more extensively used for programming, both on the machine and in the planning department. Rotation between the two groups was frequently observed.

National differences, in turn, interact with company and batch size differences. In Germany, the similarity between organization, labour and technical practices of small and larger companies was greater than in Britain, where there appeared to be a split between pragmatically flexible small plants and organizationally more segmented larger plants. Formal engineering qualifications at various levels were relatively more common in Britain in larger plants, whereas often they were not represented in small British plants. In Germany, by contrast, formal qualifications were common to smaller and larger plants alike.

In both countries, CNC operation was generally seen as exacting less 'informatics' skills than advanced machining talents. Programming aids on the machine or in the planning departments are seen to be tools of increasing facility to control a process which has become ever more demanding from the point of view of precision, machining speeds, tools, fixtures and materials.

Socioeconomic conditions

While there was a series of 'logics' moulded by distinct historical, cultural and social developments, there was also a further thrust of CNC application which was *common to Britain and Germany*. This stemmed from the relevant existing macroeconomic factors, including competitive and marketing strategies, which interact with sociotechnical considerations. They incorporate a very broad range of factors which affect enterprises, but which they themselves attempt to influence.

Firms stated that they were keen, or perhaps being forced, increasingly to cater for small market 'niches' rather than for homogeneous mass markets, given that the most generally stated competitive situation was one in which there was static, or more commonly in Britain, negative market growth. More individualized, customized products, with a greater number of product variants were seen as appropriate.

Complex design of components may be consistent with NC application, but the greater variability of products and components was more specifically associated with CNC application. The two most important factors which lead to increasing CNC application within manufacturing were as follows:

(1) the demand for more frequent and less time-consuming machine conversion from one batch to another, arising from the increased variability of products and components;

(2) the inducement to minimize finished product stocks and work-in-progress which can be substantially reduced when the full potential for manufacturing 'families' of different, but similar, components is realized and/or advantage is taken of the opportunity to produce, in a single manufacturing cycle, those subassemblies which comprise the final product.

Thus, both market-oriented as well as financial considerations point in the direction of smaller batch sizes and more frequent conversion; this has important sociotechnical implications. It is through these developments that we can see the effect on skills. The increased variability of batches is not one which can be handled bureaucratically through a conventional increase in the division of labour. CNC operators are likely to have to deal with a greater and more frequently changing range of jobs; part of this is related to the increased sophistication of the machine control system through which more flexible changeovers and improvements of programmes can be achieved.

The crucial 'bottlenecks', however, may not be information-processing and calculating skills; experience indicates that the most crucial problems refer to tooling, materials, feeds, speeds, faults and breakdowns. Skills in handling these problems are most directly developed on the machine. Thus, while programming skills are required, increasing emphasis must be placed on the maintenance of 'craft' skills on the shop floor, rather than the converse (as implied by the 'pessimists' writing on the subject).

The justification of CNC *ex ante* was however made in rather general terms, often intuitive, through lack of information about skills requirements, down-time, maintenance and service quality, equipment reliability, organizational implications and so on. *Ex ante and ex post*, NC uses in general were justified on the basis of the complexity of the geometric design of the components, and the required cutting cycles and sequences. Economic justifications were also discussed by respondents, but were often presented as secondary considerations. This may or may not be the case. Arguments for CNC, however, emphasized increased flexibility coupled with potential increases in productivity in manufacturing. In the past, productivity increases had involved the production process becoming less flexible and less capable of handling variations in component specifications, small batch size, batch conversion and so on.

Differences also existed in the kind of production technologies selected in the German and British plants. It was, for example, rather more difficult in the case of both small and bigger plants to find examples of large batch processing involving CNC equipment in Britain. This may have been because of the recession reducing batch sizes. We do not interpret this wholly as a problem of matching: it would seem that there are different nationally specific overall socioeconomic trends in operation which affect the application of CNC both quantitatively (more CNC turning in Germany) and qualitatively.

Organizational structure

Work shop and operator planning were more widespread in German small batch turning, whereas British plants still followed the more traditional NC organizational view where programmes are made in the programming department and proved on the shop floor by both programmer and the operator. This difference between the two countries also applied to a lesser extent in CNC milling, particularly where automatic tool change facilities were absent. It was however less visible with respect to machining centres, the stronghold of traditional NC philosophy in both countries.

Organizational differences as far as CNC were concerned could be very strong even on the same site. They were related to production engineering practices, but cannot be explained fully without recourse to factors of personal influence and departmental tradition. This was visible in the German large plant–large batch categories – where differences between turning and milling were striking. In the latter case, programming was centrally performed, whereas in turning a high degree of shop floor autonomy was retained.

Within all the cases, British companies used CNC in such a way as either to maintain planning department control and/or autonomy, or to segregate the NC operations from the other sections. In Germany, CNC organization was fashioned so that it linked foremen, chargehands, workers and planners around a common concern. This was particularly apparent in the case of foremen, who in Germany were deeply involved in CNC expertise, but whom in Britain had been largely by-passed, resulting in deskilling of their jobs.

In both countries, the attractiveness of programming-related functions lay less in a formalized language, than in the instrumental value of programming as a means whereby the more demanding part of the task, that of metal cutting, could be more effectively planned and controlled. This diffusion of control to the shop floor depended on the ability to make programmes on the machines without losing too much machining time but more importantly on the accepted view of operator competence in metal-removal technology. This view was more partial to physically locating programme-related functions in the hands of planners in Britain, whereas it was more likely to prefer programming by operators as experts in cutting and setting in Germany.

Training and employee qualifications

Rather than any one deterministic outcome, there was a striking variety in the training and qualification patterns under CNC application, which was closely related to differences in the organization of the work process.

In both countries, using more skilled people as CNC operators was strongly linked to the integration of the operating and machine-setting activities. Where this existed, operators were more likely to be skilled;

where it was absent, operators were unskilled. This relationship, however, was also a function of batch size, and thus it followed that there was also a *differentiation* between *semiskilled* operators and skilled setters where *large batches* were run. However, more *skilled* operators, who also performed setting functions, prevailed in *small batch* processing (see Figure 1).

In small plants, setting and programming-related functions were intimately combined. In the small plant–small batch classification, there was a large overlap of operating, setting and programming-related functions. In the small plant–large batch category, setting and programming were still closely combined, but they were differentiated from the normal running of a batch. Was the policy to encourage – and this is important in the context of the deskilling debate – the development of operator expertise in the programming-relating functions? Would this be an end in itself since the operator was uniquely placed, via the CNC 'electronics' to control and, if necessary, modify the cutting process?

Operators frequently reported that they were more concerned with the problems of metal removal, that is, feeds and speeds, tool selection, tip wear and quality, tool life and so on, than with programming difficulties. Some limitations with machine programming were nevertheless encountered; these were usually associated with the question of reduced machine utilization that might result from the direct programming of geometrically difficult components.

The most pervasive national differences occurred in the small batch–large plant category. German companies allied CNC operations to tradesman status and experience, but British companies did not necessarily do so. Where this happened, however, it was usually recently introduced and

| | | Plant size: bureaucratization of programming → | |
		small	large
Batch size polarization of functions and skills	small	Programming with larger involvement of operators, and small or nonexistent programming department	Programming more in the hands of a programming/planning department, but skills of operators and planners overlap
	large	Programming done by setters, foremen or programmer; programming and setting closely integrated	Programming in the hands of a programmer/ planning department, and differentiation between operating, setting and programming skills

Figure 1   Skills polarization and bureaucratization of programming under CNC
Source: *Sorge* et al., 1983

not uncontested within the company. There was a consistently held view in Britain that CNC tended to deskilling and that CNC operation was more routine. The same companies are however planning to introduce skilled tradesmen onto their NC and CNC machines in an attempt to increase machine utilization and, to a lesser extent, improve quality.

There were very different views of the qualifications needed for planning and programming. The higher the skills required of the CNC operators, the less was the demand placed upon the planners–programmers. This proposition held both within countries and between countries, and was linked as before for the setting and operating functions, with the effect of batch size. Thus, the application of CNC fitted with equal ease into strikingly different qualification structures; at one extreme, it could be found, within a polarized qualification pattern with unskilled operators, skilled setters and technician-planners neatly differentiated; at the other, it could be located close to conventional machining, even craftsman activity, involving a skilled, homogeneous work force.

Did this leave national differences between Britain and Germany in vocational training unaffected? There are perhaps even greater differences under CNC, although there was no direct causal connection. One of the more striking was the training of industrial workers. In Britain, there was a split at the beginning of working life, between technician and worker apprenticeships, between programming and planning as opposed to machining occupations. Training for these operated in parallel, whereas in Germany the latter set of skills was an essential prerequisite for the former. This difference appeared likely to persist and in Britain to sustain the more polarized qualification structure; whilst in Germany CNC was more visibly used to reduce training differentials between technical staff and the shop floor personnel, to increase the tradesman's status, to encourage even greater flexibility in production and to reduce the 'decision-making over-load' on top management.

Personnel structure

Personnel structure, seen in terms of proportions of white-collar staff and worker categories resulted from a complex set of influences which did not necessarily originate from CNC. There was a general tendency in most of the companies investigated, particularly the small British but including most of the German companies, for a merging or blurring of status between the shop floor and the planning functions. The personnel structure remained more or less constant through the action of the status-bound grading policy, and any shift in the distribution of actual functions had limited influence. This merging of status may have even helped advance the utilization of the CNC process.

However this was not the complete story; this process took on very different forms. When craft qualifications were more frequently held on the shop floor, and particularly on the part of the operators, programming and planning functions were often delegated to these workers. Conversely, the weaker the craft qualifications, the stronger was the tendency to reclassify shop floor personnel as white-collar staff. Both effects could be observed in Germany; but their respective incidence depended on batch size which, in turn, was strongly related to operator qualifications. In short batches, operators progressed onto programming tasks without changing status, whilst with large batches white-collar status was extended into the works.

Status up-grading in Britain took place in small companies, but occurred less in larger factories where there was a more consistent distinction between works and staff. This was despite the more legalistic character of the definition of workers and white-collar employees in Germany. Where figures are comparable, Britain used craft workers more *sparingly* than Germany. These differences in personnel structure were related to the degree to which production engineering, work planning and production control tasks were set away from the shop floor. The relationship between white-collar employees in programming and production control, and the shop floor, changed very little even with intensive CNC use. In Germany this was more visible on the shop floor, and the associated functions moved to meet it on the 'home' ground. In Britain, the programming, planning and control staff were seen as the 'proprietors' of such knowledge, but this philosophy is being questioned more and more critically, as in Germany, if not for the same reasons.

Concluding Remarks

All our data lead us to stress the extreme malleability of CNC technology; it is inadequate to consider a technology as given and to observe the effects which follow from it. But this is not to say that technology is unimportant; its significance unfolds through a continuous series of 'piecemeal' modifications. As part of a complex pattern of sociotechnical design and improvement, such changes interact with organizational and manpower innovations. Thus, in the company context, the detailed technical specifications of the CNC system adopted may reflect the specific influence of corporate and departmental strategies, and the existing production, engineering and organizational procedures and the current manpower policies, all of which vary *within* a country and *between* countries and societies when we are considering the 'labour process' in general.

In the study, we found that solutions to CNC applications were alternatively organizationally simple or complex; some stressed functional differentiation, while others emphasized functional integration within positions or departments; in some cases there was a strong element of skill polarization, but with skill enrichment at the shop floor level. None of these contrasting policies can be said to be more 'advanced' from the technical point of view, yet at the same time it cannot be said that the application of CNC was haphazard or the subject of accidental initiatives.

Companies, particularly in Germany, are increasingly seeing the merits of stressing craft skills, as a viable option, when implementing the new technology. This is not because it is a necessary consequence, but because CNC has been developed in a context which links economic success with this process. There is a striking kinship between the increasing use of CNC and the renewed interest by companies in training and employing skilled workers.

Increased programming or programme changing in the workshop may further blur status boundaries for blue- as well as white-collar workers. It would, however, be misleading to interpret this as another step towards the 'postindustrial society', as 'information-processing' work or as a 'service' function, as so often happens. Whilst it is true that workers are dealing with increasingly sophisticated information technology, this may only concern the *tools of their trade* rather than their *working goals*.

Notes and References

We gratefully acknowledge the support of the Anglo–German Foundation for the Study of Industrial Society for this study, as well as the help of CEDEFOP, the European Centre for Vocational Training. For professional dialogue and criticism, we have the pleasure to thank Derek Allen, Georges Dupont, Michael Fores, Donald Gelwin, Jonathan Hooker and Marc Maurice, and many others too numerous to mention.

1. Braverman, H., 1978 *Labour and Monopoly Capital; The Degradation of Work in the Twentieth Century*. Monthly Review Press, New York
2. Prais, S.J., 1981 *Vocational Qualifications of the Labour Force in Britain and Germany*. National Institute of Economic and Social Research, London, Discussion Paper 43
3. Rada, J., 1980 *The Impact of Microelectronics*. International Labour Office, Geneva. p. 105
4. Information on research methods is given in Sorge, A., Hartmann, G., Warner, M., Nicholas, I.J., 1983 *Microelectronics and Manpower in Manufacturing Applications of Computer Numerical Control in Great Britain and West Germany*. Gower Press, Aldershot
5. Windolf, P., 1981 Strategies of enterprises in the German labour market. *Cambridge Journal of Economics*, 5 (4): 359
6. Wood, S., (ed.) 1982 *The Degradation of Work?* Hutchinson, London, ch. 1, pp. 1–10

Information Systems and Organizational Change

Peter G. W. Keen

Introduction

This reading discusses long-term change in organizations in relation to information systems. The aim is to explain why innovation is so difficult and to point towards effective strategies for managing the process of change. Many commentators have drawn attention to the problems of implementation that result in systems being technical successes but organizational failures.[34,8,13,6] Their analyses stress the complexity of organizational systems and the social inertia that damps out the intended effects of technical innovations.

The growing body of research on implementation deals mainly with tactical issues. How to create a climate for changing, building and institutionalizing a specific system.[14] This reading focuses on strategic questions:

(1) What are the causes of social inertia?
(2) What are the main organizational constraints on change?
(3) What are the mechanisms for effecting change?

Effective implementation relies on incremental change, small-scale projects and face-to-face facilitation. A strategy for long-term change and large-scale innovation requires a broader strategy; the conceptual and empirical work on implementation, both within MIS and OR/MS and in political science, provides few guidelines and some very pessimistic conclusions. The main argument of this reading is that information systems development is an intensely political as well as technical process and that organizational mechanisms are needed that provide MIS managers with authority and resources for negotiation. The traditional view of MIS as a staff function ignores the pluralism of organizational decision making and

Source: from *Communications of the ACM*, 24 (1), 1981. ©Association for Computing Machinery, Inc.

PP 24-33

the link between information and power. Information systems increasingly alter relationships, patterns of communication and perceived influence, authority and control. A strategy for implementation must therefore recognize and deal with the politics of data and the likelihood, even legitimacy, of counterimplementation.

The Causes of Social Inertia

'Social inertia' is a complicated way of saying that no matter how hard you try, nothing seems to happen. The main causes of inertia in relation to information systems seem to be:

(1) Information is only a small component of organizational decision processes;
(2) Human information processing is experiential and relies on simplification;
(3) Organizations are complex and change is incremental and evolutionary; large steps are avoided, even resisted;
(4) Data are not merely an intellectual commodity but a political resource, whose redistribution through new information systems affects the interests of particular groups.

Computer specialists generally take for granted that information systems play a central role in decision making. Descriptive studies of managers' activities suggest this is often not the case. In general, decision processes are remarkably simple – what has worked in the past is most likely to be repeated. Under pressure, decision makers *discard* information, avoid bringing in expertise and exploring new alternatives, they simplify a problem to the point where it becomes manageable. Almost every descriptive study of a complex decision process suggests that formal analysis of quantified information is, at best, a minor aspect of the situation.[28,3] Negotiations, habit, rules of thumb and 'muddling through'[23] have far more force. This may seem an extreme assertion but there is little if any empirical evidence to challenge it. The point is not that managers are stupid or information systems irrelevant but that decision making is multifaceted, emotive, conservative and only partially cognitive. Formalized information technologies are not as self-evidently beneficial as technicians presume. Many descriptive models of decision making imply that 'better' information will have virtually no impact.[23,4,11]

Simon's concept of bounded rationality stresses the simplicity and limitations of individual information processing. There has long been a conflict between the normative perspective of OR/MS and MIS, which defines tools based on a rationalistic model of decision making, and the descriptive, largely relativistic position of many behavioural scientists who argue that that conception is unrealistic. Mitroff's study of the Apollo moon scientists is perhaps the best supported presentation of this position.[26] Regardless of one's viewpoint on how individuals *should* make decisions, it seems clear that the processes they *actually* rely on do not remotely approximate the rational ideal. This gap between the descriptive and prescriptive is a main cause of inertia:

(1) There is little evidence to support the concept of consistent preference functions
(2) Managers and students (the traditional subjects of experiments) have difficulty with simple trade-off choices
(3) Perceptions are selective
(4) There are clear biases and personality differences in problem-solving 'styles' that may even lead individuals to reject accurate and useful information
(5) Even intelligent and experienced decision makers make many errors of logic and inference
(6) Managers prefer concrete and verbal data to formal analysis

All in all, human information processing tends to be simple, experiential, nonanalytic and on the whole, fairly effective. Formalized information systems are thus often seen as threatening and unneeded. They are an intrusion into the world of the users who see these unfamiliar and nonrelevant techniques as a criticism of themselves.

Leavitt's classification of organizations as a diamond, (figure 1) in which task, technology, people and structure are interrelated and mutually adjusting, indicates the complex nature of social systems.[21] When technology is changed, the other components often adjust to damp out the impact of the innovation. Many writers on implementation stress the homeostatic behaviour of organizations and the need to 'unfreeze the status quo'. (This term is taken from the Lewin–Schein framework of social change, discussed below.)

Information systems are often intended as coupling devices that coordinate planning and improve management control. Cohen and March's view of many organizational decision processes as a garbage can[4] and Weick's powerful conception of 'loose coupling'[35] imply, however, that signals sent from the top often get diffused, defused and even lost, as they move down

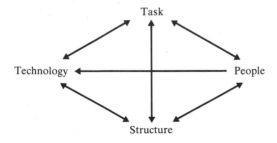

Figure 1   The Leavitt 'Diamond' components of the organization

and across units whose linkages are tenuous. The more complex the organization, the less likely the impact of technical change; homeostatic, self-equilibrating forces in loosely coupled systems are a major explanation for the frequency of failure of large-scale planning projects.

The characteristics of individuals and organizations listed above suggest that dramatic change rarely occurs in complex social systems. Lindblom's well-known concept of muddling through reinforces that view.[23] He points out the value of incremental, remedial decision making and rejects the 'synoptic ideal'. Wildavsky[37] similarly disdains formalized planning and recommends an avowedly political process based on partiality and incremental analysis. He contrasts political and economic rationality. The latter looks for optimal solutions through systematic methodologies. Compromise is pathological since by definition it represents a retreat from rationality (one might expect that few people would espouse this position in so pristine a form – until one listens to a faculty full of microeconomists). Political (or social) rationality looks only for feasible solutions and recognizes that utopian change cannot be assimilated by complex systems composed of individuals with bounded rationality. Only small increments are possible and compromise, far from being bad, is an essential aspect of the implementation process.

The final cause of inertia is less passive than the others. Data are a central political resource. Many agents and units in organizations get their influence and autonomy from their control over information. They will not readily give that up. In many instances new information systems represent a direct threat and they respond accordingly. We now have adequate theories of implementation. We have less understanding of counterimplementation, the life force of more than a few public sector organizations and a hidden feature of many private ones. This issue is discussed in more detail later.

All these forces towards inertia are constraints on innovation. They are not necessarily binding ones. Implementation *is* possible but requires patience and a strategy that recognizes that the change process must be explicitly managed. Only small successes will be achieved in most situations. These may, however, be strung together into major long-term innovations. 'Creeping socialism' is an instance of limited tactical decisions adding up to strategic redirection; no one step appears radical.

Overcoming Social Inertia: A Tactical Approach

There are several well-defined tactical models for dealing with inertia. They are tactical in the sense that they apply largely to specific projects. They recommend simple, phased programmes with clear objectives and facilitation by a change agent or a 'fixer', an actor with the organizational resources to negotiate among interested parties and make side payments. The Lewin–Schein framework and an extension of it, Kolb and Frohman's model of the consulting process,[19] have been used extensively by researchers on OR/MS and MIS implementation,[7,14] both in descriptive studies and prescriptive analysis. This conception of the change process (see figure 2) emphasizes:

(1) The immense amount of work needed prior to design; change must be self-motivated and based on a 'felt need' with a contract between user and implementer built on mutual credibility and commitment;
(2) The difficulty of institutionalizing a system and embedding it in its organizational context so that it will stay alive when the designer/consultant leaves the scene;
(3) The problem of operationalizing goals and identifying criteria for success.

This tactical approach is 'up-and-in' rather than 'down-and-out'.[22] DO is based on direction from the top, lengthy design stages and a formal system for planning and project management. UI relies on small groups, with

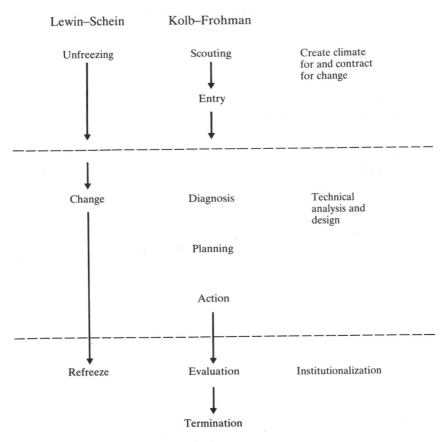

Figure 2   Tactical model for describing and/or managing change

face-to-face involvement and participative management. The design evolves out of the entry process.[19]

Leavitt and Webb point out that UI works well for small projects. However, large-scale change requires an engineering approach to design that quickly encounters social inertia. The dilemma is that UI limits itself to feasible, incremental change while DO, the broader strategic process, is rarely successful. The tactical model needs extension; facilitation is not enough and social inertia is dangerously close to social entropy.

No formal effective strategic model exists. If it did, one might expect to find it in political science, which frequently reconstructs the processes underlying efforts to deliver major social, technical or political programmes. Political science deserves the label of the 'dismal' science far more than economics, which after all believes in the eventual triumph of rationality; most studies in this field deal with failures. (Saplosky's analysis of the Polaris project[31] is a rare example of a success.) They identify as forces impeding change not only social inertia but also pluralism and counterimplementation – overt moves, often made by skilled actors, to prevent a disruption of the status quo. Counterimplementation is most likely to occur when outsiders bring in threatening new technologies.

Information systems are exactly that in many cases.

Pluralism: The Need to Mobilize

Political science views organizations mainly as groups of actors, often with conflicting priorities, objectives and values. The management literature generally assumes far more commonality of purpose. The down-and-out approach relies on this. Up-and-in evades the problem by limiting the scope of the project and hence the number of actors involved; it fails completely if consensus is not possible. The more the organization is viewed as a set of loosely coupled units where joint action rests on negotiations, the more any strategy for implementation must emphasize the need to mobilize coalitions, to provide the necessary support for an innovative proposal. Obviously, that process is based on political rather than economic rationality. The corollary of this argument is that lack of attention to the constraints on change imposed by pluralism in organizations will result in failure.

Many writers who attack the rationalist tradition on which OR/MS and MIS are based stress the legitimacy of pluralism and hence of incremental decision making. Lindblom sees the use of social interactions instead of analysis and planning as analogous to reliance on a market system to simplify the process of resource allocation.[24] Strauss argues that 'social order' and decision making in any organization are predominantly based on negotiations:

> ... when individuals or groups or organizations work together to 'get things done' then agreement is required about such matters as what, how, when, where, and how much. Continued agreement itself may be

something to be worked at ... negotiations pertain to the orderin articulation of an enormous variety of activities.[33]

In many instances, pluralistic perspectives view formal information systems as either ethically dangerous in that they impose a false rationality, naive or simply irrelevant. They also deny their value as coupling devices that help coordinate planning and communication; pluralists see *merit* in disorder and redundancy. Weiner and Wildavsky, commenting on federalism, summarize this argument: What is needed is '... planning with a different aim; to foster choice through careful structuring of social interaction'.[36]

These viewpoints are obviously not shared by most proponents of analytic methodologies. Since they are mainly based on studies of public policy issues, one may argue that business organizations are more tightly coupled and less dominated by pluralism and incrementalism. This may be true in particular instances: there are many companies whose planning systems are effective in establishing and communicating goals, involving managers in the decision process and creating a climate for innovation. Even so, most case studies of complex decisions suggest that companies are far more pluralistic than we conveniently assume. Pettigrew's analysis of a decision to purchase a computer, for example, reveals innumerable territorial disputes, manoeuvring for position, conflict over goals and irreconcilable differences in perspective among organizational units.[28] Believers in pluralism do not find that surprising but most computer specialists do.

The point is not to justify pluralism. It seems clear, however, that it is a main cause of inertia. 'Getting things done', whether down-and-out or up-and-in, requires the careful building of coalitions based on complex negotiations. The larger the scope of a project and the more strategic its goals, the truer this will be, because of the '... geometric growth of interdependencies ... whose implications extend over time'. Pressman and Wildavsky,[29] suggest some organizational mechanisms that can provide information systems developers with the authority and resources to resolve these complexities.

Counterimplementation

Believers in rationalism generally view resistance to change and protection of vested interests as faults to be ignored or suppressed. The tactical approach to implementation sees resistance as a signal from a system in equilibrium that the costs of change are perceived as greater than the likely benefits. The bringers and sellers of change – academics, computer specialists and consultants – assume that what they offer is good. In practice, there are many valid reasons to go beyond passive resistance and actively try to prevent implementation. Many innovations are dumb ideas. Others threaten the interests of individuals and groups by intruding on their territory, limiting their autonomy, reducing their influence, or adding to their workload. While we all may try to act in the 'corporate' interest, we often have very different definitions of exactly what that is. (Dearborn and Simon point out that even senior executives adopt the perspective of their department.[5])

Obviously there is a fine line between honest resistance to a project one feels is misguided and selfish sabotage of a necessary innovation. The difference is a matter for conscience and self-scrutiny. In both cases, the response is political, whether 'clean' or 'dirty' politics.

Bardach defines implementation as a game and outlines some of the moves and countermoves by which actors: (1) divert resources from a project; (2) deflect its goals; (3) dissipate its energies.[2] A central lesson to be learned from examples of successful counterimplementation is that there is no need to take the risky step of overtly opposing a project. The simplest approach is to rely on social inertia and use moves based on delay and tokenism. Technical outsiders should be kept outside and their lack of awareness of organizational issues encouraged. ('Why don't you build the model and we'll deal with the people issues later; there's no need to have these interminable meetings.') If more active counterimplementation is needed, one may exploit the difficulty of getting agreement among actors with different interests by enthusiastically saying, 'Great idea – but let's do it properly!' adding more people to the game and making the objectives of the venture broader and more ambitious and consequently more contentious and harder to make operational.

This author has found examples of most of the tactics Bardach identifies, in an ongoing study of the implementation of information systems and models for educational policy analysis in state government. Before discussing them, it is important to examine what is perhaps the single most important cause of counterimplementation in information systems development – the politics of data.

The link between control over information and influence has often been noted. 'Information is a resource that symbolizes status, enhances authority and shapes relationships.'[37] 'Information is an element of power.'[9]

Computer systems often redistribute information, breaking up monopolies. Building a database then becomes a political move; sometimes it is equivalent to a declaration of war. The system designer needs to ask:

(1) Who owns the data?
(2) Who will share it?
(3) What will be the perceived impact of redistribution on:
    (a) evaluation;
    (b) influence and authority;
    (c) communication?

He or she should then get ready to deal with counterimplementation.

A key step in the tactical approach to implementation is to convert the general impetus for change which is usually based on broad goals and rallying cries, into operational objectives and a specific contract.[19,7] Any project is very vulnerable to counterimplementation until this is done.

Programmes that have unclear goals or ambiguous specifications and that rely on continuing high levels of competence and coordination are easy targets for skilled game players.

Conclusion: A Strategic Perspective on Change

Counter-counter-implementation (CCI) is largely defensive, whereas the facilitative tactical approach is proactive. To an extent, CCI involves containing and doing the opposite of counterimplementers, whose strategy may be summarized as:

(1) Lay low;
(2) Rely on inertia;
(3) Keep the project complex, hard to coordinate and vaguely defined;
(4) Minimize the implementers' legitimacy and influence;
(5) Exploit their lack of inside knowledge.

The tactical model addresses some of these issues:

(1) Make sure you have a contract for change;
(2) Seek out resistance and treat it as a signal to be responded to;
(3) Rely on face-to-face contracts;
(4) Become an insider and work hard to build personal credibility;
(5) Co-opt users early.

A strategic model for change needs to resolve some additional concerns.

(1) What happens when consensus is impossible?
(2) How can large-scale projects evade social inertia?
(3) What authority mechanisms and organizational resources are needed to deal with the politics and data and counterimplementation?
(4) What is the role of management?

Some points are obvious from the analysis so far. Whether we like it or not, we can only hope for incremental change except, as Ansoff points out, in situations of mild crises, where the status quo is no longer satisfactory, and organizations rethink their goals and are more willing to think 'rationally'.[1] This reality suggests that systems designers must always aim for simplicity of design and precise objectives. However, if they are to go beyond tactical innovations based on up-and-in, they need down-and-out directional planning; they must establish the *direction* of change and evolve complex systems out of phased components. This requires nontechnical resources such as (1) a meaningful steering committee and (2) authority.

The analysis in this reading indicates that information development must be spearheaded by a general, not coordinated by aides-de-camp. It must be defined as part of the information function of the organization, instead of being a staff service labelled data-processing or management science. The issues of negotiations seem central. To position a system one must clarify objectives, respond to resistance, adjust other components of the Leavitt

Diamond (task, technology, people, structure) and block off counterimplementation. The politics of data (and of software engineering) make it essential that negotiations be handled by a fixer, well-linked into senior managements' decision making. Large-scale change is a process of coalition building; this cannot be done by staff analysts, who are too easily caught in the middle with no formal powers.

The strategy for managing social change is based on acceptance of the political nature of information systems development and the need for suitable authority. Many organizations have moved in this direction. Neal and Radnor and their colleagues conclude that OR/MS groups with formal charters (budgets, senior job titles for their managers, and the right to turn down user requests) are more successful than ones that are a corporate service unit.[27,30] The few Grand Old Men in the information systems field who have risen to senior positions in large companies have built up organizational mechanisms that provide them with authority and strong links with top level planning in the organization. There is perhaps an almost Darwinian process of natural selection. Where the MIS group adopts a purely technical focus or cannot obtain authority for negotiations, it becomes merely a data-processing service limited to routine applications and subject to all the forces of inertia and counterimplementation discussed here.[15,13]

It is not the aim of this reading to define a specific strategy for implementation. The outline seems clear:

(1) A senior level fixer must head the information function; he or she must have full authority and resources to negotiate with or between users and with those affected by information systems.

(2) There must be some policy planning or steering committee which includes senior line managers; it will delegate to technical staff responsibility for projects that do not have significant organizational impact but will be actively involved with ones that are part of the politics of data (the policy committee also provides a negotiating table);

(3) The planning process will require substantial time and effort in the *predesign* stages, where objectives are made operational and evolution of the larger system is defined by breaking it into clear phases;

(4) Formal contracts will be needed, in which commitments must be clearly made.

(5) 'Hybrid' skills must be developed in systems staff; they cannot dismiss organizational and political issues as irrelevant or not their responsibility, but must be able to operate in the manager's world and build credibility across the organization.[16]

(6) With the umbrella provided by the fixer's authority and the steering committee, the tactical approach remains an excellent guide to managing the implementation process for a given project.

The simple, central argument presented here is that information systems development is political as well as, sometimes far more so than, technical in nature. When that is accepted, the organizational mechanisms follow naturally. Unfortunately, 'politics' have been equated with evil, corruption and worst of all, blasphemy in the presence of the Rational Ideal, but politics are the process of getting commitment, or building support or creating momentum for change; they are inevitable.

The final comments to be made here concern research. There have been few studies of the political aspects of information systems development. The topic is rarely discussed in textbooks and even the literature on tactical implementation deals with it only peripherally. Yet when one tries to reconstruct or observe the progress of any major project, this is an obvious and important feature. It is absurd to ignore it or treat it as somehow an unsuitable subject for study or for training MIS specialists. There is some fragmented research available: Pettigrew's observation of a computer purchase decision.[28] Laudon's, *Computers and Bureaucratic Reform,*[43] and the work done by the Urban Information Systems Research Group at the University of California at Irving.[17,18] Greenberger *et al.* also provide some vivid illustrations of the political nature of computer models in public policy making.[9] Most of this work is based on case studies. Politics are hard to study. They involve many hidden agenda (counterimplementers do not boast about their triumphs) and in most instances a skilled observer has to ferret out and interpret what has happened. In political science, the work on implementation is almost entirely narrative and descriptive. A political perspective on information systems is needed in research. It will of necessity be based on comparative field studies that illustrate theoretical concepts.[25,25a] It will not fit the statndard mould for behavioural research. It can immensely add to our understanding both of the implications of information technology and the dynamics of effective implementation. For a long time the word 'implementation' was not included in the index to literature on OR/MS and MIS. It is to be hoped that 'politics', 'negotiations' and 'authority' be increasingly found in the titles of papers on information systems. That the papers will often be case studies does not mean they are not 'legitimate' research. We badly need more understanding of these issues which are of fundamental importance to the effective exploitation of computer technology.

References

1. Ansoff, H.I., 1968 *Business Strategy*, Penguin, London
2. Bardach, E., 1977 *The Implementation Game: What Happens After a Bill Becomes a Law.* MIT Press, Cambridge, Massachusetts
3. Bower, J., 1970 *The Resource Allocation Process.* Irwin, New York
4. Cohen, M.R., March, J.G., 1974 *Leadership and Ambiguity.* McGraw-Hill, New York
5. Dearborn, D.C., Simon, H.A., 1976 The identification of executives. In *Administrative Behavior*, 3rd ed. Simon, H.A. (Ed.) Free Press, New York, pp. 309–14

6. Drake, J.W., 1972 *The Administration of Transportation Modelling Projects.* Heath, Lexington, Massachusetts, pp. 14–17
7. Ginzberg, M.J., 1975 A process approach to management science implementation. Unpublished Ph.D. Dissertation. Sloan School of Management, M.I.T., Cambridge, Massachusetts
8. Grayson, C.J., 1973 Management science and business practice. *Harvard Business Rev. 51* (4) 41–48
9. Greenberger, M., Crenson, M.A., Crissey, B.L., 1976 *Models of the Policy Process*, Russell Sage Foundation, New York
10. Hall, W.K., 1973 Strategic planning models: Are top managers really finding them useful? *J. Business Policy 3* (3) 19–27
11. Hirschman, A.O., 1958 *The Strategy of Economic Development.* Yale Univ. Press, New Haven, Connecticut
12. Hoos, I.R., 1972 *Systems Analysis in Public Policy*, University of California Press, Berkeley, California
13. Keen, P.G.W., 1976 Managing organizational change: The role of MIS. in *Proc. 6th and 7th Ann. Conf. of the Soc. for Management Infor. Syst.*, White, J.D., (ed.) University of Michigan, Ann Arbor, Michigan, July 1976, pp. 129–34
14. Keen, P.G.W., 1977 Implementation research in MIS and OR/MS: Description versus prescription. Stanford Business School Research Paper No. 390, Stanford, California
15. Keen, P.G.W., Gerson, E.M., 1977 The politics of software engineering. *Datamation 23* (11) 80–86
16. Keen, P.G.W., Scott Morton, M.S., 1978 *Decision Support Systems: An Organizational Perspective.* Addison-Wesley, Reading, Massachusetts
17. Kling, R., 1978 Information systems in policy making. *Telecomm Policy 2* (1): 3–12
18. Kling, R., 1980 Social analyses of computing: Theoretical perspectives in recent empirical research. *Comptng Survey 12* (1): 61–110
19. Kolb, D.A., Frohman, A.L., 1970 An organizational development approach to consulting. *Sloan Management Rev. 12* (1): 51–65
20. Laudon, K.C., 1974 *Computers and Bureaucratic Reform: The Political Functions of Urban Information Systems.* Wiley, New York
21. Leavitt, H.J., 1965 Applying organizational change in industry: Structural, technological and humanistic approaches. *Handbook of Organizations,* March, J.G. (ed.) Rand McNally, Chicago, Illinois
22. Leavitt, H.J., Webb, L., 1978 Implementing: Two approaches. Stanford Univ. Research Paper 440, Stanford, California
23. Lindblom, C.E., 1959 The science of muddling through *Public Administration Rev. 19* (2): 79–88
24. Lindblom, C.E., 1977 *Politics and Markets.* Basic Books, New York
25. Mintzberg, H., 1973 *The Nature of Managerial Work.* Harper and Row, New York
25(a) Mintzberg, H., 1977 Policy as a field of management theory. *Academy of Management Review*, January (Unnumbered paper)
26. Mitroff, I.I., 1974 *The Subjective Side of Science: A Philosophic Enquiry into the Psychology of the Apollo Moon Scientists.* Elsevier, New York
27. Neal, R.D., Radnor, M., 1973 The relation between formal procedures for pursuing OR/MS activities and OR/MS group success. *Operations Res. 21*: 451–74
28. Pettigrew, A.M., 1973 *The Politics of Organizational Decision Making.* Tavistock, London

29. Pressman, J.L., Wildavsky, A., 1973 *Implementations*. University of California Press, Berkeley, California
30. Rubenstein, A.H., Radnor, M., Baker, N., Heiman, D., McCoy, J., 1967 Some organizational factors relative to the effectiveness of management science groups in industry. *Management Sci. 13* (8): B508–18
31. Saplosky, H.M., 1972 *The Polaris System Development*, Harvard University Press, Cambridge, Massachusetts
32. Simon, H.A., 1957 A behavioral model of rational choice: In *Models of Man*, Simon, H.A., (ed.) Wiley, New York, pp. 241–60
33. Strauss, A., 1978 *Negotiations: Varieties, Contexts, Processes, and Social Cries*. Jossey-Bass, San Francisco, California
34. Urban, G.L., 1974 Building models for decision makers. *Interfaces 4*, (3): 1–11
35. Weick, K., 1969 *The Social Psychology of Organizing*. Addison-Wesley, Reading, Massachusetts
36. Weiner, S., Wildavsky, A., 1978 The prophylactic presidency. *Public Interest 52* (52): 3–19
37. Wildavsky, A., 1974 *The Politics of the Budgetary Process*. 2nd Ed. Little, Brown, Boston, Massachusetts

6.6

Supervisors and New Technology

Sheila Rothwell

The introduction of computer technology, whether through the automation of production processes in factories or of clerical processes in offices, affects the total management information system and has implications, direct or indirect, for the work of people at all levels. Much of the literature discussing the effects of this has focused on the effects on lower level employees, or, to a slightly lesser extent, on managers. The impact on supervisors – the forgotten men in the middle – is often ignored; yet they may be the people who are most vital to the successful implementation of change.

The apparent neglect arises largely from the very ambiguities and lack of definition of their role, which has been well-documented in much previous research.[1,3] If they are described as 'first-line' managers, why are they so often regarded as 'employees', by middle and senior managers, across a wide range of industries and organizations? This problem of 'status' is also seen as their chief 'problem' by supervisors themselves and their reactions to this insecurity may well have implications for their attitude to change, particularly change on the present scale.

Research at Henley over a two-year period, supported by the Manpower Services Commission, has explored the effect of the implementation of new technology on the management of people and the organization of work in a wide range of companies and industries.[2] Over 20 cases have been studied, in various degrees of detail, of the application of information technology to materials and stock control, order processing, warehousing, quantity control, packing, confectionery making, customer service, accounting and production control. Obviously supervisory roles and functions differed in each case studied – and even varied in different sections of the same office or factory – but from an examination of the part they played, if any, in the design and implementation of change (at whatever stage it had reached) it is possible to see some patterns emerging and to draw some conclusions about supervisory roles and the impact of new technology.

Source: from *Employment Gazette*, January, 1984.

The processes of management decision making involved in investing in new technology from the preliminary feasibility study to the decision to go ahead, followed by the system design specification, rarely included any supervisory involvement. These decisions were usually made at very senior levels with the advice of systems specialists and some middle managers of the departments affected. The type of system chosen and the criteria for decision making in design and implementation tended to relate to the initial motivation for change which in most cases was market related: to improve service and respond more flexibly to changes in demand. Improved cost-effectiveness was another major aim, with the desire to cut production and labour costs, achieve economics of scale and improve productivity, profitability and efficiency, also being mentioned in many cases. That these two major strategic aims could at times conflict or create short-term operational difficulties – particularly for supervisors – was rarely acknowledged.

The development of application programmes or the modification of bought-in packages to operating requirements usually involved more user contact and discussion, so the likelihood of supervisors being involved at this stage was slightly greater, as well as in the design of operating procedures and tasks and preparation of the system manuals.

Decision making tends to centre around technical issues and broad manning plans. The need to get the system working then means that emphasis seems to be placed on those who will use the system most directly. Supervisors, if they are not necessarily primary users, are often one of the last groups whose role is considered in detail, particularly in offices and in automation of manufacturing processes. They are more likely to be involved in the material and production planning control type of management information systems projects. In these cases they have a valuable input to make about how work is actually done in their area or department. In addition where supervisors became accountable for the cost-effectiveness of their section or department their involvement was quite extensive in design, implementation and training.

This depended more on the style and approach of the senior project or implementation team leader and the extent to which they were 'technically', 'hierarchically' or 'people' oriented – rather than whether it was a 'line' or 'systems' manager in charge of the project. It could also relate to the reactions of the supervisors themselves and the extent to which they sought, or avoided, involvement.

Thus, at electronics, many production superintendents and supervisors of functions were involved in application teams at all stages, with the encouragement of senior management as well as systems specialists and senior line supervisory managers. At power tools supervisors became involved, largely at their own request, once systems specifications had been produced, in going through them 'line by line' in order that amendments could be made at an early stage. They were also critical of the project task force for not involving them at an earlier stage still.

In a few cases project leaders claimed to have involved supervisors from the beginning, but supervisors only saw this as general consultation as part of user groups. In one case however they were given a detailed role in designing operating procedures in their sections.

On the other hand, emphasis on achieving a technical best, for example, in quantity control applications at liquid food and tablets meant that systems analysis or project leaders had little contact with the supervisors; similarly in the mail order and distribution companies, in addition to the systems specialist, the office and depot managers played a central part but virtually ignored the supervisors whom they treated as senior clerks. In these cases supervisors were mainly women, but their work was in some respects less directly affected by the systems change than in the materials and production control applications.

Implementing Change

In some cases, supervisors' main functions in implementation were in keeping the old system running and production going while the new system was being phased in, then keeping the two running in tandem until implementation was complete, so that there was slightly less technical need for their consultation and involvement.

In other instances, supervisors may have had a slight say in the later stages of design, but were not really involved until commissioning – as at packet foods where supervisors were recently recruited to a new automated warehouse and at confections where the technical problems of installing and obtaining the smooth running of a new automated production line had already been experienced in an earlier unsuccessful innovation. In these cases and at a small company making printing components by a new plastics process the supervisor had considerable technical involvement in ensuring the smooth running of the new system.

At liquid food accounting supervisors and clerks were closely involved in implementation because the two establishments studied were pilot sites and users' views were seen as valuable in preparing for implementation in the remaining 31 establishments.

In several of the order processing and materials management cases, whether office or warehouse only, or factory related as well, and in some of the cases involving major changes in manufacturing processes, supervisory involvement in training and communications was significant. Responsibility for operator or clerical training was largely left to supervisors in packet foods where a training consultancy firm had proved inadequate. In electronics and to some extent in confections and power tools, supervisors were responsible for some of the instruction and practice by subordinates.

This implied, of course, that supervisors were themselves given sufficient training in advance in order to be able to instruct their sections. In some cases, where trainers played a larger role, supervisory training was separate from, though largely parallel to, that of operators. In others, supervisory training was subsequent to that of operators or almost ignored. In the order processing section of photo products this delay was later realized to have been a mistake as supervisors then lacked the skills and credibility. In the factory site, training was available but supervisors were often reluctant to accept it, partly because they felt under pressure to maintain production output but also for other reasons related both to the nature of the change and of the training provided. These issues were tackled more directly at engines, where change to self-managing group working methods on the assembly line meant that roles had changed, but supervisors were sent on the team-building and other related courses before the assembly workers and were being assisted to adjust to the new situation of 'stand back and observe', rather than take part in team activity, until they could work out a new 'contract' with both their team and their managers.

The content of training is probably therefore as relevant as its timing and extent. Training in technical skills of fault-finding and repairing was obviously needed for maintenance, engineering and, to some extent, production supervisors, while technical knowledge and skills related to inputting data, adjusting programmes (engines), correcting errors, bringing up files and obtaining information was needed by many others. Beyond this, however, training in additional knowledge about how the system affecting their section related to other parts of the organization, how to be accountable for costs, how to use information to generate new activities or write reports and how to manage people differently was probably even more important. The extent to which this was necessary related to the change, if any, in supervisors' roles and functions and the clarity with which this was perceived by both supervisors and their managers.

Erosion of Roles and Functions

Erosion of the supervisory functions by technology, managers and work groups hypothesized by previous writers were all found in some form in these studies.

The *technology* itself reduced the areas of discretion in very many instances: goods received could only be booked in one way, warehouse picking or packing might be only according to schedule, manufacture of new orders could not be begun until the system indicated, discretionary switching of materials components or of manufacturing batches could be inhibited. In short, a whole range of discretionary decisions which had been normally made by supervisors and which were once essential to keep things running smoothly (as distinct from what the formal system appeared to dictate or what management thought was happening), became 'programmed' system-made decisions.

Emphasis on ensuring that the correct disciplines of entering every transaction into the system were mastered and followed to avoid the chaos that could ensue from inaccurate data meant that conflicts inevitably arose – at least in the short run – with the old principle of getting it 'out the door' at all costs. The supervisors' skills of fixing and by-passing the formal system through a mixture of experience, cunning, personal contacts and trading of favours or indulgences, could appear to count for nought overnight. Even more straightforward expertise in planning and scheduling work, acquired through years of experience, could now be available to anyone who could operate the system.

It was acknowledged in some instances that the job was more boring but accepted that this was as inevitable as 'progress' with which the technology was equated.

Computerized materials management systems are therefore more likely to erode certain technical aspects of the supervisor's role – the progress-chasing and materials scheduling aspects in particular. In order processing, customer service and accounting systems, too, much of the technical expertise of the supervisor in dealing with the problem cases, the alternative sources of supply or the customer idiosyncracies is diminished.

Some of the people aspects of the supervisory role are also likely to be eroded by the new technology. For example, manning levels may be determined by the system, and work allocated by it – functions that were previously supervisory. But this is not inevitable.

The *work group* has traditionally been seen as a major 'competitor' in eroding the supervisory role. As workers become better educated, more content and competent as well as more unionized they may also have greater expectations of autonomy and job satisfaction and greater resentment of authority and supervision. Our cases showed that these trends tended to be reinforced by the new technology in several ways, both directly and indirectly. Directly:

(a) in that these supervisory 'allocation and requisitionary' functions incorporated in the system were now exercised and utilized by employees themselves – who were often performing a wider range of tasks, and

(b) that 'system' authority did not appear to be resented in a way that personal authority might have been in, for example, the designation of 'picking' routes for fork-lift operators, or the scheduling of work priorities for clerks.

(c) This seems to have meant the reduction of interpersonal friction and of grievances between worker/worker and supervisor/worker. While the potential use of information technology for monitoring worker effort and errors is resented by trade unions, in practice, the facility does not appear to have been used extensively, other than to identify sources of error. Moreover, workers appear to have been less resentful of errors unambiguously attributable to them than in being blamed for omissions which were largely the responsibility of others.

Indirectly the technology has assisted work-group erosion of the supervisor role, in that a major aim of introducing a management information system was in some cases that of reducing labour costs. This was achieved in some instances by reducing the numbers of supervisory and indirect workers. Decisions to move to group working rarely appear to have been supervisory led but mostly came from higher management and, in several instances, from the US parent company.

Coordination of activities of different sections, or between the supervisor's own section and another may also have been taken over by a computer system that relates transport, warehouse, customer orders, materials and production capacity or one that schedules the date of a service visit according to customer and engineer availability, orders and parts and prepares the invoice.

The supervisor's *status* in managerial eyes as a source of specialist information may also be weakened if data are as readily available in the line manager's office as the supervisor's, and usable by other specialists in finance, personnel or planning.

That the supervisory functions were to some extent being usurped by the development of *more specialist management functions* – personnel, work study and cost control specialists, production engineers – has been a feature of industrial development over the past 30 years. The adoption of 'new' technology is only reinforcing this trend to the extent that management services and 'DP' specialists are also intervening and either directly, or indirectly through the systems they devise, reshaping supervisory functions.

Moreover the likelihood of the supervisor's *promotion* to manager may be further diminished, not only by the increasing number of graduate specialists but by the fact that he or she is no longer acquiring the necessary understanding of the business, in the old way, by progressing through different departments. If special training actuarial and business practice has to be given, as at insurance, for example, it can be given to graduate recruits more easily than to older supervisors.

Increased Scope

On the other hand, supervisory functions can in some instances be enhanced by the new technology. Depending on the implementation and training process, they may have acquired more understanding of the system than other operators or line managers and become as expert in utilizing it to achieve their ends as they were in the previous informal one. The extra and up-to-date information available – for example on average weight (confectioner) – can enable them to take corrective action more quickly when faults occur, before a whole batch is 'lost', to reschedule a new order or to diagnose where bottlenecks and delays are occurring.

Freedom from routine planning, progress chasing and paperwork should enable them to give more attention to the important 'people' components of the job, although this may have to be more subtly exercised than before, with reduced opportunities for 'telling' and chasing output. If clerks or operators are more independent or more 'group related', then supervisors need to develop more team-building ability and 'indirect' motivational skills in order to maintain both output levels and group cohesion. Involvement in 'training' may increase supervisory authority in the sense of technical expertise, although instructional skills may prove difficult for some to acquire (and something for which they themselves may need training).

Supervisory authority and credibility may also be enhanced when they are given a key role in communications relating to implementation – particularly when video is used and they are 'on the screen' explaining what is going on.

Where the group is given responsibility for its own output and quality control the supervisors need to motivate through the group and keep morale (and attendance) high, although it is difficult for them to adjust in this way. Where the supervisors retain accountability for output but the working group is given considerable autonomy supervisors retain traditional responsibilities in one sense, but also have to find new ways of group motivation to achieve them and experience some problems of responsibility without power.

In some cases, however, supervisors needed to play less of a 'people motivator' role and their 'technical' function became more important – in getting the new equipment to work at all on a day-to-day basis, as at print, confectioner and packet foods. In the latter case, where an automated warehouse was opened on a greenfield site, ex-RAF men were recruited as craftsmen and then quickly promoted to supervisors on account of their technical expertise.

At photo products too as the numbers of team supervisors decreased, and maintenance arrangements were adjusted, the role of the engineering supervisors became particularly important.

In several instances, relations between supervisors were altered by technical change – liaison becoming as important between themselves, as up or down the line, although each reported to different heads. In so far as this was successful it helped to overcome some of the major problems of 'new technology' which for its benefits to be fully realized needs greater functional integration than is normally feasible in British work places, which are highly functionally oriented and occupationally specialized with competitive rather than cooperative relationships tending to predominate. In itself, the technology also facilitates cooperation at times – for example, at tablets, where relationships between the packing and warehouse supervisors improved considerably when she was able to order materials from him through the system, which provided an unambiguous record of requests and responses.

Greater managerial skills were also being demanded in several cases where supervisors were given cost-accountability and expected to use the new information available to them to identify sources of improvement and to generate management reports. They found this difficult at photo products however and tended to ignore the computer printouts. Nor did management appear to have found ways of ensuring that supervisors used this information, although they tended to complain that supervisors were unwilling to receive training.

Supervisors in this case, however, in so far as their attitudes could be judged from questionnaire responses and conversation tended to feel that there was already too much paperwork required of them and that they would have welcomed training in report writing. Most felt that their responsibilities had increased considerably with technological change and that they needed more assistance in acquiring new problem-solving skills. In this case several of the production supervisors were relatively young and they had not necessarily worked their way up from the line – one, for example, had previously trained as a trainer. Engineering supervisors had followed more traditional routes.

Numbers of supervisors tended to have diminished in several cases although this was not always easy to identify because of variations in occupational title between manager, foreman, supervisor and chargehand. Nor in all cases was the reduction directly attributable to the technology if it took place in advance of implementation (electronics) or could be justified in terms of general contraction or reduction of indirects. Nevertheless it was not uncommon to find two levels of supervision merged into one although numbers reduced were not large and the change was usually said to have been achieved on a voluntary or early retirement basis. At electronics, where several were affected, some foremen were given the alternative option of returning to 'the bench', but most left rather than accept that.

Conclusions

Changes in supervisory positions could therefore be more clearly observed in factory rather than office situations and successful implementation (from the point of view of supervisors, and from an observer's view of the change process as a whole) tended to relate mainly to the extent to which management foresaw, planned and managed the changes. This was often inadequate where managers themselves were too highly involved in planning the technicalities of the new system or only interested, as a result of schedule slippages, in enabling operators to get it running. Thus supervisors, unless they played a critical part in its operation, were often left to cope with keeping things going and adjust as best they could. Some, as always, managed to adapt, to see what was needed and to perform in such a way as to overcome the failures and teething troubles of the new system. Others tended, as might be predicted, to experience the ambiguities and insecurities of their position more intensely and to withdraw from involvement and commitment, or at least to fluctuate between the two. By the time management realized this, certain patterns of resistance to change or, at the least, cynicism about its benefits might well have set in and spread to others.

Speculations on the future role of the supervisor are not new, but the nature of the current changes in systems of working implied by the new technology may in some areas be such as to require a complete rethink of the position of supervisors. Is there a need for such a role at all? Is it provided through the system? Certainly some organizations have already reduced numbers and this trend seems likely to continue (especially in engineering and elsewhere in manufacturing) as, on the one hand, operators obtain greater autonomy and responsibility through the system and, on the other, managers become more professional. Materials management systems in particular reduce the need for supervisors to 'chase' materials and work-in-progress to the same extent as previously.

Yet if computer systems lead to greater centralization, then 'flatter' hierarchies are more appropriate and there is still scope for autonomy at lower levels; thus the new technology can give scope for enhancing the role of those supervisors who remain.

Alternative arguments can be advanced for the need for more supervisors – or more levels of supervisor. If operators acquire more responsibility, and the machine or the system becomes the 'operator', then operators become virtual supervisors themselves. Or alternatively, if motivation of operators relies less on 'carrot and stick' approaches and more on team-working and/or career progression techniques, then there will be more 'teamleader' or other specialist function appointments. If hierarchies thus lengthen, or branch out, there may be a danger of increased sectionalism, unless coordinating mechanisms, through individuals or systems, are effective. 'Gate-keeper' supervisors can also become 'gate-blockers', particularly if their traditional responsibilities and status diminish.

Definition of an appropriate supervisory role clearly needs to start with a positive analysis of what supervisors are really required for and whether they are being used most effectively. This needs a critical analysis of the *work situation* – as a system, or part of a larger system, in terms of the really critical flows of work and linkages between plant, materials and information, and between people both vertically and horizontally. Such an analysis may show that 'coordinators' more than supervisors are needed, or 'technical' or systems engineers, or primarily facilitators and people motivators. In some cases the particular range of skills and planning abilities called for is such that genuine first line *managers* are needed and they should be designated and treated as such. Human relations and communications skills seem the most critical factor common to most of the roles, followed closely by some 'problem-solving' expertise, since technical breakdowns of some sort still tend to create the biggest headaches for supervisors.

Analysis of the work situation must also be accompanied, however, by analysis of the individual styles, personalities and abilities of the supervisors to ensure a 'fit'. This is unlikely to be instantaneous but it is not always possible (or desirable) to get rid of one lot (usually the older) supervisors and recruit new ones. Even where this is done over a period, or on a greenfield site, it is not necessarily successful unless they are given *individual* roles and training. Especially if the generic category of 'supervisor' no longer still exists, but rather a variety of roles depending on the new technology and the way it is applied.

The question of title is a difficult one since it confers 'status' in itself and it is often easier to redesign a function if the title remains (or *vice versa*) because the individual feels less threatened. A cosmetic issue can however become a substantial one if others in the hierarchy do not adjust their expectations and demands appropriately. Confusion between 'foreman', 'supervisor', 'leadinghand' titles still abound, and some would advocate a role and title akin to the German 'meister' as appropriate to many manufacturing positions.

People need time and training to develop into a role but unless specific coaching is given by their own managers, this is unlikely to happen. Our research found that managers themselves were often the real problem – in that either they were too busy with the technicalities of the new system and its implementation to plan and handle the 'people' side adequately; or that they did not understand the implications of the change sufficiently them-selves (sometimes resenting it or 'opting out' mentally) to cope: or else that their own traditional prejudices (about the status of women supervisors for example) or fears (of industrial relations repercussions) prevented them from tackling issues and developing or reorganizing their supervisors in such a way as to realize the potential of individuals and of the new system. The rare promotion of an occasional supervisor may demonstrate the existence of a potential ladder, but it is hardly seen by the rest as a substitute for adequate career development. Moreover, if managers are unable to achieve the new patterns of interdepartmental coordination required by more integrated computer systems, then the necessary func-tional coordination between sectional supervisors and other specialists, is also unlikely to be achieved.

Is the implication of all this that as numbers diminish and supervision becomes less of a shop floor job, then the utility of the role will disappear? Such a radical change obviously needs careful consideration, but in the new technology, as often in the old, the problem of supervision is the problem of management.

References

1. Child, J., Partridge, B., 1982 *Lost Managers*, Cambridge
2. Rothwell, S., Davidson, D., 1982 New technology and manpower utilization. *Employment Gazette* June, pp. 252–54 and July, pp. 280–81
3. Thurley, K., Wirdenius, H., 1973 *Supervision: A Reappraisal*. Heinemann

Uncertainty and the Innovation Process for Computer-integrated Manufacturing Systems: Four Case Studies

D. Gerwin and J.C. Tarondeau

How can we understand what goes on during the process in which major equipment innovations in manufacturing are adopted and implemented? Can we explain adoption decisions? Do the behaviours exhibited during implementation conform to certain patterns? This is a tall order. Purely economic explanations are not sufficient because they don't deal with the high degree of uncertainty that managers must face. Adoption decisions must be made even though information on future benefits and costs is unavailable, and implementation is plagued by unanticipated consequences which are difficult to quantify. In this reading the concept of uncertainty is made the cornerstone of our attempt to study major process innovations in manufacturing.

Conceptual Framework

Perhaps the most influential work in organization theory, in the past 15 years is that of Thompson.[10] He viewed organizations as open systems faced by ambiguity and uncertainty, but requiring clarity and certainty in order to function in a rational manner. Every organization has a technical core devoted to efficient performance of some processing function. In manufacturing firms this of course involves the transformation of raw materials into finished products. Technical core managers perceive at least three sources of uncertainties: the nature of the transformation process, the behaviour of human resources and the requirements of environmental elements such as suppliers and customers. However, to the extent that a

Source: reprinted with permission from the *Journal of Operations Management*, the American Production and Inventory Control Society, Inc., pp. 87–100.

growing number of managers view labour not as a source of uncertainty in the production process, but as a source of ideas on how to improve the production process, the second factor will become less salient. In Thompson's theory management's role is to reduce existing uncertainties so that the core may operate as efficiently as possible. Uncertainty reduction is achieved through the application of several strategies including buffering, smoothing, adapting and rationing.

Nowhere, however, does Thompson mention innovations in process technology as a possible strategy. We believe that this is a significant omission. Manufacturing management considers new equipment to be an important means of controlling uncertainty. Technological innovation in mass production and process production has contributed to the high degrees of predictability associated with these methods of manufacturing. It has led to standardization of inputs and repeatability of outputs in the transformation process. It has reduced human variability through the rationalization of work roles and the substitution of capital for labour. And, it may handle environmental contingencies as when a firm decides to manufacture in-house a component previously purchased outside. Abernathy showed how process improvements in the auto industry led to uncertainty reduction and high efficiency, but at the expense of new product development.[1] Noble explained how record playback control lost out to numerical control (NC) because the latter made it possible for managements to have greater control over production, even though at the time the cost-effectiveness of NC was not firmly established.[9]

Why did such an astute organizational theorist as Thompson avoid dealing with technological innovation? Perhaps the reason was that it raises some perplexing questions for his theory. If in one sense process innovations can be used to reduce uncertainty, and are thus compatible with his views, in another sense they introduce uncertainty into the technical core during adoption and implementation, and are thus to be avoided. Innovating organizations must grapple with whether: sufficient political support will be generated, consequences can be accurately forecasted, new equipment will work in the prescribed manner, new equipment will prove to be economical and whether the social and political fabric of the core will be altered. One specific example is the mounting evidence that, due to lack of information, only a veneer of objectivity surrounds the adoption decision for major manufacturing innovations.[5,6,8] It is little wonder that a recent critical review identified managing the uncertainties created by innovation as one of the major concepts pervading the innovation literature.[11]

How can Thompson's theory be revised to make it compatible with the dual nature of innovation? In general, coping with uncertainty needs to be substituted for uncertainty reduction. We need to distinguish between long-run uncertainties, nagging issues that have had and threaten to continue to have a prolonged impact and short-run certainties, issues it is believed may arise but can be readily resolved. Process innovation is undertaken to reduce long-run uncertainties, but only at the expense of introducing what it is hoped will be short-run uncertainties. Manufacturing management's role is to consider the trade-off between these two factors when adoption decisions are made. This is likely to introduce a great deal of heuristic as opposed to analytical thinking into the decision process. Further, management must devise strategies that will reduce the new contingencies associated with the innovation. To a large extent, successful innovations are those for which managers have developed successful coping strategies. Since some new contingencies may turn out to be of the nagging variety, one typical strategy is likely to be additional innovation.

Nature of the Study

At this stage in our thinking we want to explore and develop our ideas rather than test them. For this purpose we have chosen to study the introduction of computer-integrated manufacturing systems (CIM) into batch production. Batch manufacturing, as opposed to mass and process, is characterized by a relatively high degree of uncertainty. In our view CIM has the potential for handling certain variables that many managers perceive as contingencies. As the latest step in the development of numerical control, CIM possesses such NC characteristics as repeatability of outputs. In addition, the systemic nature of CIM contributes to uncertainty reduction. The first generation systems, known as direct numerical control (DNC), consist of a battery of NC or CNC machines connected to a central computer which handles their operation, control and reporting. DNC facilitates management's control of shop operations by providing a centralized source of information. The second generation, under the name of flexible manufacturing systems (FMS), combines DNC capabilities with automated transfer of parts to machines in the system. This latest development, which increases the predictability of material handling, represents our closest approximation to the automated job shop.[2,3]

Since our purpose is to explore and develop our ideas, and because the number of operational CIMs in the world is small, it was decided to study a few firms in depth. In order to obtain the broadest possible insights, and due to problems of access, these firms differ in terms of culture, task environment and stage of CIM development. It was also believed that the international significance of and interest in CIM technology warranted the selection of companies in different countries. Governments and private industries in at least the United States, France, Great Britain, West Germany, East Germany, Czechoslovakia, Japan and Norway are currently involved in the development of CIMs. Due to these criteria and constraints our findings are based on data collected during 1979 in four companies; one each in the United States, Great Britain, West Germany and France.

Reasons for Adoption

In the US, declining sales and profits led the division to propose a new product line and modernized shop facilities. Adoption of an FMS was one of many choices that had to be made as part of a major capital improvement programme. Risk reduction was the major factor instrumental in the FMS's selection. The merchandizing department's sales forecasts for the new product line were steadily decreasing during the evaluation period under pressure from a hard-to-convince management. It began to be felt that estimated production quantities were falling out of the realm of economical transfer line operations. In addition, the forecast changes implied that merchandizing was not particularly confident in its projections. The result was a change in objectives from developing an optimal programme to developing one which was most likely to prevent disaster. The FMS fit into this thinking because if the new product line, or some segment of it, failed to appeal to customers, the system could be adapted to product design changes with little added cost as compared to the other machining possibilities under consideration.

Several years ago the company in Great Britain was purchased by a large British corporation. The new corporate office encouraged its acquisition to replace antiquated equipment and increase capacity in all of its shops. In the medium-sized motor and generator shop an individual was available to conduct a study of long-term needs. Consequently, the DNC concept was part of a wider plan for the entire shop. A primary factor in the decision to adopt the system was its potential for giving GB a competitive edge in the industry. Approximately six firms compete in the British motor and generator market. Since they all have roughly equal technical competence, price and delivery date are critical factors. It was estimated that a DNC system could lower price by at least doubling output per man-hour. With respect to delivery date it was anticipated that the new system would have a 24-hour throughput time for prime components. With existing equipment it took about two weeks to get these parts through the shop of which perhaps ten hours involved machining. Most time was spent in waiting lines or in materials handling. A second consideration in adopting the DNC system was the skilled labour shortage in British industry. The estimated increase in productivity would mean a labour saving of perhaps 50 individuals. Finally, the DNC's ability to provide manufacturing management with a centralized information source was seen as an important advantage. Running through the shop to collect data on several large and expensive components was considered to be very inefficient.

West Germany began to manufacture prototypes for two new airplanes in the early 1970s. Twelve stand-alone NC machines were installed for this purpose. Numerical control is particularly well-suited to the complicated metal work and special light alloys used in the aerospace industry.[8] When full-scale production was approved around 1974 it was necessary to make a large capital investment to increase the capacity of the main machine shop. Since the existing experience with NC was satisfactory various studies were conducted to determine what kinds of additional NC equipment should be purchased. Reducing costs through maximizing machine utilization was an important consideration which led to an integrated system. A work piece analysis revealed that there were a lot of small-sized parts and a few large-sized parts to be machined. It was then determined what would be the part size below which it would be economical to use automated machinery. With this selection of parts machining time would be greater than set up time only on the average. In other words, for some parts setting up the next piece would take longer than machining the current piece. In order to minimize the resulting idle time it was decided to perform machining and set up operations in different locations and place a buffer storage between them. Then when a machine is about to become idle an already set up part could be quickly sent to it via an automated transport system. A second consideration was that the highly automated nature of an integrated system reduced labour requirements. Owing to Germany's codetermination policy, the company had in the past found it difficult to lay off workers when production dropped. Since it was felt there would be only about ten years of continuous production for the new equipment as low a labour requirement as possible was desired.

In France the starting point of the FMS adoption process was the creation of a new product line of industrial vehicles. The FMS concept was adopted because it was seen as a means of achieving both productivity and flexibility. Nonintegrated production systems need a very high level of buffer stocks and typically have a low utilization rate. The company could have tried to improve the inventory control system and to increase the utilization rate of a conventional production system. Instead, it is believed, in part due to simulation results, that the FMS will lower work-in-process inventories, insure a high utilization factor and reduce delivery time. The capability of on-line control of the entire system should permit the simultaneous attainment of these objectives.

The industrial vehicle industry is faced with a highly competitive world market and cyclical demand. Both conditions necessitate that manufacturing processes have good adaptation capabilities. The FMS concept can deal with the uncertainty of product demand because the investment is not rigidly tied to a given production requirement. It offers the opportunity to modify either the products or the mix without large new investments.

Adoption Strategies

In the US the division took the initiative in preparing a proposal for a new product line and associated manufacturing processes. Corporate staff personnel were not involved. The head of manufacturing engineering appointed a five-person innovation group from his department in order to see whether it was feasible to build the new line. The group worked closely with a liaison from purchasing who helped locate potential vendors, and a member of accounting who developed financial analyses.

The innovation group's task was exceedingly difficult. It was faced with a large number of highly technical decisions that had to be made in the face of considerable uncertainty due to the facts that a new product line and some innovative manufacturing processes were being considered. Consequently, detailed financial analyses were made only for the programme as a whole rather than for each individual decision. While the exact values of these numbers could not be considered very reliable, they satisfied the corporate guidelines for rationality in capital budgeting, and provided some indication of whether the minimum limits on acceptability were likely to be exceeded.

Acquisition cost assumed a disproportionate role in decision making for two reasons. The difficulty in making accurate forecasts of future net revenues led the innovation group to deal with the certain here and now rather than the uncertain future. Corporate headquarters was unable to invest a large sum in the division, and the division for its part did not want its proposal rejected on this basis. The FMS had a relatively small acquisition cost compared to its rivals and this represented a useful selling point to corporate management.

The company had a close relationship with the vendor of the FMS. They are located nearby each other in the same city and have a history of dealing with each other. Here was one instance in which some reliable information existed. As a result, the innovation group was convinced that the vendor could develop a system along the lines it was promising. The vendor's professional sales effort, which projected a capable image, reinforced the company's thinking.

At least four hierarchical levels above manufacturing engineering had to be taken into consideration in selling the proposal. At these levels, there could not be much inquiry into individual machine tool decisions. These were considered as technical choices which were best left to manufacturing engineering. If, on the other hand, a single machine tool is to be requested for an existing product line, justification would have to include sophisticated financial analyses which would be intensively reviewed. Thus, it can be inferred that the FMS decision, in spite of its magnitude and uniqueness, lost visibility by being one among many choices which were highly technical in nature. Ultimately, the corporate president decided to approve the programme. By this time almost three years had elapsed and over 30 revisions in the proposal had been made.

In Great Britain, plans for modernizing the shop were devised by a single individual, the retired chief of manufacturing development for the plant concerned. His work was facilitated by his being freed from operating pressures, and by his taking a trip to survey manufacturing installations in foreign countries. In addition, he worked within a supportive environment at the plant and corporate levels. His plan for what the shop should look like in ten years was more like an artist's sketch than a finished painting. The most critical element was the DNC system for which a facilities layout, a simple list of estimated costs and benefits and a description of the steps in implementation, were produced. After approval by local and corporate management the plan became a goal toward which the company is working.

In West Germany it is typical for individual plants to select new manufacturing processes with the advice of staff departments in corporate headquarters. However, the magnitude of this particular investment dictated that the central office would play a more influential role than usual. Top management set up a special *ad hoc* task force, consisting of an equal number of plant and corporate people, for the purpose of selecting manufacturing processes. Representing the plant were the managers of production, production preparation and planning. The corporate personnel included an individual concerned with NC machining from the department of production techniques. The process planning, and buildings and grounds departments were also represented.

The task force made its equipment selection within three kinds of important constraints. First, choices had to be made under considerable time pressure. Second, product designs were expected to govern the selection of manufacturing processes. There was no opportunity for two-way influence as is the situation in US aircraft companies where production volumes are much larger. Third, it was necessary to meet capacity objectives within a tight, immutable financial constraint. The machine tool choices made within these constraints were formalized as much as possible using decision-making techniques developed by Kepner and Tregoe.[7] However, there were problems in dealing with subjective aspects such as gaining agreement on the weights to be assigned to various criteria.

The plant's accountants conducted a comprehensive financial analysis for the programme as a whole. Some uncertainty was removed from the calculations because prices and order quantities were specified in the contracts. Financial analyses were not made for the individual equipment choices. Top management's review of the task force's proposals concentrated upon its financial aspects. At this level no consideration was given to the technical issues involved.

In France, the decision to purchase an FMS in concept was made by corporate management. It then set up a task force consisting of manufacturing managers and manufacturing engineers for design purposes. It is important to note that the institutional proximity of the user and the vendor was a risk minimization factor in the adoption of the new system. Both companies belong to the same overall firm. The task force will also be responsible for implementation and control.

Implementation Strategies

In the US, implementation of the FMS was facilitated by installing it in stages. However, several other factors unintentionally reduced the chances of a completely smooth introduction. The corporate office expected returns on its investment in the short run, which meant that the vendor had to work within a shorter delivery time than is usually expected, and that daily production was expected soon after installation. The vendor found it impractical to conduct in-house capability studies of such a large system, hence tests had to be conducted after installation. At the same time, sales of the new product line turned out to be much greater than anticipated, compelling productive use of the FMS on a round-the-clock basis. While division personnel had some experience with numerical control, about 12 of the shop's 500 machine tools were NC, the new system represented a great deal more uniqueness and complexity. In order to achieve desired production levels in as short a time as possible, a team of highly skilled personnel led by the manufacturing engineering department initially operated the system. This minimized difficulties during the breaking-in period but created problems for the manufacturing people when they took over the system later.

In the company in Great Britain, the plan's developer gave explicit attention to the process by which his ideas would be implemented. He was particularly concerned with developing an infrastructure in the shop that could support advanced manufacturing technology. There was very little experience with NC machine tools at the site. No NC equipment existed in the shop which was to be modernized. A few years before an NC lathe had been installed in a nearby shop and there had been considerable difficulty in getting it to operate properly.

In order to develop the necessary infrastructure, three steps were taken. First, facilities had to be brought up to the point where they could accommodate the new technology. A critical decision was made to stop using corporate's mainframe computer and to begin developing a local computer centre with a minicomputer. Currently, the minicomputer is being used for process planning. Ultimately, it will run the DNC system.

Second, it was necessary to change employees' skills and attitudes. Wherever possible the shop's technical people are being assigned to handle problems with the new machines rather than outside contractors or consultants. Training courses, especially in electrical maintenance which is normally a problem for NC machines, are being offered. However, it has been found that training increases a skilled person's marketability. Workers have been encouraged to learn a variety of skills so that they can more easily adapt to the imminent changes. While the most skilled operators will prove out the first generation of new equipment, the union has accepted having semiskilled employees operate the machines afterwards. However, every skilled machinist has been guaranteed a job in an area as close as possible to his specialty. Finally, the company is trying to change the union's attitude toward added shift work, since high utilization rates are necessary to defray the costs of the new equipment.

Third, the DNC system will be installed in an evolutionary manner so that employees have a chance to digest each step before going on to the next. Machines are being purchased sequentially over a five-year period. Each one will be tested and then operated autonomously using its own CNC computer. When all the machines and their accompanying handling systems are operating the movement to DNC will commence. Of course, installing in phases also spreads out the capital investment.

In the company in West Germany, implementation was facilitated by the following strategies. Top management selected the particular plant to be affected: the company's main machine shop. Consideration had been given to building a new plant but this was ruled out. The main shop already had an infrastructure to support the new equipment. Moreover, this infrastructure had already accumulated experience in dealing with stand-alone NC machine tools. Introduction of advanced manufacturing technology including an FMS would not call for radical changes in employees' skills and attitudes. For example, a policy decision had previously been made to rely mainly on in-house maintenance for NC because vendors tended to be located far away from the plant. Training and recruiting, particularly in electronics, were part of this policy.

It was decided to install the FMS in two stages. Requiring that each machine have CNC capabilities allows them to be productive before the complete system is put into operation. Production supervisors and foremen were involved in development and installation. The best operators of the stand-alone NC machines were selected to operate the new system. In spite of all the precautions taken, there were still problems in bringing the first phase machines up to desired production levels. The vendor and company did not have a great deal of experience in dealing with such complexity. In addition, the workers' representatives would not agree to going to three shifts. Currently, the policy is for two shifts plus overtime, and discussion is underway on running the machines during lunch hours and breaks.

In France, one implementation issue that must be dealt with is a lack of NC experience. Problems of acceptance of this new technology came from a lack of confidence in the skills of the company in the fields of electronics and computer science. This is particularly true for maintenance personnel. A one-year training programme has been set up jointly by the company and the vendor to overcome these difficulties. Shop management will also have to become more qualified. On the other hand, few problems are anticipated with workers and unions since automation is not new to the company. There is considerable transfer line experience. The number and skills of the FMS operators will approximate those for a transfer line.

The very close relationship between the company and the vendor should avoid some implementation problems faced by other innovators. First of all, the vendor has responsibility for the development, construction and implementation of the entire system even if aspects of it are acquired from several suppliers. Product designs have been worked out in close collaboration between the two companies. Each machine can be operated by its own CNC system so that implementation can be in stages. Implementation plans are now under development by the company and the vendor.

Implementation Problems

The firm in the US is the only one of the firms studied which has had a complete CIM in operation. It provides the most detailed insights into implementation problems and solutions. Problems with quality control and accounting will be discussed here.

When the division's quality control unit became involved with the FMS it found a great deal of uncertainty. It was important to quickly identify the occurrence of defects in order to prevent damage to as few as possible of the expensive housings. However, quality checks could be made easily only at the end of the machining process or between machining sequences. It could take up to two hours after occurrence to identify the first defective part. It was also necessary to readily discover and correct the causes of defects. Otherwise, the system or part of it would have to be shut down to avoid further damage. However, defects could arise due to any number of factors in the machine tools, computers, loading and material handling system, as well as in the parts themselves.

In order to deal with this situation, the DNC computer was programmed to shut off machines at certain times so the operators could conduct inspections of the parts being processed. While defects are found more quickly, machine utilization is reduced. The long-run solution is automated continuous monitoring which depends upon state of the art advances being made. It was also decided to switch from outside suppliers of raw castings to the division's foundry in order to gain more control over some of the factors affecting quality. Finally, the original theoretical objectives for quality were lowered when they proved difficult to attain in practice.

FMS implementation had a profound effect upon the cost accounting standards against which performance is judged. First, the standard cost of machining a part could no longer be expressed in terms of direct labour hours because this quantity does not vary with the cost of the part being processed. It was decided to express FMS cost standards in terms of machining hours while the rest of the shop remained with direct labour hours. One side effect was that manufacturing managers could not rely upon the informal procedures they had previously used, based on direct labour hours, for controlling the operations of the new system. Control passed out of their hands into the hands of staff personnel.

Second, none of the typical sources of data for circulating standard cost parameters were available due to the uniqueness of the manufacturing process and the newness of the product line. Neither the shop nor facilities elsewhere in the country could provide pertinent historical data. Consequently, the values of standards have been based on intuitive estimates. Experience has shown that total planned costs for the FMS are a fairly reliable benchmark, but that planned values of some major cost components such as rework and maintenance are not very accurate. Currently, the problem remains because past performance data for the FMS have been affected by starting up conditions and a period of abnormally high demand.

In Great Britain, problems were noted in accounting, maintenance and production scheduling. A machine hour basis had been previously employed for large conventional machine tools so attention focused on starting to charge costs at as accurate a rate as possible. It was decided to wait until the initial machining centre was completely tested to develop accounting controls. Only then would it be known whether each part would pass through every operation, or different combinations of operations would pertain to different parts. In addition, more reliable estimates of utilization, maintenance, machining times and related quantities could then be obtained. There was also concern about being able to identify indirect factors such as reductions in materials handling due to combining several operations in a single machine, and increases in the number of programmers.

At the present time the company relies upon vendors and contractors for major repairs of NC machines and computers. As the equipment's reliability is not as high as desired, it is thought that these companies do not have highly qualified people. It is also difficult to have repairmen arrive when they are needed. For one particular machine breakdown, the foreign vendor could not send anyone for about three months. The company also has problems with its own maintenance men. Owing to the large capital investment the new machines are run on second and even third shifts, but it is hard to provide maintenance coverage at these times. One partial remedy is more planned maintenance but only weekends are available and skilled people do not want to work then either. The entire situation is complicated by the fact that any single individual is unlikely to have all the answers for repairing such complex units.

Production scheduling has become a more difficult task at this firm due to the new NC machines. In the past, the existence of machining alternatives allowed production control to load machines without much regard to capacity. Introduction of the new machines caused the infinite capacity assumption to be invalidated. Design of the products is built around this equipment so it is no longer easy to subcontract or use the shop's conventional machines as alternatives. Breakdowns of NC machines cause scheduling problems for the same reasons, especially when several operations are combined in one.

In West Germany the respondents talked about problems in quality control, accounting, production scheduling and production preparation. Currently, there are no serious quality control problems, presumably due to the parts' relatively simple shapes and low costs. However, a large number of inspectors are needed because the system has not been completely checked out yet, and because of the large volume of parts being produced. Control is exercised after parts come off the FMS rather than within the system. Attempts to develop a within-system quality check have so far not been successful. For example, it is believed that having the computer stop the machines would lower machine utilization too much. A completely automated quality control system is viewed as the long-run solution for guaranteeing that the first time a defect occurs it will be detected.

The accounting department, owing to its experience with the first generation of stand-alone NC machine tools, already employed a machine hour basis for computing cost standards when the first FMS equipment arrived. However, there was difficulty in finding reliable information to estimate cost parameters. Energy usage, maintenance and tool usage among other items were assumed to be quite different than for other shop facilities. The procedure was to develop initial cost estimates from the plans for the system and to improve stepwise each time new performance data become available. Now that costs for the first stage of implementation are fairly reliable, they can be used to help predict second stage costs.

The computerized system which does weekly production scheduling for the entire shop cannot be used for the FMS. It does not allow for setup and machining in parallel on different machines rather than sequentially on each machine. A separate real-time scheduling system is being developed for the FMS which must be interfaced with the global system. It is planned that weekly information from the latter will be input for the former, which in turn will provide feedback data, but integration has not yet been achieved.

Production preparation had disagreements with manufacturing concerning which parts should be moved from the stand-alone NC machines to the FMS. Manufacturing wanted a large number of parts moved due to the FMS's higher productivity. It hoped that the FMS's capacity would be increased, and there would be less of a chance of diverting parts to outside subcontractors. Production preparation, on the other hand, had to be concerned with whether conversion costs might outweigh productivity gains.

The French system won't begin to be installed until next year so little can be reported on implementation. Nevertheless, the company and the vendor anticipate quality control problems. A provision for some automatic quality inspection during machining will contribute toward early defect detection. In addition, certain parts have been designed so that they can be recycled automatically through the system if they are subsequently found to be defective.

Discussion

In order to determine whether our ideas concerning uncertainty are compatible with the interview data it is useful to identify common patterns and differences among the four companies. Some commonalities do exist in the main reasons for adopting CIMs. A starting point for innovation in the US and Great Britain was modernization of facilities, a need which is undoubtedly linked to reducing uncertainties such as equipment breakdowns. However, new product development and increasing capacity were starting points in three companies. This raises the obvious but important point that uncertainty reduction for existing facilities, as implied in our conceptual framework, is not the only occasion for technological innovation. Manufacturing management must be concerned with reducing uncertainties in planned facilities as well.

The criteria used in making adoption choices were relatively diverse reflecting each company's attempt to seize upon different CIM advantages. Some of these criteria do reflect attempts to cope with uncertainty. Both the US and French companies selected CIMs because of their flexibility in dealing with market uncertainties. This was the critical factor in the American company's choice. In Great Britain, uncertainty reduction was reflected in desiring a centralized source of production information and wanting to avoid the vagaries of the skilled labour market. The experience in France suggests that one coping strategy, technological innovation, can be used to substitute for a less effective one, in-process inventories. To be sure, other criteria were also employed by the firms. Each of the following were mentioned by two firms: productivity improvement, utilization rate improvement and reduction in delivery times.

The companies differed with respect to adoption strategies in the degree of local versus corporate influence in initiating the adoption process. In the US the initiative was with the division, and it took three years to gain corporate's approval. In Great Britain, corporate level let it be known that it wanted an initiative which was subsequently developed at the local level. In West Germany and France the initiative was mainly from corporate personnel who then brought local people into the process. While adoption occurred with all of the above models, it appears that in those companies in which the critical uncertainty of corporate commitment was resolved early the length of the adoption phase was shortest.

In the US, Great Britain and West Germany a special task force or individual had the main responsibility for adoption. In France it was corporate management. In at least three of the firms (no information is available for France), equipment was selected for a large capital improvement programme. A CIM turned out to be the main ingredient of each programme. In at least the same three companies the equipment selectors had the responsibility for making technical decisions which were not intensively reviewed by general management.

Of particular interest are the strategies employed to cope with the uncertainties of the adoption decision process. In at least the US and West German companies the task forces prepared comprehensive financial analyses but only for their programme as wholes. These were evaluated in detail by their superiors. However, the exact values of the numbers used could not be considered very reliable because of the uncertain future. At the level of individual decisions, financial analyses were largely qualitative, especially for the CIMs, due to the lack of available information on consequences. In the US and France, a premium was placed on having a vendor with which a satisfactory working relationship already existed; in other words, one for which some certain knowledge was available. Apparently, this strategy pays off. Ettlie found that a good working relationship was associated with successful NC implementation.[4] However, our reading suggests that the strategy has its limitations. Even a trusted vendor will not have answers for all user problems when the equipment is novel. In the American firm the known quantity, acquisition cost, played a more important role than the uncertain estimates of future net earnings.

It has been possible to identify four strategies used to deal with implementation uncertainties. Perhaps the most critical is to develop a human and technical infrastructure which will be able to support the innovation. If this support is not available the equipment will present contingencies with which the technical core will be unable to cope. We believe that lack of an infrastructure is one significant factor impeding the diffusion of CIMs.

Possessing a support system is closely related to the particular stage of NC development in which a company is to be found. The more highly developed NC is within a firm, the more its support activities will be able to cope with additional NC advances. The infrastructure will have experience that can be used in solving new problems. And, its previous solutions to problems will facilitate NC advances in the future; for example, having already converted from labour hours to machine hours. These observations are supported by Ettlie's study which found that successful NC implementation was associated with a company having prior exposure to it.[4] However, our study suggests that there may be a qualitative leap in uncertainty from stand-alone NC machines to an integrated system. Experience with just the former cannot completely prepare a firm for the complexities of the latter.

The infrastructure's significance was highlighted by the West German company's decision to install its FMS and other new equipment in an existing factory. The already established support systems eased implementation problems here, and to some extent in the US. In the British company, great care is being taken to develop the necessary support before conversion to a CIM takes place. The French company has invested in training programmes.

The second strategy is to install complex manufacturing equipment in stages to limit the problems that must be dealt with at any one point in time, and to contribute to the development of experience in solving problems. All four companies installed their CIMs in phases. At least two approaches are available, the choice in part depending upon a firm's prior exposure to NC. Firms starting from scratch may want to prove out each machine one at a time. Taking advantage of CNC capabilities on individual machines will facilitate this strategy. Companies with prior NC experience may want to install an integrated system in modules. Of course, other factors such as the redundancy built into the system will also influence this choice.

Having operating managers participate in adoption and implementation is the third strategy. This decreases uncertainty by increasing their information about the innovation. Of course, it also tends to increase commitment. The companies studied differed in the degree to which manufacturing managers shared in adoption and implementation. The US company had difficulties when its manufacturing people took over the CIM because they had not been much involved previously. The British and West German companies had relatively fewer problems. In the former, the individual responsible for renovating the plant sought out the opinions of operating personnel. In the latter, manufacturing people were involved during adoption, development and installation. France had manufacturing managers participate in design, and plans to have them take part in implementation.

Finally, Great Britain and West Germany used the most skilled operators in the shop to help prove out the new equipment; a strategy which facilitates coping with initial operating uncertainties. However, this objective must be traded off against other criteria. The US has used and France will employ semiskilled workers; an approach which reduces wages costs and decreases the chances that operators will believe they have been deskilled. Britain avoided the trade-off by negotiating with its union to employ semiskilled workers after the testing phase is completed. All four companies would like to have adequately skilled operators and service people available for second and third shifts. However, they are faced with a good deal of reluctance on the part of their workers.

The most frequently cited implementation problems were quality control, accounting, equipment maintenance and production scheduling. These are all examples of the maintenance and control activities that manufacturing management relies upon to provide information that will reduce uncertainty in the technical core. It is therefore significant that the complexities of CIMs have introduced contingencies into most of these functions.

In the US company, up until recently, there had been uncertainty in discovering and correcting quality defects. Consequently, this company tried to reduce uncontrolled variability in the quality and receipt of raw materials by using its own foundry as supplier. The American, West German and French companies also spoke of quality problems in terms of the need to have earlier detection and diagnosis of defects. Otherwise, rework, scrap and inspection costs become too high. Automated in-process inspection is a likely solution but is not yet commercially suited to these companies' needs.

In the accounting area, the American, British and West German companies all had problems with uncertain cost standards due to lack of available data. The US company's experience has been that it takes several years to compile a reliable performance history. One maintenance issue was the uncertain support by outside sources due to lack of information about the equipment or scheduling difficulties. However, the most frequently mentioned maintenance problem was the reluctance of skilled people to work on second or third shifts. The issues pertaining to production scheduling were not directly related to uncertainty.

The fact that many of these problems are easing as companies are gaining experience in dealing with them, illustrates their short-term nature.

Concluding Remarks

Coping with uncertainty may be a useful concept for organizing observations of the innovation process for manufacturing equipment. Major aspects of the reasons for adoption, adoption and implementation strategies and implementation problems, were subsumed under our framework. Clearly, it was not the only factor operating. Some data could not be explained by the framework, and other data were accounted for by other concepts as well.

The applicability of our approach for companies in different cultures and task environments was especially encouraging. We believe that these two variables will tend to affect the nature of the specific uncertainties and strategies of a firm rather than the basic theme of coping with uncertainty. Future research involving a more rigorous study with a larger sample is needed to verify this claim. A broad range of process innovations also needs to be investigated. It is hoped that the ideas generated here will stimulate further research in an area of growing international concern.

Notes and References

The authors are indebted to CERESSEC, Cergy, France; the Institute for Social Research in Industry, Norwegian Technical University, Trondheim, Norway; and the Management Research Center, University of Wisconsin-Milwaukee for financial support. We also wish to thank Elizabeth Williams for her editorial assistance.

1. Abernathy, W.J., 1978 *The Productivity Dilemma*. The Johns Hopkins University Press, Baltimore
2. Burge, A.J., Goforth, R.E., 1976 Survey of numerical control equipment application in Texas manufacturing plants. *Industrial Productivity Program*, Texas Engineering Experiment Station, Texas A. & M. University System, May
3. Cook, N.H., 1975 Computer Managed Parts Manufacture. *Scientific American* 232: 22–29
4. Ettlie, J., 1973 Technology transfer – from innovators to users. *Industrial Engineering* 16: 16–23
5. Gerwin, D., 1981 Control and evaluation in the innovation process: The case of flexible manufacturing systems. *IEEE Transactions in Engineering Management,* forthcoming
6. Gold, B., Rosegger, G., Boylan, M.G. Jr., 1980 *Evaluating Technological Innovations*. Lexington Books, Lexington, Massachusetts
7. Kepner, C.H., Tregoe, B.B., 1965 *The Rational Manager*, McGraw-Hill, New York
8. Nabseth, L., Ray, G.F., 1974 *The Diffusion of New Industrial Processes*. Cambridge University Press, London
9. Noble, D.F., 1980 Social choice in machine design: The case of automatically-controlled machine tools, and a challenge for labor. In: Zimbalist, A., (ed.), *Case Studies on the Labor Process*. Monthly Review Press, New York
10. Thompson, J.P., 1967 *Organizations in Action*, McGraw-Hill, New York
11. Tornatzkey, L.G., *et al.*, 1980 *Innovation Processes and Their Management: A Conceptual, Empirical, and Policy Review of Innovation Process Research*. Division of Policy Research and Analysis, National Science Foundation

6.8

Advanced Manufacturing Systems in Modern Society

J. Hatvany, Ø. Bjørke, M.E. Merchant, O.I. Semenkov and H. Yoshikawa

Skill and Training

The principle of numerical control involves a radical change in working method, which has consequences for the training of skilled workers. At the beginning of this century we went through a correspondingly radical change which changed the operator's main function from energy generation to control. These tasks demanded manual skills and technological know-how. Thus the main objective of training skilled workers was to develop these manual skills. The step we now have to take is the substitution of manual control by verbal control based on supplying a written message, specifying the path to be followed. A consequence of this is that manual control skills will be replaced by verbal skills. This results in making the operator's job more like that of the technician or engineer, working in the methods office.

Integrated manufacturing systems go much further in this direction. In fact the acceptance of microprocessors, robots, highly automated and intelligent apparatus and its related software, causes serious impacts upon the quality and the quantity of the demands made on workers. Generally speaking, there is a greater demand for workers with a thorough, fundamental schooling in the sciences and in modern technologies, such as electronics and computers. This trend coincides well with that of the new labour supply in most industrialized countries, where the percentages of secondary, technical college and university graduates have all been growing in recent years. Nevertheless, a number of problems remain to be solved.

Source: from *Computers in Industry*, 4, 1983. A selection of chapters from the work in the title is reproduced here.

One general feature of CAD/CAM developments to date, has been the necessity for extant industrial employees at all levels to accommodate changes in their roles. This involves a massive adult education effort, training skilled workers to become programmers, electricians to handle electronics, technicians to communicate with data-bases, process planning and scheduling programmes and engineers to understand the overall nature of the new systems. The didactic, psychological and organizational tasks of such a broad retraining programme require very careful planning and execution, for if they are incorrectly implemented they can ruin all the benefits expected from the introduction of advanced manufacturing systems.

Training good maintenance personnel for these systems is especially difficult, since they have to have a very generalistic, multidisciplinary grounding to avoid the familiar phenomenon of pushing a fault over into the other man's domain. This is a point where a well-conceived computer-based system can itself offer much help, both by including a wide range of diagnostic services and also by assisting maintenance personnel through tutorial-style diagnostic and repair databases.

The hardest training task of all, however, is to educate the engineers who will design, implement and support the systems. So far, these people are very scarce everywhere. They are mostly innovative types, more scientist than engineer, whereas the proliferation of the systems requires that they be increasingly treated not as scientific experiments, but as sound, high-level engineering products. Technical universities in the FRG are now running engineering courses with a major CAD/CAM content. Computer science courses at French universities have been doubled. In the USSR, the universities at Moscow, Leningrad, Kiev and Minsk have established new faculties for training experts in robot systems, in computer-aided design and in computer-aided process planning. Project work at these faculties is closely linked to industry, and they are also running retraining courses for graduate engineers and awareness courses for managers. In Japan, departments called 'Precision Engineering' of universities and colleges have taught manufacturing technology for a long time. During the past ten years, these departments have become most advanced from the viewpoint of computer applications. Almost all the Precision Engineering departments have started CAD and CAM undergraduate courses. In the University of Tokyo, for example, 30 per cent of the graduate students are engaged in CAD/CAM research projects.

Particularly significant has been the establishment of small, but advanced experimental CAD/CAM systems in universities at Trondheim (Norway), Budapest (Hungary), West Berlin and Provo (Utah, USA), which have served as powerful focal points for training all levels of CAD/CAM personnel. These pilot plants have been utilized for advanced research, for graduate and postgraduate project work, for undergraduate demonstrations, for training workers of plants which are to be equipped with CAD/CAM and for tutorial demonstrations to politicians and top management.

Settlement Patterns

The biggest cities of our times have evolved around industry and the related sevices, transport, trade and finance. Many of these huge settlements have, indeed, grown up around one single factory, which needed tens of thousands of workers to man its machinery. It now appears that in fact industrial activity might well undergo a kind of polarization. The really big mass production plants (particularly in the continuous process industries) will form one pole, the extremely flexible, locally based small-scale manufacturing facilities will form the other. In the case of the first, the major change to be brought about by intensive computerization will be that since the labour force required to run them can be reduced to a few tens of people for, say, a multimillion ton steel works, the plants can mostly be located in areas remote from centres of human habitation and the crews to man them can be transported in, from a distant city (if they wish). This obviously makes for better environmental protection and secondarily obviates the load on communal services (transport, electricity etc.) which any large plant would otherwise impose.

At the other end of the scale, the highly concentrated, completely closed facilities for the flexible production of one-off or small batch quantities of a very great variety of goods, with their own carefully controlled antipollution measures, can well be sited in the midst of otherwise residential areas. Here they will be near their labour force, their markets and their fellows. These two factors may in due course reverse the hitherto concentrative effect of industry on the settlements of the advanced nations.

In actual fact, two major models have emerged for the settlement patterns of advanced manufacturing systems. One of these originated in Norway, the other in Japan.

Norway is a relatively sparsely populated country with high living standards, a high cost of labour and – due to oil – adequate investment resources. A large portion of the Norwegian population lives in small, scattered communities along the country's very long, highly serrated coastline. Traditionally employment in the provinces has been in agriculture and fishing. Reduced demands for workers in these occupations have led to a weak economic structure and encouraged the population to leave these areas. Northern Norway is an extreme example of this situation. Generally it is accepted in Norway that the small rural communities ought to be conserved as far as is reasonably possible. The means to obtain this is to strengthen the basis of employment in the provinces. In trying to fulfill these goals, the Production Engineering Laboratory of NTH-SINTEF has developed the following manufacturing system concept.

The size of firms should be such that they can cope with the problems of product development and marketing for the world market. These and other functions which can only be performed by large concerns are kept at a central location. However, production facilities can be divided into cells and be spread geographically around the provinces. Each cell unit should have about 10 production workers together with additional indirect workers. Such units are of a size which could be put into most local societies without creating a stressful situation. If more jobs are required, multiple cells would be located under the same roof. Thus industry is moved towards where the people are living, instead of the contrary, which is the rule today. Advantages include a stable and motivated work force, and all the positive effects associated with the 'small is beautiful' concept. Disadvantages might include absenteeism and the additional transportation of products. Today, however, it is not unusual for a person living in the provinces to incur home/work travel costs of 25 Nkr daily. Consequently cells could incur at least 250 Nkr additional product transport costs per day and still be in balance.

In order to handle the data flow between the cell and the parent company, a well-developed telecommunication system is needed. The tariff for telecommunication services should also be adjusted to fit an industrial expansion of this type.

A typical cell would have five functions:

(1) Manufacturing
(2) Planning
(3) Preparatory functions
(4) Supplementary functions
(5) Stocking functions

The preparatory function would cover the manual processing prior to manufacturing. This could include cutting of blanks, machining of clamping surfaces, palletizing of blanks etc. The supplementary functions would also be of a manual nature and include depalletizing of machined parts, deburring, heat treatment, assembly etc.

Actual manufacturing is performed in the cell-core, which is where added value accrues. A cell-core is a number of numerically controlled machine tools. In the pilot plant the cell-core consists of four NC machine tools plus an automatic workpiece handling device (an industrial robot). Each machine tool, including the robot, is CNC-controlled and fed with data by a DNC-system. The robot moves workpieces both between machines and between machines and the corestock. The corestock has the capacity to handle the number of workpieces machined during two shifts. Thus, the cell-core is working around the clock unmanned, while the cell as such is manned only during daytime. This makes possible the rapid amortization of the capital invested in the cell-core.

The planning function covers all cell-internal planning. This includes detailed technological planning as well as production planning. Consequently the cell crew is delegated the responsibility to utilize the cell resources. There is thus plenty of scope for local initiative and pride of achievement.

The Japanese model is very different. Japan is very densely populated along the narrow strips of utilizable seashore that are easily accessible and inhabited. Industrial plants of varying sizes are closely interwoven with residential settlements throughout these areas and consequently serious problems have arisen with environmental and scenic pollution. The goal is therefore to *concentrate* production in a smaller number of highly efficient units, liberating land areas for residential and recreational use. These concentrated industrial plants are required to have extremely strict environmental control (zero pollution) and to operate around the clock with a minimum number of night-shift workers. They must also be attractive places of employment for highly educated people, so that no dirty or heavy work has to be performed by humans.

Taking these (and similar) premises as their point of departure, Japanese planners in 1974 elaborated a project for an 'unmanned metal working factory'.[2] It was planned that the plant – a floor area of 20,000–30,000 sq. metres – would be staffed by a control crew of 10, compared with a normal complement of 700 to 800 workers. It would be directly linked to a remote technological data bank and have the following major components:

(1) control centre
(2) operation department
(3) machining department
(4) pollution disposal department
(5) warehouse department

All but the first of these would be fully automated. Human activities in such a factory would be limited to the functions of top management, CAD, programming, equipment supervision and maintenance, data input, outward liaison and external transport.

Since 1974, almost all these goals have been realized. Fujitsu FANUC was the first to construct a complete factory along these lines.[3] Located in a 20,000 sq. metre building, this plant has 29 cell-like work stations. Seven of them are equipped with robots; 22 are equipped with automatic pallet changers with pallet tools. These stations are connected by unmanned vehicles guided by electromagnetic and optical methods. The plant has two automatic warehouses, one for material and another for finished parts and subassemblies. The vehicles transport material from the warehouse to the unmanned machining stations. Robots or automatic pallet changers load materials onto the stations from the vehicles. Finished parts are transferred again automatically by the vehicles to the second warehouse.

The plant has an assembly floor, where workers work during the day. Transportation between the parts warehouse and this floor is also by unmanned vehicles. In the daytime, 19 workers are working around machining stations, mainly for palletizing, and 63 workers are in the assembly section. (At a later stage it is planned to automate most of the assembly work, also using robots.) Thus, at present 82 workers are in the plant during the day, but at night there is only one. The assembly floor is closed, and the machining floor is operated without any workers. Every station is equipped with a monitoring device with TV camera, and one person sits in the control room to monitor all the working stations. He observes the working status of stations through the camera, without touring the factory. The monitoring device also records the spindle motor current, calculating the cutting force and time, to evaluate the cutting conditions.

The factory is located in a scenic area (the Mt. Fuji National Park), where very strict conditions were stipulated. It is built in a hollow and concealed from view by trees. It is a zero-pollution factory – the effluent water is as clean as the mountain brook entering the plant. Heating and hot water are supplied by solar energy converters covering the entire roof.

Many other Japanese machine tool manufacturers have also recently completed commercial systems with monitoring devices which can operate nearly unmanned night shifts, and are building factories along similar lines. Toshiba Machinery Co., Toyoda Machine Works Co., Makino Co., Yamazaki Iron Works Co. and Niigata Iron Works Co. are the leaders in this area.

Having considered these two, widely differing models, what should be the engineer's conclusions? Is one model 'right' and the other 'wrong'? Previously, an engineer was successful when the technology he introduced showed economic benefits and was reasonably adapted to human and environmental needs. Today, economic benefits are in themselves no longer acceptable, unless technology is also seen to be actively beneficial from a human and environmental point of view. Environmental acceptance is no longer a question of morals, but of survival.

The social value systems which determine the settlement patterns of modern industry are, however, changing not only as a function of time in a single country. They differ also from country to country, and even from region to region. These differences may be due to the mentality and traditions of the people or to different levels of educational, economic and industrial development, or even to geographic and climatic factors. A good illustration of the latter is to be found in the Soviet Union. The principal raw material resources of the country are located in the sparsely populated, remote and inhospitable regions of the North, Siberia and the Far East. In these areas huge territorial industrial and energy complexes are being built, but labour and transport costs are extremely high. This is obviously a case for highly productive, highly concentrated plants, employing relatively few people – close to the 'Japanese model'. The most densely populated European regions of the USSR have a developed infrastructure and advanced educational facilities. Their settlement pattern is a multitude of small- and medium-sized towns. Here the present strategy is to establish many small, decentralized workshops that are organizationally parts of large concerns – mainly in the instruments and allied industries. This pattern is more akin to the 'Norwegian model'.

We may conclude, therefore, that there is no universally optimal settlement pattern for all countries and all regions. What advanced manufacturing systems do offer, is the *possibility* to adapt flexibly and efficiently to the social, economic, geographic and environmental criteria of a wide variety of strategies.

Intelligent People

Adaptivity and satisfaction

The willingness of people to adapt themselves to the new conditions brought about by advanced automation appears to differ widely according to a number of factors. Age, social status, the educational level, the employment situation, incentive systems, consultative and participative arrangements, planning procedures and the circumstances of retraining opportunities all play important roles. It is obvious for instance, that a middle-level shop foreman who is over 50 and has spent two-thirds of his life in the classical metalworking environment will rarely be among the enthusiastic supporters of computerization. A highly skilled milling machine operator who is given a dismissal notice in a city with 23 per cent unemployment and no subsidized retraining courses, is also likely to be a bitter opponent, especially if no previous consultation or warning has taken place. These are familiar problems to all students of the sociology of automation and have received detailed treatment in many publications. Some of these writings are enthusiastically optimistic, others cooly analytical, while some are quite firmly opposed to the trends we have been discussing in this reading. Of the latter category, which is perhaps less familiar to the international engineering community. Blomberg and Gerwin provide a good example in this abstract:[1]

Computer-aided manufacturing technology supposedly holds one of the keys to productivity growth through its potential for revolutionizing the production of discrete parts. Unfortunately, complex, large-scale, integrated systems may be beyond the capabilities of most companies to deal with. Data from four companies in three countries suggests that management, staff specialists and workers experience great difficulty in coping with the cognitive and motivational problems emerging from the acquisition and use of this new technology. Greater emphasis needs to be placed on investment in small-scale, decentralized, less capital intensive, manufacturing systems.

(It should be added that the 'three countries' did not include Japan and the 'four companies' had some of the earliest experimental systems.) [Blumberg and Gerwin report on similar work in reading 4.7.]

This occasion is not an appropriate one to add to the voluminous literature on the general social implications and problems of automation, since few areas have solutions that are universally applicable to the various backgrounds and environments represented by the authors. Rather we choose to discuss two factors of adaptivity to the new manufacturing methods, which have a direct bearing on the engineer's activities. One of these is the innovative process, the other is job satisfaction.

The receptivity of Japanese people to new technologies has always been very great. They welcomed and transformed traditional Chinese technology, they welcomed and adapted readily the very same Western technologies which had elicited considerable resistance in their countries of origin. There are two reasons for this. One is that the new technology introduced into Japan had previously been well-tried in the West and reformed into an acceptable format. The second, and more important reason lies in the different innovative thought processes of Japan.

Japanese culture is based on the ideal of harmonization with the natural environment and careful, piecemeal improvement by small steps.

Let us consider the example of a teacup. Originally, many thousands of years ago, the teacup was just a broken piece of wood or stone, with an appropriate shape for drinking. In the Western world, this was subjected to functional analysis and radical innovation. The morphology of the existing teacup was put aside, a handle was conceived for holding, a cylindrical part for containing, a smooth contact surface for the lips, and these were combined into a brand new system. Japanese people did not split function and morphology, they tried to change the original shape as little as possible. Morphological developments were confined within infinitesimal deformations. The conduct of the teadrinker was adapted to the extant teacup. Since it has no handle, it is difficult to use when it is full – consequently Japanese manners dictate that a teacup is 'beautiful' if it is half full. The innovative approach is therefore one of harmonization with the environment by reforming oneself, rather than reforming the environment. By virtue of this characteristic, Japanese people accept new technologies without intense resistance. They are confident of reforming the new technology gradually and of developing new behaviour patterns for utilizing it. These considerations in the approach to innovation have important lessons for us all.

For the workers, technicians and engineers who operate advanced manufacturing systems, the prime criterion determining their readiness to adapt to the new conditions lies in the degree of satisfaction they obtain from their jobs. It has been estimated that productivity may easily be changed by a factor of two as a consequence of satisfactory or bad motivation. One of the most frequently emphasized conclusions in this respect, is that the previous fragmentation of workers' tasks (apostrophized as Taylorism) should be reversed. The increasing educational level of workers requires vertical job enlargement under which workers should be delegated the responsibility to control their own working conditions. Such responsibility can evidently not be delegated to every individual, since in such a situation the activities of a workshop would be impossible to coordinate. Responsibility should instead be given to a group of workers, implying a given set of resources. The cellular (maybe even geographically decentralized) structures of the new, advanced manufacturing systems should facilitate such an approach. With cell-level operation planning and part programming, routine maintenance etc. delegated to a cell crew, it is to be expected that their sense of achievement, their scope for original, creative action, their collective loyalty and independent group posture towards management will do a great deal to raise job satisfaction and counter the image of subservience to machinery, of button-pushing drudgery and of the alienation of the individual which has been gloomily forecast for the future.

Interfaces and intelligence

We have witnessed over the past years an increasing rejection by workers of the classical numerical control technology. In this technology part programming was moved from the shop floor into an office. The technologists in the office plan the machining process, clerical personnel punch the tape, all the worker has to do is to mount the tape, push the button and watch the process. So far most part programmers have been recruited from among the best operators, but where will they come from in the future, when the damage of reducing operators to 'workpiece handlers' and 'button pushers' starts to have effect? Both the human and the long-term recruiting problems demonstrate that we have to give back to operators the control of their own machines or, in the case of a highly automated system, of their groups of machines. The 'workpiece handler' and 'button pusher' jobs must be eliminated.

Similar problems have arisen at the engineering level. Early concepts of computer-aided design envisaged a design job as a sequentially deterministic set of actions, embodied in an algorithm. There was 'a correct way' to design a product – this was embedded in the CAD system, and the designer was told at each step what data he must enter. Evidently a well-trained, innovative engineer, a creative intellectual, will find such an environment most frustrating.

In both cases the reasonable approach appears to be to design systems where man is always in the decision-making role and the computerized system is always in the slave role. However, beyond a certain degree of system complexity this relationship is very difficult to maintain with the technical means at present available. As was recently pointed out, the time seems opportune to reconsider shop-floor and design office man–machine communications in the light of these problems.[4] The new tools which are becoming available for this are:

(1) functional system design techniques
(2) new computer graphics applications
(3) natural language communication
(4) intelligent dialogue techniques

Recent results in artificial intelligence research have already been experimentally applied to such problems as the presentation of strategically selected information, the analysis of characteristic event sequences, anticipative decision queries, etc. The essence of these efforts is to create man–machine interfaces where the human operator is not swamped by irrelevant information, but is shown those facts which he needs at that moment, where *choices* are always made by the human operator, but their *consequences* are presented for his consideration, where the human master of the system can impose communication procedures and processes on an essentially permissive and obedient system.

The degree of intelligence of such systems and interfaces may be measured by the degree of intelligence which they allow people to exercise in their activities.

References

1. Blumberg, M., Gerwin, D., Coping with advanced manufacturing technology, in: *Quality of Working Life in the Eighties* (CCQWL, Toronto, 1981)
2. Bulletin of MEL. 1974 Methodology for unmanned metal working factory. 13 Tokyo
3. Inaba, S., An experience and effect of FMS in machine factory, in: *Preprints*, IFAC 8th Triennial World Congress, CS-6 (IFAC, Kyoto, 1981)
4. Nemes, L., Hatvany, J., Design criteria and evaluation methods for man–machine communications on the shop floor, in Sata, T. and Warman, E.A.(eds.), *Man–machine communication in CAD/CAM* 1981, North-Holland, Amsterdam

Index